D0947783

The Science of Mechanics

The Science of Mechanics

Account of Its Developmen

TRANSLATED BY THOMAS J. McCORMACK

SIXTH EDITION • WITH REVISIONS THROUGH

THE OPEN COURT PUBLISHING COMPAN

A Critical and Historical

y Ernst Mach

EW INTRODUCTION BY KARL MENGER

HE NINTH GERMAN EDITION

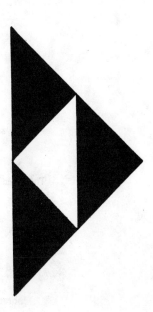

A SALLE • ILLINOIS •

Copyright 1893, 1902, 1919, 1942

By

The Open Court Publishing Company

© 1960

THE OPEN COURT PUBLISHING CO.

Third Paperback Edition 1974

Library of Congress Catalog

Card Number: 60-10179

ISBN 0-87548-202-3

INTRODUCTION TO THE
SIXTH AMERICAN EDITION, 1960

Ernst Mach's book *Die Mechanik in Ihrer Entwicklung Historisch-Kritisch Dargestellt*, one of the great scientific achievements of the last century, remains a model for the presentation of the development of ideas in any field. In its own domain, the work is still full of vitality. It is an inspiration to philosophers of science, a valuable source of information for historians of physics, and a splendid help to teachers of mechanics. Its first half is a most stimulating introduction of unsurpassed clarity and depth for beginners.

The book follows the development of mechanics up to the turn of the century.[1] As the title indicates, the work is historical and critical.

That the *historical* presentation of a branch of science is the most penetrating approach to the subject matter and leads to the deepest insight was one of Mach's general methodological ideas. Nor is anything more conducive to creative thinking than an exposition of ideas as they have developed, of notions abandoned long since, and of the role that historical accidents have played in the genesis of current concepts—techniques that Mach introduced and superbly developed in his *Science of Mechanics*. Mach also applied the historical method of presentation to the theory of heat and to

[1] Nine German editions of this book have been published. Seven of them appeared during Mach's lifetime (1838-1916), in 1883, 1888, 1897, 1901, 1904, 1908, and 1912. The eighth and ninth German editions appeared in 1921 and 1933. English translations were published by the Open Court Publishing Company in 1893, 1902, 1915, 1919, 1942, and 1960.

parts of optics and the theory of electricity.[2] But the method has further potentialities. Other parts of mathematics, especially algebra, would profit from a similar treatment; and, most of all, the science of mechanics itself might gain if its development since 1900, including the theory of relativity, wave mechanics, and quantum mechanics, were presented *à la* Mach.

The *critical* parts of Mach's book culminate in his analysis of the concept of mass and his examination of Newton's ideas of absolute space and absolute time. The latter critique is quoted in almost every presentation of the theory of relativity. "This book," Einstein wrote in his autobiography[3] about Mach's *Science of Mechanics*, "exercised a profound influence upon me . . . while I was a student."

*   *   *

In this century, the analysis of the mass concept and, even more, physicists' views on space and time, have advanced beyond Mach. Yet his original discussions remain classics not only of physics but also of *philosophy of science*.

Against Newton's definition of mass as a quantity of matter, Mach raised the objection that it was of no help in actual operations with masses; and he formulated a new definition, based on Newton's Third Law— a definition that made it possible to measure masses. Mach's treatment initiated a method that was later greatly elaborated and applied to the philosophy of

---

[2]E. Mach, *Prinzipien der Wärmelehre* (Leipzig: 1896);
———, *Principles of Physical Optics* (Leipzig: J. A. Barth, 1921; London: Methuen Co., 1926; New York: Dover Press, 1953.)
———, "On the Fundamental Concepts of Electrostatics," in *Popular Scientific Lectures* (5th ed., La Salle, Illinois: Open Court, 1943), pp. 107-136.

[3]"Albert Einstein: Philosopher-Scientist," in *Library of Living Philosophers*, ed. P. A. Schilpp (Evanston: 1949), pp. 20-21.

physics by P. W. Bridgman[4] under the name of *operationalism.*

Mach rejected absolute space and time because they are unobservable. More generally, he proposed to eliminate from science notions which lack counterparts that are actually or at least potentially observable. He thereby became one of the initiators of *antimetaphysical positivism.*

A third point that Mach stressed over and over again was his view that science had the purpose of saving mental effort. General laws are shorter and easier to grasp than enumerations of specific instances. Simpler theories are preferable to more complicated ones. His theory of "economy of thought" is Mach's main point of contact with R. Avenarius who, in his *empirio-criticism,*[5] regarded philosophy as thinking about the world with minimum effort.

A fourth point of philosophical importance in Mach's program was the replacement of causal explanations by functional connections. In this respect, Newton had been a shining model. Without entering into the question that was uppermost in the minds of his contemporaries—the question as to *why* bodies attract each other—Newton was satisfied with formulating the specific connection of the attractive force between two bodies with their masses and with the distance between them. In the 18th and 19th centuries, countless attempts, now all but forgotten, were made to *explain* gravitation. Physicists hypostasized vortices, or tensions in media, or bombardments of the bodies by particles traversing space at random and driving, for instance, a stone toward the earth because the latter shields the stone against the particles coming from

[4] P. W. Bridgman, *The Nature of Physical Theories* (Princeton: 1936; New York: Dover Press, 1936).

[5] R. Avenarius, *Philosophie als Denken der Welt gemäss dem Princip des kleinsten Kraftmasses* (Leipzig, 1876).

below. But the real triumphs of human insight into gravitation—Newton's deduction of Kepler's Laws, the prediction of new planets which were subsequently observed, and, in the present age of experimental astronomy, the control of the motion of artificial satellites—these triumphs are independent of any explanation of gravitation and are entirely anchored in Newton's law that the attractive force between two bodies is proportional to their masses and inversely proportional to the square of the distance between them. Similarly, Hertz' prediction of electromagnetic waves is based on Maxwell's equations connecting the fundamental electric and magnetic quantities with each other, without aiming at an explanation of phenomena.

Mathematics has the tremendous creative power of evolving hidden consequences from assumptions about the observable universe and thus prompting predictions of previously unobserved phenomena. Not that the universe is under any obligation to conform to those predictions! But if these consequences are verified, then mathematics has led to new discoveries; and if they are not borne out by observations, then mathematics has necessitated a revision of the underlying general assumptions.

Mach's emphasis on functional connection raises two questions: What are functions? And what is it that functions connect?

Once the logarithm of any positive number has been explained as the exponent to which 10 must be raised to give that number, mathematicians define the logarithmic function by pairing, to every positive number, its logarithm. So, e.g., to 10, they pair 1; to 100, 2; to .1, —1. The *logarithmic function* (the result of this definition) is the class of all pairs of numbers thus obtained—a class including in particular the pairs (10,

1), (100, 2), (.1, —1). More generally, mathematicians say that a function has been defined if, to every number or to every number of a certain kind, a number somehow has been paired. The *function* (the result of this definition) is the class of all pairs of numbers thus obtained.

The traditional answer to the second question is: Functions connect variable quantities or, briefly, *variables*. A thorough analysis has revealed that the term variable is used in several totally discrepant meanings.[6] For instance, it is applied to the letters x and y in the mathematical statement:

(1) $x + \log y = \log y + x$ for any number x and any positive number y. Variables in this sense are used according to the following rules: (a) In the formula, according to the accompanying legend, the letters x and y may be replaced with numerals, say, —3 and 10, each such replacement yielding a specific formula, e.g., $-3 + \log 10 = \log 10 - 3$. (b) Without any change in the meaning of the statement (1), any two unlike letters may be used as variables; e.g., one may write:

$x + \log b = \log b + x$ for any number x and any positive number b; or x and y may even be interchanged as in:

$y + \log x = \log x + y$ for any positive number x and any number y.

The so-called variables that are functionally connected in physical laws, however, are of a completely different nature. Suppose, e.g., that, in the course of a process of a certain type, the work w in joules done by changing the pressure p in atmospheres of a gas

[6]K. Menger, "On Variables in Mathematics and in Natural Science," *British Journal for Philosophy of Science*, V., 1954, pp. 134-142. An elementary presentation of the concepts of variable, function, and fluent is contained in K. Menger, *Calculus. A Modern Approach:* (Boston: Ginn & Co., 1955). See particularly chapters IV and VII.

from its initial value p_0 is connected with p/p_0 by the logarithmic function—in a formula:

(2) $w = \overline{w} \log p/p_0$.

The contrast between p and w in (2) on the one hand, and x and y in (1), on the other, could hardly be greater than it is. (a) The letters p and w must not be replaced with just any two values of pressure and work. For instance, the work done by compressing the helium in one tank on Monday is not in general the logarithm of the pressure of oxygen in another tank on Wednesday. Nor is the formula (2) accompanied by any legend authorizing any replacements. (b) The meaning of (2) changes completely if, instead of w, say, the designation v of the gas volume is introduced in the formula or if p and w are interchanged. (In fact, the formulae thus obtained are, in general, false.) For, whereas x and y in (1) stand for *any* number and do not designate anything specific, p and w do; they designate pressure and work.

But exactly what is gas pressure in atmospheres? One way of defining it is by pairing, to each state S of a gas, the pressure in atmospheres $p(S)$ of the gas in the state S. *Gas pressure* (the result of this definition) is the class of all pairs $(S, p(S))$ thus obtained. Similarly, work in joules is a class of pairs $(S, w(S))$. The traditional formula (2) is nothing but an abbreviation of the following law:

(2_s) $w(S) = \overline{w} \log p(S)/p_0$ for any state S of any gas undergoing a process of the type under consideration.

In a strictly positivistic and operationalist spirit, one would take a step beyond the preceding definition of pressure in atmospheres and define *observed pressure in atmospheres* by pairing, to each act A of reading a pressure gauge calibrated in atmospheres, the number $p^*(A)$ that is read as the result of the act. The result of this definition is the class of all pairs

(A, $p^*(A)$)) thus obtained.[7] Similarly, one can define w^*, the *observed work in joules* as a class of pairs (B, $w^*(B)$) for any act B of measurement of a certain other kind. The formula (2) then is an abbreviation of the following more articulate formulation of the physical law: (2*) $w^*(B) = \log p^*(A)/p_0$ for any two acts of reading pressure gauges and work meters, respectively—acts simultaneously directed to the same gas sample undergoing a process of the type under consideration.

According to their definitions, p^*, w^*, p and w are classes of pairs, yet not only, as such, fundamentally different from the variables x and y in (1) but also basically different from functions. For while the latter e.g., the logarithmic functions, are purely logico-mathematical objects, the definitions of pressure, observed pressure, and the like include references to physical states or observations. The work, which is the logarithm of the pressure, bears to the logarithmic function a relation similar to the relation that a yard, which is equal to three feet, bears to the number three.[8] In order to distinguish objects of scientific studies such as p and p^* both from variables and from functions, the writings quoted in [6] and [8] have revived the term coined by Newton for time, distance traveled, gas pressure, work and the like—the term *fluents*. During the 18th century, not only did this term fall into almost

[7] An extreme positivist might question whether, beyond the *observed* pressure p^*, there is any (so to speak *objective*) pressure p. How, indeed, is the pressure $p(S)$ of a gas in the state S ascertained? It is derived (by more or less arbitrary averaging processes) from values of the observed pressure, namely, by somehow averaging numbers $p^*(A_1)$, $p^*(A_2)$, ... that result from acts A_1, A_2, ... of reading various guages—acts that various observers direct to the gas in the state S. In the case of helium in the corona of the sun or the terrestial atmosphere a million years ago, the pressure is ascertained by what Bridgman calls pencil-and-paper-operations.

[8] Detailed discussions of these and related points are contained

complete oblivion, being replaced by the word *variable,* but the underlying concept was confused with that of a variable in the logico-mathematical sense of a letter meant to be replaced with any numeral or, more generally, with the designation of any element of a certain class of objects.

The questions raised by Mach's emphasis on functional connections thus can be answered as follows: Functions are certain classes of pairs of numbers. The objects that, in physical laws, are connected by functions are fluents—classes of pairs, each of which results from pairing a number to a physical state or an act of observation.

The fifth and last philosophical point to be mentioned here (which is only briefly touched on at the end of the present book) is Mach's emphasis on the immediate sense data. He called them *elements* and used them as building blocks in constituting *complexes* such as the idea of the various things surrounding us as well as the idea of ourselves. He refused to look for objective causes of the phenomena or data.

"As several ideas imprinted on the senses are observed to accompany each other, they become marked by one name and so to be reputed as one thing. Thus, for example, a certain color, taste, smell, figure and consistence having been observed to go together, are accounted one distinct thing, signified by the name apple. Other collections of ideas constitute a stone, a tree, a book and the like sensible things." This passage, which precisely renders the first part of the contention,

in the book quoted in footnote 6, and in the following articles by the same author: K. Menger, "Mensuration and Other Mathematical Connections of Observable Material," in *Measurement: Definitions and Theories,* ed. C. W. Churchman and P. Ratoosh (New York: Wiley & Sons, 1959), pp. 97-128; and ————, "An Axiomatic Theory of Functions and Fluents," in *The Axiomatic Method* ed. Henkin, Suppes, and Tarski (Amsterdam: North-Holland Publishing Co., 1959), pp. 454-473.

might well have been written by Mach; actually, however, it is a quotation from the very first section of the
*Treatise concerning the Principles of Human Knowl-
edge* written by Bishop Berkeley in 1710. In developing
these thoughts further, however, Mach diverged altogether from Berkeley, who assumed extra-physical,
spiritual causes of sense data in which, because he was
a theologian, he was greatly interested. Mach shunned
a search for causes of data, in particular, for extra-
physical and spiritual causes, and confined himself to
the phenomena. His fear of being identified with
Berkeley's spiritualism probably explains why Mach
did not mention the development of what he called
the theory of elements and complexes in the first part
of Berkeley's classical *Treatise*—a book that can hardly
have escaped Mach's attention.[9] In contrast to Berkeley's idealistic metaphysics, Mach's philosophy of science has often been called *phenomenalism*.

*   *   *

Mach's operationalist, antimetaphysical, anticausal
views and his ideas on economy of thought pervade
his presentation of the science of mechanics. Yet an
introduction to the present book seems to call less for
a discussion of some moot points of Mach's philosophy[10] than for an appraisal of its author as a scientist.

Mach repeatedly emphasized that admiration for a
great physicist of the past should not keep historians
from discussing the master's limitations. In keeping

[9]On the occasion of the 200th anniversary of Berkeley's death,
K. R. Popper published "A Note on Berkeley as a Precursor of
Mach," *British Journal for Philosophy of Science*, IV, 1953, including an amazing list of quotations from lesser known writings
by Berkeley wherein also some of Mach's scientific ideas (connected with the concept of force, the critique of absolute space,
time, and motion and with the economy of thought) are clearly
anticipated.

[10]Debates have centered on the question whether references to
basically unobservable entities can be, and should be, eliminated

with this admonition, our respect for Mach should not prevent us from observing that, in the course of this century, three limitations of Mach himself as a scientist have become apparent.

Even though Mach is generally recognized as one of the principal precursors of the theory of relativity, he himself not only ignored that theory in the editions of the present book[1] that he published after the appearance of Einstein's first paper in 1905, but actually underlined his aloofness. Remarks to that effect are included in the Preface to his book *Principles of Physical Optics;*[2] and his son, Ludwig Mach, quoted the following passage from papers left by his father: "I do not consider the Newtonian principles as completed and perfect; yet, in my old age, I can accept the theory of relativity just as little as I can accept the existence of atoms and other such dogma."

This leads to the second point where the actual development of physics has completely diverged from Mach's views. Mach seems to have been unimpressed by Boltzmann's triumphs in the kinetic theory of gases as well as by Perrin's experiments on Brownian motion; at least, it appears from the quoted passage, he was not sufficiently impressed to attribute physical significance to the assumption of atoms. We can only speculate as to how he would react to science of the mid-twentieth century, which is completely dominated by atomic physics. Would he admit that the phenomena in a Wilson cloud chamber make a granular structure

only from the final statements of a theory or also from all intermediate steps and from the basic assumptions. Other discussions have dealt with the concept of the simplicity of a theory. On questions of this type there is not only disagreement between various scientists, but some of them, in the course of their lives, have changed their own opinions. For instance, Einstein mentions in his autobiography[3] that, while always admiring Mach as a physicist, at a later age he abandoned many of Mach's philosophical views which had impressed and influenced him greatly in his youth.

of matter and electricity almost visible? Would he perhaps, while admitting granular structure, question the precise equality of all grains of like type and point to galaxies, which also have a granular structure without all grains of like type (e.g., all red giants) having precisely equal masses or sizes? How would he react to the discovery of ever more types of elementary particles?

Strangely, Mach, who had such a sharp eye for the difficulties of atomism, did not seem to appreciate that the idea of matter continuously filling space leads to other conceptual problems and perhaps to even greater difficulties. One can imagine advances in the technique of experimentation and observation that will make statements about hitherto unobservable particles verifiable. Certain statements about the behavior of matter continuously filling space, however, are fundamentally unverifiable.

The writer of this Introduction strongly believes that some of the difficulties common to *all* theories of microphysical phenomena have a common cause—the lack of an adequate microgeometry. The current views on geometry are still essentially Euclid's. The various non-Euclidean geometries developed in the 19th century differ from Euclid's geometry only with regard to assumptions such as the parallel postulate which, when applied to nature, are reflected in properties of space *in the large,* wherefore the main domain of application of those geometries is astronomy or cosmology. *In the small,* all the 19th century non-Euclidean geometries are indistinguishable from Euclid's geometry. Any two points are assumed to have an exact distance from one another and (at least if they are not too far apart) to be joined by exactly one straightest line. And even the sum of the angles in any small non-Euclidean triangle is, according to the assumptions, indistinguishable from two right angles. Only rather recent work has abandoned the conception that the same laws are valid or even the same general notions are applicable in the very small that everyone knows from the geometry of larger regions. In fact, the very ideas of numerical distance and of points have been challenged.[11] And this is what microphysics probably needs: a microgeometry built on assumptions that are completely different from Euclid's; a theory of lumps rather than of points, and of distance distributions rather than of exact numerical distances; that is to say, given any two lumps and any interval of numbers, all that can be determined is the probability that the distance between the two lumps belongs to the said interval.

Mach's third limitation lies in his neglect of logic and of the critique of language—even the prelogical critique developed by his contemporary F. Mauthner[12] who, unfortunately, has been sadly neglected not only in his own day but by his successors as well. The way in which a man combines immediate sense data or elements in constituting complexes is profoundly influenced by others: mainly by those who, long since, taught him to speak; then by his teachers and educators; and finally by people with whom he exchanges information and views. He is, in other words, strongly influenced by language and all the wisdom and all the folly which, since time immemorial, his ancestors have stored in that means of communication.

Connected with Mach's alogical orientation are two

---

[11]See K. Menger, "Theory of Relativity and Geometry," in *Albert Einstein: Philosopher-Scientist*,[3] especially pp. 472-74; ———, "Statistical Metric," *Proceedings National Academy of Science*, XXVIII, 1942, p. 535 et seq.
———, "Probabilistic Geometry," *ibid*, XXXVII, 1951, p. 226.
B. Schweizer and A. Sklar, "Statistical Metric Spaces." *Pacific Journal of Mathematics*, X, 1960, p. 313 et seq.
[12]F. Mauthner, *Beiträge zu einer Kritik der Sprache* (Stuttgart: 1901-2), I, II, III.

other aspects of his work: a certain lack of precision in the formulation of some philosophical ideas and strictly empiristic views on mathematics. Mach considers even arithmetic as entirely based on experience. While in recent times empiristic views in mathematics have been decidedly underemphasized and, therefore, still seem to present unexplored potentialities, they unquestionably are one-sided and in definite need of a complementation by logic, by the logical analysis of language and, perhaps, by some of the ideas advanced by H. Poincaré.[13]

*   *   *

Of the many scientists who have been influenced by Mach, only Einstein and Bridgman have been mentioned on the preceding pages; of the kindred philosophers, only Avenarius and Poincaré. The 1920's witnessed the constitution of a group of philosophers of science who may be considered as direct successors and continuators of Mach—even in a geographical sense: they taught at the two schools with which Mach had been connected: the Universities of Vienna and Prague.[14] This group has become widely known under the name of *Wiener Kreis or Vienna Circle.**

In its beginnings, the Vienna Circle was altogether Mach-oriented. The philosopher M. Schlick emphasized[15] that the postulates of a theory are implicit definitions of its basic concepts—an idea which, in the

[13]H. Poincaré, *Science and Hypothesis* (New York: Dover Press, 1952). First French edition, 1912.

[14]Mach, who was born in Turas, Moravia, was professor at the University of Prague from 1867 to 1895, and at the University of Vienna from 1895 until his retirement in 1901. He died near Munich in 1916.

*The writer of this Introduction was a member of the Vienna Circle (remark by the editor of the 6th American Edition, 1960).

[15]M. Schlick, *Allgemeine Erkenntnislehre* (2d ed.; Berlin: Springer, 1923).

field of mechanics, goes back to Mach's use of New-
ton's Third Law as a definition of mass. In 1927, O.
Neurath founded an Ernst Mach Society in Vienna;
and the first meeting of the philosophers of science in
Prague stood, as P. Frank emphasized, in the sign of
Mach.

It was the mathematician H. Hahn who first directed the interest of the Vienna Circle to logic by his
detailed presentation of the ideas of B. Russell and
the *Principia Mathematica*.[16] In this way, L. Wittgenstein's *Tractatus*[17] became a topic—and in the years
1925-27 the dominant topic—of the discussions in the
Circle. The interest shifted from Mach's elements and
complexes to the ways of talking about observations
and of formulating laws; from the analysis of sensations to the analysis of language.[18]

Even in this respect, and in spite of his limited
interest in logic, Mach was a precursor. His antimetaphysical attitude, exemplified in his views on
absolute space and time, anticipated the statement that
only verifiable propositions are meaningful or, to put
it somewhat less dogmatically, the positivistic postulate
that extra-logical propositions should be verifiable.
Moreover, long before Wittgenstein and Carnap, Mach
had used, if only on the base of common sense and
unsystematically, the terms *Scheinprobleme* (pseudo—
or apparent problems) and *meaningless questions*. "Re-
fraining from answering questions that have been
recognized as meaningless," Mach wrote in the *Anal-*

[16]B. Russell, *Introduction to Mathematical Philosophy* (London: 1919); and A. N. Whitehead and B. Russell, *Principia
Mathematica* (2d ed.; Cambridge: Cambridge University Press,
1925-27), I, II, III.

[17]L. Wittgenstein, *Tractatus Logico-Philosophicus* (London:
1922).

[18]A more detailed account of this transition is given in R.
von Mises, *Positivism. A Study in Human Understanding* (Cambridge: Harvard University Press, 1951).

ysis of Sensations, "is by no means resignation. It is, in the presence of the enormous material that may be meaningfully investigated, the only reasonable attitude of scientific investigators." Later, on the basis of linguistic analysis, R. Carnap tried to eradicate pseudo-problems systematically. The philosophy of the Vienna Circle developed into *logical positivism.*

A description of Mach's influence would be incomplete without a mention of the impact of his ideas in the first decade of this century on philosophers in Russia. That development seems to have culminated in 1908, when an outline of philosophy with empirio-critical contributions by A. Bodgdanov and A. Luna-charsky was published in St. Petersburg. The man who was to shape the future of Russia, however, was opposed to Mach. Even in letters from his exile in Siberia in the first years of this century, Lenin had been critical of "Machism." In 1908, he went to London for extensive studies of the philosophical literature and, as their result, in the following year, he published a violent attack on Mach, Avenarius, Poincaré,. and related thinkers in a book *Materialism and Empirio-Criticism. Critical Notes on a Reactionary Philosophy.*[19] Lenin begins by emphasizing the strong similarity of Mach's constitution of the idea of objects with Berke-ley's. He then quotes several passages from Mach (e.g., references to the world of which we form pictures) which actually are at variance with Mach's own general views, whose presentation here and there lacks precision. Lenin further criticizes some statements that were also abandoned by logical positivists—two decades later, but on the basis of an analysis of language, whereas Lenin's critique is based on dialectical ma-

[19]An English translation with a Foreword by A. Deborin describing the book's background was published as volume 13 of Lenin's *Collected Works* (New York: International Publishers, 1927).

terialism. Lenin asserts, for instance, that "there is nothing in the world but matter in motion; and matter can not move save in space and time"; and "on the basis of relative conceptions we arrive at the absolute truth"—statements that logical positivists analyzing language find just about as unacceptable as the (it goes without saying, un-Machian) opposite statements that one used to hear from idealistic and relativistic philosophers: "there is nothing but ideas and minds wherein they exist; and ideas cannot exist without having a cause"; and "all truth is relative; there is no absolute truth." Lenin then proceeds to identify Mach's philosophy with Berkeley's idealistic and theological views, from which Mach definitely kept aloof; and he finally condemns the author of the *Science of Mechanics* because of the pietistic utterances of some of Mach's minor followers. In the Soviet Union, Lenin's views on Mach's philosophy became authoritative.

*   *   *

Mach's life, his son Ludwig once wrote, was dominated by a fundamental impulse toward personal clarity. He was a champion of mass education and progress and always a fearless advocate of truth as he saw it. "I see Mach's greatness," Einstein wrote,[3] "in his incorruptible skepticism and independence." In the prosperous but nationalistic and militaristic atmosphere of Central Europe in the late Victorian and Edwardian era, Mach seems to have felt a strong affinity to the English speaking world. Special ties connected him with Paul Carus and the Edward Hegeler family, who founded the Open Court Publishing Company in La Salle, Illinois. To them Mach dedicated his last book, *The Principles of Physical Optics,* in gratitude for their help in disseminating his ideas. In that dedication Mach expressed the wish that in discussions of his work mention should be made of their names.

The first English translation of Mach's *Science of Mechanics* was made and published by Open Court in 1893, and it is indeed appropriate that the first book to appear in a new series of quality paperbound books now being published by Open Court is this new edition of Mach's great work on *The Science of Mechanics*.

Karl Menger

Illinois Institute of Technology
Chicago, March, 1960

PREFACE TO THE FIRST GERMAN EDITION

The present volume is not a treatise upon the application of the principles of mechanics. Its aim is to clear up ideas, expose the real significance of the matter, and get rid of metaphysical obscurities. The little mathematics it contains is merely secondary to this purpose.

Mechanics will here be treated, not as a branch of mathematics, but as one of the physical sciences. If the reader's interest is in that side of the subject, if he is curious to know how the principles of mechanics have been ascertained, from what sources they take their origin, and how far they can be regarded as permanent acquisitions, he will find, I hope, in these pages some enlightenment. All this, the positive and physical essence of mechanics, which is of greatest and most general interest for a student of nature, is in existing treatises completely buried and concealed beneath a mass of technical considerations.

The gist and kernel of mechanical ideas has in almost every case grown up in the investigation of very simple and special cases of mechanical processes; and the analysis of the history of the discussions concerning these cases must ever remain the method at once the most effective and the most natural for laying this gist and kernel bare. Indeed, it is not too much to say that it is the only way in which a real comprehension of the general upshot of mechanics is to be attained.

I have framed my exposition of the subject agreeably to these views. It is perhaps a little long, but, on the other hand, I trust that it is clear. I have not in every case been able to avoid the use of the abbreviated and precise terminology of mathematics. To do so would have been to sacrifice matter to form; for the language of everyday life has not yet grown to be sufficiently accurate for the purposes of so exact a science as mechanics.

The elucidations which I here offer are, in part, substantially contained in my treatise, *Die Geschichte und die Wurzel des Satzes von der Erhaltung der Arbeit* (Prague, Calve, 1872). At a later date nearly the same views were expressed by KIRCHHOFF (*Vorlesungen über mathematische Physik: Mechanik,* Leipzig, 1874) and by HELMHOLTZ (*Die Thatsachen in der Wahrnehmung,* Berlin, 1879), and have since become commonplace enough. Still the matter, as I conceive it, does not seem to have been exhausted, and I cannot deem my exposition to be at all superfluous.

In my fundamental conception of the nature of science as Economy of Thought,—a view which I indicated both in the treatise above cited and in my pamphlet, *Die Gestalten der Flüssigkeit* (Prague, Calve, 1872), and which I somewhat more extensively developed in my academical memorial address, *Die ökonomische Natur der physikalischen Forschung* (Vienna, Gerold, 1882)—I no longer stand alone. I have been much gratified to find closely allied ideas developed, in an original manner, by DR. R. AVENARIUS (*Philosophie als Denken der Welt gemäss dem Princip des kleinsten Kraftmasses,* Leipzig, 1876). Regard for the true endeavor of philosophy, that of guiding into one common stream the many rills of knowledge, will not be found wanting in my work, although it takes a determined stand against the encroachments of speculative methods.

The questions here dealt with have occupied me since my earliest youth, when my interest for them was powerfully stimulated by the beautiful introductions of LAGRANGE to the chapters of his *Analytic Mechanics,* as well as by the lucid and lively tract of JOLLY, *Principien der Mechanik* (Stuttgart, 1852). If DÜHRING's estimable work, *Kritische Geschichte der Principien der Mechanik* (Berlin, 1873), did not particularly influence me, it was that at the time of its appearance, my ideas had been not only substantially worked out, but actually published. Nevertheless, the reader will, at least on the negative side, find many points of agreement between Dühring's criticisms and those here expressed.

The new apparatus for the illustration of the subject, here figured and described, were designed entirely by me and constructed by Mr. F. Hajek, the mechanician of the physical institute under my control.

In less immediate connection with the text stand the facsimile reproductions of old originals in my possession. The quaint and naïve traits of the great inquirers, which find in them their expression, have always exerted upon me a refreshing influence in my studies, and I have desired that my readers should share this pleasure with me.

E. MACH.

PRAGUE, May, 1883.

AUTHOR'S PREFACE TO THE TRANSLATION

Having read the proofs of the present translation of my work, *Die Mechanik in ihrer Entwickelung,* I can testify that the publishers have supplied an excellent, accurate, and faithful rendering of it, as their previous translations of essays of mine gave me every reason to expect. My thanks are due to all concerned, and especially to Mr. McCormack, whose intelligent care in the conduct of the translation has led to the discovery of many errors, heretofore overlooked. I may, thus, confidently hope, that the rise and growth of the ideas of the great inquirers, which it was my task to portray, will appear to my new public in distinct and sharp outlines.   E. Mach.

Prague, April 8th, 1893.

PREFACE TO THE SECOND EDITION

In consequence of the kind reception which this book has met with, a very large edition has been exhausted in less than five years. This circumstance and the treatises that have since then appeared of E. Wohlwill, H. Streintz, L. Lange, J. Epstein, F. A. Müller, J. Popper, G. Helm, M. Planck, F. Poske, and others are evidence of the gratifying fact that at the present day questions relating to the theory of cognition are pursued with interest, which twenty years ago scarcely anybody noticed.

As a thoroughgoing revision of my work did not yet seem to me to be expedient, I have restricted myself, so far as the text is concerned, to the correction of typographical errors, and have referred to the works that have appeared since its original publication, as far as possible, in a few appendices.

E. MACH.

PRAGUE, June, 1888.

PREFACE TO THE SEVENTH GERMAN EDITION

When, forty years ago, I first expressed the ideas explained in this book, they found small sympathy, and indeed were often contradicted. Only a few friends, especially Josef Popper, the engineer, were actively interested in these thoughts and encouraged the author. When, two years later, Kirchhoff published his well-known and often-quoted dictum, which even today is hardly correctly interpreted by the majority of physicists, people liked to think that the author of the present work had misunderstood Kirchhoff. I must decline with thanks this, as it were, prophetical misunderstanding as not corresponding either to my faculty of presentiment or to my powers of understanding.

However, the book has reached a seventh German edition, and by means of excellent English, French, Italian, and Russian translations has spread over almost all the world. Gradually some of those who work at this subject, like J. Cox, Hertz, Love, MacGregor, Maggi, H. von Seeliger, and others, gave voice to their agreement. For them, of course, only details in a book meant for a general introduction could be of interest.

In this subject, I could hardly avoid touching upon philosophical, historical, and epistemological questions; and by this the attention of various critics was aroused. I took special joy in the recognition which I found with the philosophers, R. Avenarius, J. Petzoldt, H. Cornelius, and, later, W. Schuppe. The apparently small concessions which philosophers of another tendency, like G. Heymans, P. Natorp, and Aloys Müller, have granted to my characterization of absolute space and absolute time as misconceptions suffice for me; indeed, I do not wish for anything more. I thank Messrs. L. Lange and J. Petzoldt not only for their agreement in certain details, but also for their active and fruitful collaboration. In an historical respect, the criticisms of Emil Wohlwill, whose death, I regret to say, has just been announced to me, were valuable and enlightening, especially on the period of Galileo's youthful work; further, critical remarks of P. Duhem and G. Vailati have also been valuable. I am very grateful to Mr. Philip E. B. Jourdain of Cambridge for his critical notes that unfortunately, for the most part, came too late for inclusion in *this* edition, which was already nearly finished. P. Duhem, O. Hölder, G. Vailati, and P. Volkmann have taken part in the epistemological discussions with vigor, and their remarks have been helpful to me.

At the end of the last century my disquisitions on mechanics fared well as a rule; it may have been felt that the empirico-critical side of this science was the most neglected. But now the Kantian traditions have gained power once more, and again we have the demand for an *a priori* foundation of mechanics. Now, I am indeed of the opinion that all that can be known *a priori* of an empirical domain must become evident to mere logical circumspection only after frequent surveys of this domain, but I do not believe that investigations like those of G. Hamel* do any harm to the subject. Both sides of mechanics, the empirical and the logical side, require investigation. I think that this is expressed clearly enough in my book, although my work is for good reasons turned especially to the empirical side.

I myself—seventy-four years old, and struck down by a grave malady—shall not cause any more revolutions. But I hope for important progress from a young mathematician, Dr. Hugo Dingler, who, judging from his publications,† has proved that he has attained to a free and unprejudiced survey of *both* sides of science.

This edition will be found somewhat more homogeneous than the former ones. Many an ancient dispute which today interests nobody any more is left out and many new things are added. The character of the book has remained the same. With respect to the monstrous conceptions of absolute space and absolute time I can retract nothing. Here I have only shown more clearly than hitherto that Newton indeed spoke much about these things, but throughout made no serious application of them. His fifth corollary‡ contains the only practically usable (probably approximate) *inertial system.*

* "Über Raum, Zeit, und Kraft als apriorische Formen der Mechanik." *Jahresber. der deutschen Mathematiker-Vereinigung,* xviii, 1909; "Über die Grundlagen der Mechanik," *Math. Ann.,* lxvi, 1908.

† *Grenzen und Ziele der Wissenschaft,* 1910; *Die Grundlagen der angewandten Geometrie,* 1911.

‡ *Principia,* 1687, p. 19.

ERNST MACH.

VIENNA, February 5th, 1912.

TABLE OF CONTENTS

CHAPTER I

THE DEVELOPMENT OF THE PRINCIPLES OF STATICS

CHAPTER II

THE DEVELOPMENT OF THE PRINCIPLES OF DYNAMICS

CHAPTER III

THE EXTENDED APPLICATION OF THE PRINCIPLES

OF MECHANICS AND THE DEDUCTIVE

DEVELOPMENT OF THE SCIENCE

CHAPTER IV

THE FORMAL DEVELOPMENT OF MECHANICS

CHAPTER V

THE RELATIONS OF MECHANICS TO OTHER DEPARTMENTS OF KNOWLEDGE

Ernst Mach

INTRODUCTION

1. THAT branch of physics which is at once the oldest and the simplest and which is therefore treated as introductory to other departments of this science, is concerned with the motions and equilibrium of masses. It bears the name of mechanics.

2. The history of the development of mechanics is quite indispensable to a full comprehension of the science in its present condition. It also affords a simple and instructive example of the processes by which natural science generally is developed.

An *instinctive,* irreflective knowledge of the processes of nature will doubtless always precede the scientific, conscious apprehension, or *investigation,* of phenomena. The former is the outcome of the relation in which the processes of nature stand to the satisfaction of our wants. The acquisition of the most elementary truth does not devolve upon the individual alone: it is pre-effected in the development of the race.

In point of fact, it is necessary to make a distinction between mechanical experience and mechanical science, in the sense in which the latter term is at present employed. Mechanical experiences are, unquestionably, very old. If we carefully examine the ancient Egyptian and Assyrian monuments, we shall find there pictorial representations of many kinds of implements and mechanical contrivances; but accounts of the scientific knowledge of these peoples are either totally lacking, or point conclusively to a very inferior grade of attainment. By the side of highly ingenious appliances, we behold the crudest and roughest expedients employed—as the use of sleds, for instance, for the transportation of enormous blocks of stone. All bears an instinctive, unperfected, accidental character.

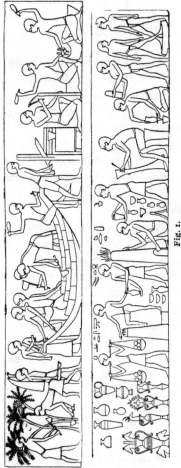

Fig. 1.

So, too, prehistoric graves contain implements whose construction and employment imply no little skill and much mechanical experience. Thus, long before theory was dreamed of, implements, machines, mechanical experiences, and mechanical knowledge were abundant.

3. The idea often suggests itself that perhaps the incomplete accounts we possess have led us to underrate the science of the ancient world. Passages occur in ancient authors which seem to indicate a pro

founder knowledge than we are wont to ascribe to those nations. Take, for instance, the following passage from Vitruvius, *De Architectura*, Lib. V, Cap. III, 6:

"The voice is a flowing breath, made sensible to the organ of hearing by the movements it produces in the air. It is propagated in infinite numbers of circular zones, exactly as when a stone is thrown into a pool of standing water countless circular undulations are generated therein, which, increasing as they recede from the center, spread out over a great distance, unless the narrowness of the locality or some obstacle prevent their reaching their termination; for the first line of waves, when impeded by obstructions, throw by their backward swell the succeeding circular lines of waves into confusion. Conformably to the very same law, the voice also generates circular motions; but with this distinction, that in water the circles remaining upon the surface, are propagated horizontally only, while the voice is propagated both horizontally and vertically."

Does not this sound like the imperfect exposition of a popular author, drawn from more accurate disquisitions now lost? In what a strange light should we ourselves appear, centuries hence, if our popular literature, which by reason of its quantity is less easily destructible, should alone outlive the productions of science? This too favorable view, however, is very rudely shaken by the multitude of other passages containing such crude and patent errors as cannot be conceived to exist in any high stage of scientific culture.

Recent research has contributed greatly to our knowledge of the scientific literature of antiquity, and our opinion of the achievements of the ancient world in science has been correspondingly increased. Schiaparelli has done much to place the work of the Greeks in astronomy in its right light, and Govi has disclosed many precious treasures in his edition of the *Optics* of Ptolemy. The view that the Greeks were especially neglectful of experiment can no longer be maintained unqualifiedly. The most ancient experiments are doubtless those of the Pythagoreans, who employed a monochord with movable bridge for determining the lengths of strings emitting harmonic notes. Anaxagoras's demonstration of the corporeality of the air by means of closed inflated tubes, and that of Empedocles by means of a vessel having its orifice inverted in water (Aristotle, *Physics*) are both primitive experiments. Ptolemy instituted systematic experiments on the refraction of light, while his observations in physiological optics are still full of interest today. Aristotle (*Meteorology*) describes phenomena that go to explain the rainbow. The absurd stories which tend to arouse our mistrust, like that of Pythagoras and the anvil which emitted harmonic notes when struck by hammers of different weights, probably sprang from the fanciful brains of ignorant reporters. Pliny abounds in such vagaries. But they are not, as a matter of fact, a whit more incorrect or nonsensical than the stories of Newton's falling apple and of Watts's tea-kettle. The situation is, moreover, rendered quite intelligible when we consider the difficulties and the expense attending the production of ancient books and their consequent limited circulation. The conditions here involved are concisely discussed by J. Mueller in his paper, "Ueber das Ex-

periment in den physikalischen Studien der Griechen,"
Naturwiss. Verein zu Innsbruck, XXIII, 1896-1897.

4. When, where, and in what manner the development of science actually began, is at this day difficult
historically to determine. It appears reasonable to
assume, however, that the instinctive gathering of experiential facts preceded the scientific classification of
them. Traces of this process may still be detected in
the science of today; indeed, they are to be met with,
now and then, in ourselves. The experiments that
man heedlessly and instinctively makes in his strug-
gles to satisfy his wants, are just as thoughtlessly and
unconsciously applied. Here, for instance, belong the
primitive experiments concerning the application of
the lever in all its manifold forms. But the things
that are thus unthinkingly and instinctively discovered,
can never appear as peculiar, can never strike us as
surprising, and as a rule therefore will never supply an
impetus to further thought.

The transition from this stage to the classified,
scientific knowledge and apprehension of facts, first becomes possible on the rise of special classes and pro-
fessions who make the satisfaction of definite social
wants their lifelong vocation. A class of this sort occupies itself with particular kinds of natural processes.
The individuals of the class change; old members
drop out, and new ones come in. Thus arises a need
of imparting to those who are newly come in, the
stock of experience and knowledge already possessed;
a need of acquainting them with the conditions of the
attainment of a definite end so that the result may be
determined beforehand. The communication of knowledge is thus the first occasion that compels distinct re-

flection, as everybody can still observe in himself. Further, that which the old members of a guild mechanically pursue, strikes a new member as unusual and strange, and thus an impulse is given to fresh reflection and investigation.

When we wish to bring to the knowledge of a person any phenomena or processes of nature, we have the choice of two methods: we may allow the person to observe matters for himself, when instruction comes to an end; or, we may describe to him the phenomena in some way, so as to save him the trouble of personally making anew each experiment. Description, however, is only possible of events that constantly recur, or of events that are made up of component parts that constantly recur. That only can be described, and conceptually represented, which is uniform and conformable to law; for description presupposes the employment of names by which to designate its elements; and names can acquire meanings only when applied to elements that constantly reappear.

5. In the infinite variety of nature many ordinary events occur; while others appear uncommon, perplexing, astonishing, or even contradictory to the ordinary run of things. As long as this is the case we do not possess a well-settled and unitary conception of nature. Thence is imposed the task of everywhere seeking out in the natural phenomena those elements that are the same, and that amid all multiplicity are ever present. By this means, on the one hand, the most economical and briefest description and communication are rendered possible; and on the other, when once a person has acquired the skill of recognizing these permanent elements throughout the great-

est range and variety of phenomena, of seeing them in the same, this ability leads to a *comprehensive, compact, consistent,* and *facile conception of the facts.* When once we have reached the point where we are everywhere able to detect the *same* few simple elements, combining in the ordinary manner, then they appear to us as things that are familiar; we are no longer surprised, there is nothing new or strange to us in the phenomena, we feel at home with them, they no longer perplex us, they are *explained.* It is a process of adaptation of thoughts to facts with which we are here concerned.

6. Economy of communication and of apprehension is of the very essence of science. Herein lies its pacificatory, its enlightening, its refining element. Herein, too, we possess an unerring guide to the historical origin of science. In the beginning, all economy had in immediate view the satisfaction simply of bodily wants. With the artisan, and still more so with the investigator, the concisest and simplest possible knowledge of a given province of natural phenomena—a knowledge that is attained with the least intellectual expenditure—naturally becomes in itself an economical aim; but though it was at first a means to an end, when the mental motives connected therewith are once developed and demand their satisfaction, all thought of its original purpose, the personal need, disappears.

To find, then, what remains unaltered in the phenomena of nature, to discover the elements thereof and the mode of their interconnection and interdependence—this is the business of physical science. It endeavors, by comprehensive and thorough description, to make the waiting for new experiences unnecessary; it seeks to save us the trouble of experimentation, by making use, for example, of the known interdependence of phenomena, according to which, if one kind of event occurs, we may be sure beforehand that a certain other event will occur. Even in the description itself labor may be saved, by discovering methods of describing the greatest possible number of different objects at once and in the concisest manner. All this will be made clearer by the examination of points of detail than can be done by a general discussion. It is fitting, however, to prepare the way, at this stage, for the most important points of outlook which in the course of our work we shall have occasion to occupy.

7. We now propose to enter more minutely into the subject of our inquiries, and, at the same time, without making the history of mechanics the chief topic of discussion, to consider its historical development so far as this is requisite to an understanding of the present state of mechanical science, and so far as it does not conflict with the unity of treatment of our main subject. Apart from the consideration that we cannot afford to neglect the great incentives that it is in our power to derive from the foremost intellects of all epochs, incentives which taken as a whole are more fruitful than the greatest men of the present day are able to offer, there is no grander, no more intellectually elevating spectacle than that of the utterances of the fundamental investigators in their gigantic power. Possessed as yet of no methods, for these were first created by their labors, and are only rendered comprehensible to us by their performances, they grapple with and subjugate the object of their inquiry, and imprint upon it the forms of conceptual thought. They that know the entire course of the development of science,

will, as a matter of course, judge more freely and more correctly of the significance of any present scientific movement than they, who, limited in their views to the age in which their own lives have been spent, contemplate merely the momentary trend that the course of intellectual events takes at the present moment.

CHAPTER I.

THE DEVELOPMENT OF THE PRINCIPLES OF STATICS.

I.

THE PRINCIPLE OF THE LEVER.

1. The earliest investigations concerning mechanics of which we have any account, the investigations of the ancient Greeks, related to statics, or to the doctrine of equilibrium. After the conquest of Constantinople by the Turks in 1453, when a fresh impulse was imparted to the thought of the Occident by the ancient writings that the fugitive Greeks brought with them, it was likewise investigations in statics, principally evoked by the works of Archimedes, that occupied the foremost investigators of the period.

Researches in mechanics by the Greeks do not begin until a late date, and in no wise keep pace with the rapid advancement of the race in the domain of mathematics, notably in geometry. Reports of mechanical inventions, so far as they relate to the early inquirers, are extremely meager. Archytas, a distinguished citizen of Tarentum (*circa* 400 B. C.), famed as a geometer and for his employment with the problem of the duplication of the cube, devised mechanical instruments for the description of various curves. As an astronomer he taught that the earth was spherical and that it rotated upon its axis once a day. As a mechanician he founded the theory of pulleys. He is also said to have applied geometry to mechanics in a treatise on this latter science, but all information as to details is lack-

ing. We are told, though, by Aulus Gellius (X. 12)
that Archytas constructed an automaton consisting of
a flying dove of wood and presumably operated by
compressed air, which created a great sensation. It is,
in fact, characteristic of the early history of mechanics
that attention should have been first directed to its
practical advantages and to the construction of automata designed to excite wonder in ignorant people.

Even in the days of Ctesibius (285-247 B. C.) and
Hero (first century A. D.) the situation had not ma-
terially changed. So, too, during the decadence of
civilization in the Middle Ages, the same tendency asserts itself. The artificial automata and clocks of this
period, the construction of which popular fancy ascribed to the machinations of the Devil, are well
known. It was hoped, by imitating life outwardly, to
apprehend it from its inward side also. In intimate
connection with the resultant misconception of life
stands also the singular belief in the possibility of
perpetual motion. Only gradually and slowly, and in
indistinct forms, did the genuine problems of mechanics loom up before the minds of inquirers. Aristotle's
tract, *Mechanical Problems* (German trans. by Poselger,
Hanover, 1881) is characteristic in this regard. Aristotle is quite adept in detecting and in formulating
problems; he perceived the principle of the parallelogram of motions, and was on the verge of discovering centrifugal force; but in the actual solution of
problems he was infelicitous. The entire tract par-
takes more of the character of a dialectic than of a
scientific treatise and rests content with enunciating
the "apories," or contradictions, involved in the problems. But the tract upon the whole very well illus-

trates the intellectual situation that is characteristic
of the beginnings of scientific investigation.

"If a thing take place whereof the cause be not
apparent, even though it be in accordance with nature, it appears wonderful. . . . Such are the instances
in which small things overcome great things, small
weights heavy weights, and incidentally all the problems that go by the name of 'mechanical.' . . . To
the apories (contradictions) of this character belong
those that appertain to the lever. For it appears contrary to reason that a large weight should be set in
motion by a small force, particularly when that weight
is in addition combined with a larger weight. A weight
that cannot be moved without the aid of a lever can be
moved easily with that of a lever added. The pri-
mordial cause of all this is inherent in the nature of
the circle, which is as one should naturally expect:
for it is not contrary to reason that something wonderful should proceed out of something else that is
wonderful. The combination of contradictory properties, however, into a single unitary product is the
most wonderful of all things. Now, the circle is actually composed of just such contradictory properties.
For it is generated by a thing that is in motion and
by a thing that is stationary at a fixed point."

In a subsequent passage of the same treatise there
is a very dim presentiment of the principle of virtual
velocities.

Considerations of the kind here adduced give evidence of a capacity for detecting and enunciating problems, but are far from conducting the investigator to
their solution.

2. ARCHIMEDES of Syracuse (287-212 B. C.) left

behind him a number of writings, of which several have come down to us in complete form. We will first employ ourselves a moment with his treatise *De Æquiponderantibus,* which contains propositions respecting the lever and the center of gravity.

In this treatise Archimedes starts from the following assumptions, which he regards as self-evident:

a. Magnitudes of equal weight acting at equal distances (from their point of support) are in equilibrium.

b. Magnitudes of equal weight acting at unequal distances (from their point of support) are not in equilibrium, but the one acting at the greater distance sinks.

From these assumptions he deduces the following proposition:

"Commensurable magnitudes are in equilibrium when they are inversely proportional to their distances (from the point of support)."

It would seem as if analysis could hardly go behind these assumptions. However, when we carefully look into the matter, this is not the case.

Imagine (Fig. 2) a bar, the weight of which is neglected. The bar rests on a fulcrum. At equal distances from the fulcrum we append two equal weights. That the two weights, thus circumstanced, are in equilibrium, is the assumption from which Archimedes starts. We might suppose that this was self-evident entirely apart from any experience, according to the so-called principle of sufficient reason; that in view of the symmetry of the entire arrangement there is no reason why rotation should occur in the one direction

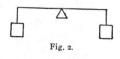

Fig. 2.

rather than in the other. But we forget, in this, that a great multitude of negative and positive experiences is implicitly contained in our assumption; the negative, for instance, that dissimilar colors of the lever-arms, the position of the spectator, an occurrence in the vicinity, and the like, exercise no influence; the positive, on the other hand, (as it appears in the second assumption,) that not only the weights but also their distances from the supporting point are decisive factors in the disturbance of equilibrium, that they also are circumstances determinative of motion. By the aid of these experiences we do indeed perceive that rest (no motion) is the only motion which can be uniquely* determined, or defined, by the determinative conditions of the case.†

Now we are entitled to regard our knowledge of the decisive conditions of any phenomenon as sufficient only in the event that such conditions determine the phenomenon precisely and uniquely. Assuming the fact of experience referred to, that the weights and their distances *alone* are decisive, the first proposition of Archimedes really possesses a high degree of evidence and is eminently qualified to be made the foundation of further investigations. If the spectator place himself in the plane of symmetry of the arrangement in question, the first proposition manifests itself, moreover, as a highly imperative *instinctive* perception—a result determined by the symmetry of our own body. The pursuit of propositions of this character is, furthermore, an excellent means of accustoming ourselves

* So as to leave only a single possibility open.

† If, for example, we were to assume that the weight at the right descended, then rotation in the opposite direction also would be determined by the spectator, whose person exerts no influence on the phenomenon, taking up his position on the opposite side.

in thought to the precision that nature reveals in her processes.

3. We will now reproduce in general outlines the train of thought by which Archimedes endeavors to reduce the general proposition of the lever to the particular and apparently self-evident case. The two equal weights 1 suspended at *a* and *b* (Fig. 3) are, if the bar *ab* be free to rotate about its middle point *c*, in equilibrium. If the whole be suspended by a cord at *c*, the cord, leaving out of account the weight of the bar, will have to support the weight 2. The equal weights at the extremities of the bar accordingly replace the double weight at the center.

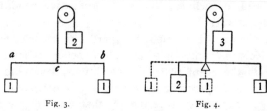

Fig. 3.   Fig. 4.

On a lever (Fig. 4), the arms of which are in the proportion of 1 to 2, weights are suspended in the proportion of 2 to 1. The weight 2 we imagine replaced by two weights 1, attached on either side at a distance 1 from the point of suspension. Now again we have complete symmetry about the point of suspension, and consequently equilibrium.

On the lever-arms 3 and 4 (Fig. 5) are suspended

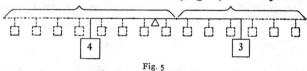

Fig. 5

the weights 4 and 3. The lever-arm 3 is prolonged

the distance 4, the arm 4 is prolonged the distance 3, and the weights 4 and 3 are replaced respectively by 4 and 3 pairs of symmetrically attached weights $\frac{1}{2}$, in the manner indicated in the figure. Now again we have perfect symmetry. The preceding reasoning, which we have here developed with specific figures, is easily generalized.

4. It is of interest to see how Archimedes's mode of view was modified by Stevinus and GALILEO.

Galileo imagines (Fig. 6) a heavy horizontal prism, homogeneous in material composition, suspended by its extremities from a homogeneous bar of the same length. The bar is provided at its middle point with a suspensory attachment. In this case equilibrium will obtain; this we perceive at once. But in this case is contained every other case. Galileo shows this in the following manner. Let us suppose the whole length of the bar or the prism to be $2(m + n)$. Cut the prism in two, in such a manner that one portion shall have the length $2m$ and the other the length $2n$. We can effect this without disturbing the equilibrium by previously fastening to the bar by threads, close to the point of proposed section, the inside extremities of the two portions. We may then remove all the threads, if the two portions of the prism be antecedently attached to the bar by their centers. Since the whole length of the bar is $2(m + n)$, the length of each half

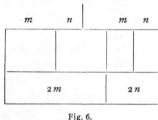

Fig. 6.

is $m + n$. The distance of the point of suspension of
the right-hand portion of the prism from the point of
suspension of the bar is therefore m, and that of the
left-hand portion n. The experience that we have
here to deal with the weight, and not with the form,
of the bodies, is easily made. It is thus manifest, that
equilibrium will still subsist if *any* weight of the magnitude $2m$ be suspended at the distance n on the one
side and *any* weight of the magnitude $2n$ be suspended
at the distance m on the other. The instinctive elements
of our perception of this phenomenon are even more
prominently displayed in this form of the deduction
than in that of Archimedes.

We may discover, moreover, in this beautiful presentation, a remnant of the ponderousness which was
particularly characteristic of the investigators of antiquity.

How a modern physicist conceived the same problem, may be learned from the following presentation of
LAGRANGE. He says: Imagine a horizontal homogeneous prism suspended at its center. Let this
prism (Fig. 7) be conceived divided into two prisms
of the lengths $2m$ and $2n$. If now we consider the

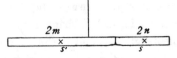

Fig. 7.

centers of gravity of these two parts, at which we may
imagine weights to act proportional to $2m$ and $2n$, the
two centers thus considered will have the distances n
and m from the point of support. This concise disposal of the problem is only possible to the practised
mathematical perception.

5. The object that Archimedes and his successors sought to accomplish in the considerations we have here presented, consists in the endeavor to reduce the more complicated case of the lever to the simpler and apparently self-evident case, to *discern* the simpler in the more complicated, or *vice versa*. In fact, we regard a phenomenon as explained, when we discover in it known simpler phenomena.

But surprising as the achievement of Archimedes and his successors may at the first glance appear to us, doubts as to its correctness, on further reflection, nevertheless spring up. From the mere assumption of the equilibrium of equal weights at equal distances is derived the inverse proportionality of weight and lever-arm! How is that possible? If we were unable philosophically and *a priori* to excogitate the simple fact of the dependence of equilibrium on weight and distance, but were obliged to go for *that* result to experience, in how much less a degree shall we be able, by speculative methods, to discover the *form* of this dependence, the proportionality!

As a matter of fact, the assumption that the equilibrium-disturbing effect of a weight P at the distance L from the axis of rotation is measured by the product $P \cdot L$ (the so-called statical moment), is more or less covertly or tacitly introduced by Archimedes and all his successors.

First it is obvious that if the arrangement is absolutely symmetrical in every respect, equilibrium obtains on the assumption of *any* form of dependence whatever of the disturbing factor on L, or, generally, on the assumption $P \cdot f(L)$; and that consequently the *particular* form of dependence PL cannot possibly be inferred from the equilibrium. The fallacy of the deduction must accordingly be sought in the transformation to which the arrangement is subjected. Archimedes makes the action of two equal weights to be the same under all circumstances as that of the combined weights acting at the middle point of their line of junction. But, seeing that he both knows and assumes that distance from the fulcrum is determinative, this procedure is by the premises unpermissible, if the two weights are situated at unequal distances from the fulcrum. If a weight situated at a distance from the fulcrum is divided into two equal parts, and these parts are moved in contrary directions symmetrically to their original point of support; one of the equal weights will be carried as near to the fulcrum as the other weight is carried from it. If it is assumed that the action remains constant during such procedure, then the particular form of dependence of the moment on L is implicitly determined by what has been done, inasmuch as the result is only possible provided the form be PL, or be *proportional* to L. But in such an event all further deduction is superfluous. The entire

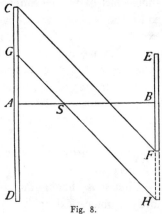

Fig. 8.

deduction contains the proposition to be demonstrated, by assumption if not explicitly.

6. HUYGENS, indeed, reprehends this method, and gives a different deduction, in which he believes he

has avoided the error. If in the presentation of Lagrange we imagine the two portions into which the
prism is divided turned ninety degrees about two
vertical axes passing through the centers of gravity
s, s' of the prism-portions (see Fig. 9), and it be
shown that under these circumstances equilibrium
still continues to subsist, we shall obtain the Huygenian deduction. Abridged and simplified, it is
as follows: In a rigid weightless plane (Fig. 9)

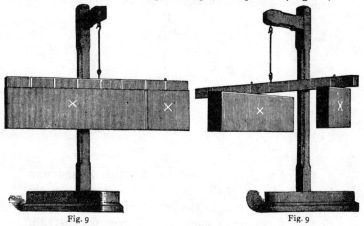

Fig. 9   Fig. 9

through the point *S* we draw a straight line, on which
we cut off on the one side the length 1 and on the other
the length 2, at *A* and *B* respectively. On the extremities, at right angles to this straight line, we place,
with the centers as points of contact, the heavy, thin,
homogeneous prisms *CD* and *EF,* of the lengths and
weights 4 and 2. Drawing the straight line *HSG*
(where $AG = \frac{1}{2}AC$) and, parallel to it, the line *CF,*
and translating the prism-portion *CG* by parallel displacement to *FH,* everything about the axis *GH* is
symmetrical and equilibrium obtains. But equilibrium

also obtains for the axis *AB;* obtains consequently for
every axis through *S*, and therefore also for that at
right angles to *AB:* wherewith the new case of the
lever is given.

Apparently, nothing else is assumed here than that
equal weights *p,p* (Fig. 10) in the same plane and at
equal distances, *l,l* from an axis *AA'* (in this plane)
equilibrate one another. If we place ourselves in the
plane passing through *AA'* perpendicularly to *l,l*, say

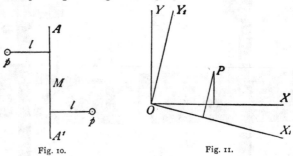

Fig. 10.   Fig. 11.

at the point *M*, and look now towards *A* and now
towards *A'*, we shall accord to this proposition the
same evidentness as to the first Archimedean proposition. The relation of things is, moreover, not altered if
we institute with the weights parallel displacements
with respect to the axis, as Huygens in fact does.

The error first arises in the inference: if equilibrium obtains for two axes of the plane, it also obtains
for every other axis passing through the point of intersection of the first two. This inference (if it is not to
be regarded as a purely instinctive one) can be drawn
only upon the condition that disturbant effects are as-
scribed to the weights *proportional* to their distances
from the axis. But in this is contained the very kernel
of the doctrine of the lever and the center of gravity.

Let the heavy points of a plane be referred to a system of rectangular coördinates (Fig. 11). The coördinates of the center of gravity of a system of masses $m\ m'\ m''$ ... having the coördinates $x\ x'\ x''$ ... $y\ y'\ y''$ ... are, as we know,

$$\xi = \frac{\Sigma mx}{\Sigma m}, \quad \eta = \frac{\Sigma my}{\Sigma m}.$$

If we turn the system through the angle α, the new coördinates of the masses will be

$$x_1 = x\ \cos\alpha - y\ \sin\alpha, \quad y_1 = y\ \cos\alpha + x\ \sin\alpha$$

and consequently the coördinates of the center of gravity

$$\xi_1 = \frac{\Sigma m\ (x\cos\alpha - y\sin\alpha)}{\Sigma m} = \cos\alpha\ \frac{\Sigma mx}{\Sigma m} - \sin\alpha\ \frac{\Sigma my}{\Sigma m}$$
$$= \xi\ \cos\alpha - \eta\ \sin\alpha$$

and, similarly,

$$\eta_1 = \eta\ \cos\alpha + \xi\ \sin\alpha.$$

We accordingly obtain the coördinates of the new center of gravity, by simply transforming the coördinates of the first center of the new axes. The center of gravity remains therefore *the self-same* point. If we select the center of gravity itself as origin, then $\Sigma m\ x = \Sigma m\ y = 0$. On turning the system of axes, this relation continues to subsist. If, accordingly, equilibrium obtains for two axes of a plane that are perpendicular to each other, it also obtains, and obtains then only, for every other axis through their point of intersection. Hence, if equilibrium obtains for any two axes of a plane, it will also obtain for every other axis of the plane that passes through the point of intersection of the two.

These conclusions, however, are not deducible if the coördinates of the center of gravity are determined by some other, more general equation, say

$$\xi = \frac{mf(x) + m'f(x') + m''f(x'') + \cdots}{m + m' + m'' + \cdots}$$

The Huygenian mode of inference, therefore, is inadmissible and contains the very same error that we remarked in the case of Archimedes.

Archimedes's self-deception in this, his endeavor to reduce the complicated case of the lever to the case instinctively grasped, probably consisted in his unconscious employment of studies previously made on the center of gravity *by the help of the very proposition he sought to prove.* It is characteristic, that he will not trust on his own authority, nor perhaps even on that of others, the easily presented observation of the import of the product $P \cdot L$, but searches after a further verification of it.

Now as a matter of fact we shall not, at least at this stage of our progress, attain to any comprehension whatever of the lever unless we directly *discern* in the phenomena the product $P \cdot L$ as the factor decisive of the disturbance of equilibrium. In so far as Archimedes, in his Grecian mania for demonstration, strives to get around this, his deduction is defective. But regarding the import of $P \cdot L$ as given, the Archimedean deductions still retain considerable value, in so far as the modes of conception of different cases are supported the one on the other, in so far as it is shown that one simple case contains all others, in so far as the same mode of conception is established for all cases. Imagine (Fig. 12) a homogeneous prism, whose axis is AB, supported at its center C. To give a graphical representation of the sum of the products of the weights and distances, the sum decisive of the disturbance of equilibrium, let us erect upon the elements of the axis, which are proportional to the elements of the weight, the distances as ordinates; the ordinates to the right

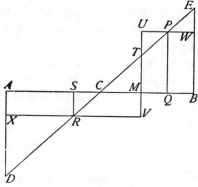

Fig. 12.

of C (as positive) being drawn upwards, and to the left of C (as negative) downwards. The sum of the areas of the two triangles, $ACD + CBE = 0$, illustrates here the subsistence of equilibrium. If we divide the prism into two parts at M, we may substitute the rectangle $MUWB$ for $MTEB$, and the rectangle $MVXA$ for $TMCAD$, where $TP = \frac{1}{2}TE$ and $TR = \frac{1}{2}TD$, and the prism-sections MB, MA are to be regarded as placed at right angles to AB by rotation about Q and S.

In the direction here indicated the Archimedean view certainly remained a serviceable one even after no one longer entertained any doubt of the significance of the product $P \cdot L$, and after opinion on this point had been established historically and by abundant verification.

Experiments are never absolutely exact, but they at least may lead the inquiring mind to *conjecture* that the key which will clear up the connection of all the

facts is contained in the exact metrical expression
PL. On no other hypothesis are the deductions of
Archimedes, Galileo, and the rest intelligible. The
required transformations, extensions, and compres-
sions of the prisms may now be carried out with perfect certainty.

A knife edge may be introduced at any point under a prism suspended from its center without dis-

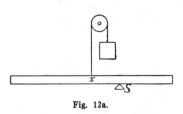

Fig. 12a.

turbing the equilib-
rium (see Fig. 12a),
and several such ar-
rangements may be
rigidly combined to-
gether so as to form
apparently new cases
of equilibrium. The
conversion and disintegration of the case of equilibrium into several other cases (Galileo) is possible
only by taking into account the value of *PL.* I cannot agree with O. Hölder who upholds the correctness of the Archimedean deductions against my criticisms in his essay *Denken und Anschauung in der Geometrie*, although I am greatly pleased with the extent
of our agreement as to the nature of the exact sciences and their foundations. It would seem as if
Archimedes (*De æquiponderantibus*) regarded it as a
general experience that two equal weights may under
all circumstances be replaced by one equal to their
combined weight at the center (Theorem 5, Corollary 2). In such an event, his long deduction (Theorem 6) would be necessary, for the reason sought follows immediately (see pp. 19, 20). Archimedes's
mode of expression is not in favor of this view.

Nevertheless, a theorem of this kind cannot be regarded as *a priori* evident; and the views advanced on pp. 19-20 appear to me to be still uncontroverted.

I must here draw my readers' attention to a beautiful paper by G. Vailati[1], in which the author takes sides with Hölder against my criticism of Archimedes' deduction of the law of the lever but at the same time he criticizes Hölder somewhat. I believe that everyone may read Vailati's exposition with profit and, by comparison with what I have said will be in a position himself to form a judgment upon the points at issue. Vailati shows that Archimedes derives the law of the lever on the basis of general experiences about the center of gravity. I have never disputed the view that such a process is possible and permissible and even very fruitful at a certain stage of investigation, and further, is perhaps the only correct one at that stage. On the contrary, by the manner in which I have exposed the derivations of Stevinus and Galileo, which were made after the example of Archimedes, I have expressly recognized this. But the aim of my whole book is to convince the reader that we cannot make up properties of nature with the help of self-evident suppositions, but that these suppositions must be taken from experience. I would have been false to this aim if I had not striven to disturb the impression that the general law of the lever could be deduced from the equilibrium of equal weights on equal arms. I had, then, to show where the experience, that already contains the general law of the lever, is introduced. Now this experience lies in the supposition emphasized on p. 19, and in the same way it lies in every one of the

[1] La dimonstrazione del principio delle leva data "la Archimede," *Bolletino di bibliografia storia delle scienze matematiche*, May and June 1904.

general and undoubtedly correct theorems on the center of gravity brought forward by Vailati. Now, because the fact that the value of a load is proportional to the arms of the lever is not directly and in the simplest way apparent in such an experience, but is found in an artificial and roundabout way, and is then offered to the surprised reader, the modern reader has to object to the deduction of Archimedes. This deduction from simple and almost self-evident theorems may charm a mathematician who either has an affection for Euclid's method, or who puts himself into the appropriate mood. But in other moods and with other aims we have all the reason in the world to distinguish in value between getting from one proposition to another, and conviction, and between surprise and insight. If the reader has derived some usefulness out of this discussion, I am not very particular about maintaining every word I have used.

7. The manner in which the laws of the lever, as handed down to us from Archimedes in their original simple form, were further generalized and treated by modern physicists, is very interesting and instructive. LEONARDO DA VINCI (1452-1519), the famous painter

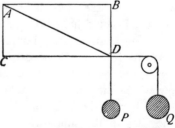

Fig. 13.

and investigator, appears to have been the first to recognize the importance of the general notion of the so-called statical moments. In the manuscripts he has left us, several passages are found from which this clearly appears. He says, for example: We have a bar *AD* (Fig. 13) free to rotate about *A,* and suspended from the bar, a weight *P*, and suspended from a string which passes over a pulley, a second weight *Q*. What must be the ratio of the forces that equilibrium may obtain? The lever-arm for the weight *P* is not *AD,* but the "potential" lever *AB*. The lever-arm for the weight *Q* is not AD, but the "potential" lever *AC*. The method by which Leonardo arrived at this view is difficult to discover. But it is clear that he recognized the essential circumstances by which the effect of the weight is determined.

Considerations similar to those of Leonardo da Vinci are also found in the writings of GUIDO UBALDI.

8. We will now endeavor to obtain some idea of the way in which the notion of statical moment, by which as we know, is understood the product of a force into the perpendicular let fall from the axis of rotation upon the line of direction of the force, could have been arrived at—although the way that really led to this idea is not now fully ascertainable. That equilibrium exists (Fig. 14) if we lay a cord, subjected at both sides to equal tensions, over a pulley, is perceived without difficulty. We shall always find a plane of symmetry for the apparatus—the plane which stands at right angles

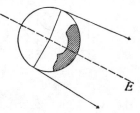

Fig. 14.

to the plane of the cord and bisects (*EE*) the angle made by its two parts. The motion that might be supposed possible cannot in this case be precisely determined or defined by any rule whatsoever: no motion will therefore take place. If we note, now, further, that the material of which the pulley is made is essential only to the extent of determining the form of motion of the points of application of the strings, we shall likewise readily perceive that almost any portion of the pulley may be removed without disturbing the equilibrium of the machine. The rigid radii that lead out to the tangential points of the string, are alone essential. We see, thus, that the rigid radii (or the perpendiculars on the linear directions of the strings) play here a part similar to the lever-arms in the lever of Archimedes.

Let us examine a so-called wheel and axle (Fig. 15) of wheel-radius 2 and axle-radius 1, provided respectively with the cord-hung loads 1 and 2; an apparatus which corresponds in every respect to the lever of Archimedes. If now we place about the axle, in any manner we may choose, a second cord, which we subject at each side to the tension of a weight 2, the second cord will not disturb the equilibrium. It is plain, however, that we are also permitted to regard

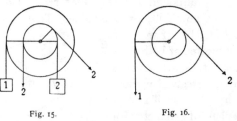

Fig. 15.   Fig. 16.

the two pulls marked in Fig. 16 as being in equilibrium, by leaving the two others, as mutually destructive, out of account. But we arrive in so doing, dismissing from consideration all unessential features, at the perception that not only the pulls exerted by the weights but also the perpendiculars let fall from the axis on the lines of the pulls, are conditions determinative of motion. The decisive factors are, then, the products of the weights into the respective perpendiculars let fall from the axis on the directions of the pulls; in other words, the so-called statical moments.

9. What we have so far considered, is the development of our knowledge of the principle of the lever. Quite independently of this was developed the knowledge of the principle of the inclined plane. It is not necessary, however, for the comprehension of the machines, to search after a new principle beyond that of the lever; for the latter is sufficient by itself. Galileo, for example, explains the inclined plane from the lever in the following manner. We have before us (Fig. 17) an inclined plane, on which rests the weight Q, held in equilibrium by the weight P. Galileo, now, points out the

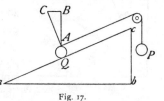

Fig. 17.

fact, that it is not requisite that Q should lie directly upon the inclined plane, but that the essential point is rather the form, or character, of the motion of Q. We may, consequently, conceive the weight attached to the bar AC, perpendicular to the inclined plane, and rotatable about C. If then we institute a very slight rotation about the point C, the weight will move in the element of an arc coincident with the inclined plane. That the path assumes a curve if the motion be continued is of no consequence here, since this further movement does not in the case of equilib-

rium take place, and the movement of the instant alone
is decisive. Reverting, however, to the observation
of Leonardo da Vinci, mentioned before, we readily
perceive the validity of the theorem $Q \cdot CB = P \cdot CA$
or $Q/P = CA/CB = ca/cb,$ and thus reach the law of
equilibrium on the inclined plane. Once we have
reached the principle of the lever, we may, then, easily
apply that principle to the comprehension of the other
machines.

II.

THE PRINCIPLE OF THE INCLINED PLANE.

1. STEVINUS, or STEVIN, (1548-1620) who investigated the mechanical properties of the inclined
plane, did so in an eminently original manner. If a

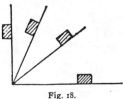

Fig. 18.

weight lie (Fig. 18) on a horizontal table, we perceive at
once, since the pressure is directly perpendicular to the
plane of the table, by the principle of symmetry, that equilibrium subsists. On a vertical
wall, on the other hand, a weight is not at all obstructed
in its motion of descent. The inclined plane accordingly will present an intermediate case between these
two limiting suppositions. Equilibrium will not exist
of itself, as it does on the horizontal support, but it
will be maintained by a weight less than that necessary to preserve it on the vertical wall. The ascertainment of the statical law that obtains in this case, caused
the earlier inquirers considerable difficulty.

Stevinus's manner of procedure is in substance as
follows. He imagines a triangular prism with horizontally placed edges, a cross-section of which *ABC* is

represented in Fig. 19. For the sake of illustration
we will say that $AB = 2BC$; also that AC is horizontal. Over this prism Stevinus lays an endless string
on which 14 balls of equal weight are strung and tied
at equal distances apart. We can advantageously replace this string by an endless uniform chain or cord.
The chain will either be in equilibrium or it will not.
If we assume the latter to be the case, the chain, since

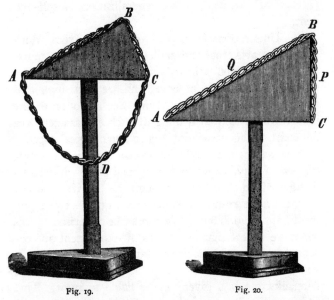

Fig. 19.   Fig. 20.

the conditions of the event are not altered by its motion, must, when once actually in motion, continue to
move forever, that is, it must present a perpetual motion, which Stevinus deems absurd. Consequently only
the first case is conceivable. The chain remains in equilibrium. The symmetrical portion ADC may, therefore, without disturbing the equilibrium, be removed.

The portion AB of the chain consequently balances the portion BC. Hence: on inclined planes of equal heights equal weights act in the inverse proportion of the lengths of the planes.

In the cross-section of the prism in Fig. 20 let us imagine AC horizontal, BC vertical, and $AB = 2BC;$ furthermore, the chain-weights Q and P on AB and BC proportional to the lengths; it will follow then that $Q/P = AB/BC = 2$. The generalization is self-evident.

2. Unquestionably in the assumption from which Stevinus starts, that the endless chain does not move, there is contained primarily only a *purely instinctive* cognition. He feels at once, and we with him, that we have never observed anything like a motion of the kind referred to, that a thing of such a character does not exist. This conviction has so much logical cogency that we accept the conclusion drawn from it respecting the law of equilibrium on the inclined plane without the thought of an objection, although the law if presented as the simple result of experiment, or otherwise put, would appear dubious. We cannot be surprised at this when we reflect that all results of experiment are obscured by adventitious circumstances (as friction, etc.), and that every conjecture as to the conditions which are determinative in a given case is liable to error. That Stevinus ascribes to instinctive knowledge of this sort a higher authority than to simple, manifest, direct observation might excite in us astonishment if we did not ourselves possess the same inclination. The question accordingly forces itself upon us: Whence does this higher authority come? If we remember that scientific demonstration, and scientific criticism generally can

only have sprung from the consciousness of the individual fallibility of investigators, the explanation is not far to seek. We feel clearly, that we ourselves have contributed *nothing* to the creation of instinctive knowledge, that we have added to it nothing arbitrarily, but that it exists in absolute independence of our participation. Our mistrust of our own subjective interpretation of the facts observed, is thus dissipated.

Stevinus's deduction is one of the rarest fossil indications that we possess in the primitive history of mechanics, and throws a wonderful light on the process of the formation of science generally, on its rise from instinctive knowledge. We will recall to mind that Archimedes pursued exactly the same tendency as Stevinus, only with much less good fortune. In later times, also, instinctive knowledge is very frequently taken as the starting-point of investigations. Every experimenter can daily observe in his own person the guidance that instinctive knowledge furnishes him. If he succeeds in abstractly formulating what is contained in it, he will as a rule have made an important advance in science.

Stevinus's procedure is no error. If an error were contained in it, we should all share it. Indeed, it is perfectly certain, that the union of the strongest instinct with the greatest power of abstract formulation alone constitutes the great natural inquirer. This by no means compels us, however, to create a new mysticism out of the instinctive in science and to regard this factor as infallible. That it is not infallible, we very easily discover. Even instinctive knowledge of so great logical force as the principle of symmetry employed by Archimedes, may lead us astray. Many of my readers will recall, perhaps, the intellectual shock they experienced when they heard for the first time that a magnetic needle lying in the magnetic meridian is deflected in a definite direction away from the meridian by a wire conducting a current being carried along in a parallel direction above it. The instinctive is just as fallible as the distinctly conscious. Its only value is in provinces with which we are very familiar.

Let us rather put to ourselves, in preference to pursuing mystical speculations on this subject, the question: How does instinctive knowledge originate and what are its contents? Everything which we observe in nature imprints itself *uncomprehended* and *unanalyzed* in our percepts and ideas, which, then, in their turn, mimic the processes of nature in their most general and most striking features. In these accumulated experiences we possess a treasure-store which is ever close at hand and of which only the smallest portion is embodied in clear articulate thought. The circumstance that it is far easier to resort to these experiences than it is to nature herself, and that they are, notwithstanding this, free, in the sense indicated, from all subjectivity, invests them with a high value. It is a peculiar property of instinctive knowledge that it is predominantly of a negative nature. We cannot so well say what must happen as we can what cannot happen, since the latter alone stands in glaring contrast to the obscure mass of experience in us in which single characters are not distinguished.

Still, great as the importance of instinctive knowledge may be, for discovery, we must not, from our point of view, rest content with the recognition of its authority. We must inquire, on the contrary: Under what conditions could the instinctive knowledge in question have originated? We then ordinarily find that the very principle to establish which we had recourse to instinctive knowledge, constitutes in its turn the fundamental condition of the origin of that knowledge. And this is quite obvious and natural. Our instinctive knowledge leads us to the principle which explains that knowledge itself, and which is in its turn also corroborated by the existence of that knowledge, which is a separate fact by itself. This we will find on close examination is the state of things in Stevinus's case.

3. The reasoning of Stevinus impresses us as so highly ingenious because the result at which he arrives apparently contains more than the assumption from which he starts. While on the one hand, to avoid contradictions, we are constrained to let the result pass, on the other, an incentive remains which impels us to seek further insight. If Stevinus had distinctly set forth the entire fact in all its aspects, as Galileo subsequently did, his reasoning would no longer strike us as ingenious; but we should have obtained a much more satisfactory and clear insight into the matter. In the endless chain which does not glide upon the prism, is contained, in fact, everything. We might say, the chain does not glide because no sinking of heavy bodies takes place here. This would not be accurate, however, for when the chain moves many of its links really do descend, while others rise in their place. We must say, therefore, more accurately, the chain does not glide because for every body that could possibly descend an equally heavy body would have to ascend equally high, or a body of double the weight half the height, and so on. This fact was familiar to Stevinus, who likewise presented it, in his theory of pulleys; but he was plainly too distrustful of himself to lay

down the law, without additional support, as also valid for the inclined plane. But if such a law did not exist universally, our instinctive knowledge respecting the endless chain could never have originated. With this our minds are completely enlightened.—The fact that Stevinus did not go as far as this in his reasoning and rested content with bringing his (indirectly discovered) ideas into agreement with his instinctive thought, need not further disturb us.

Stevinus's procedure may be looked at from still another point of view. If it is a fact, for our mechanical instinct, that a heavy endless chain will not rotate, then the individual simple cases of equilibrium on an inclined plane which Stevinus devised and which are readily controlled quantitatively, may be regarded as so many special experiences. For it is not essential that the experiments should have been actually carried out, if the result is beyond question of doubt. As a matter of fact, Stevinus experiments mentally. Stevinus's result could actually have been deduced from the corresponding physical experiments, with friction reduced to a minimum. In an analogous manner, Archimedes's considerations with respect to the lever might be conceived after the fashion of Galileo's procedure. If the various mental experiments had been executed physically, the linear dependence of the static moment on the distance of the weight from the axis could be deduced with perfect rigor. We shall have still many instances to adduce, among the foremost inquirers in the domain of mechanics of this tentative adaptation of special quantitative conceptions to general instinctive impressions. The same phenomena are presented in other domains also. I may be permitted to refer in this connection to the expositions which I have given in my *Principles of Heat,* page 151. It may be said that the most significant and most important advances in science have been made in this manner. The habit which great inquirers have of bringing their single conceptions into agreement with the general conception or ideal of an entire province of phenomena, their constant consideration of the whole in their treatment of parts, may be characterized as a genuinely philosophical procedure. A truly philosophical treatment of any special science will always consist in bringing the results into relationship and harmony with the established knowledge of the whole. The fanciful extravagances of philosophy, as well as infelicitous and abortive special theories, will be eliminated in this manner.

It will be worth while to review again the points of agreement and difference in the mental procedures of Stevinus and Archimedes. Stevinus reached the very general view that a mobile, heavy, endless chain of any form stays at rest. He is able to deduce from this general view, without difficulty, special cases, which are quantitatively easily controlled. The case from which Archimedes starts, on the other hand, is the most special conceivable. He cannot possibly deduce from his special case in an unassailable manner the behavior which may be expected under more general conditions. If he apparently succeeds in so doing, the reason is that he already knows the result which he is seeking, whilst Stevinus, although he too doubtless knows, approximately at least, what he is in search of, nevertheless could have found it directly by his manner of procedure, even if he had not known it. When the static relationship is rediscovered in such a manner it has a higher value than the result of a metrical experiment would have, which always deviates somewhat from the theoretical truth. The deviation increases with the disturbing circumstances, as with friction, and decreases with the diminution of these difficulties. The exact static relationship is reached by idealization and disregard of these disturbing elements. It appears in the Archimedean and Stevinian procedures as an *hypothesis* without which the individual facts of experience would at once become involved in logical contradictions. Not until we have possessed this hypothesis can we by operating with the exact concepts reconstruct the facts and acquire a scientific and logical mastery of them. The lever **and the** inclined plane are self-created ideal objects of mechanics. These objects alone completely satisfy the logical demands which we make of them; the physical lever satisfies these conditions only in measure in which it approaches the ideal lever. The natural inquirer strives to *adapt* his *ideals* to reality.

The service which Stevinus renders himself and his readers, consists, therefore, in contrasting and comparing knowledge that is instinctive with knowledge that is clear, in bringing the two into connection and accord with one another, and in supporting one upon the other. The strengthening of perception which Stevinus acquired by this procedure, we learn from the fact that a picture of the endless chain upon the prism graces as vignette, the title-page of his work *Hypomnemata Mathematica* (Leyden, 1605)* with the inscription "Wonder en is gheen wonder." As a fact,

* The title given is that of Willebrord Snell's Latin translation (1608) of Simon Stevin's *Wisconstige Gedachtenissen*, Leyden, 1605—*Trans.*

every enlightening progress made in science is accompanied with a certain feeling of disillusionment. We
discover that that which appeared wonderful to us is
no more wonderful than other things which we know

Fig. 21.

instinctively and regard as self-evident; nay, that the
contrary would be much more wonderful; that everywhere the same fact expresses itself. Our puzzle turns
out then to be a puzzle no more; it vanishes into
nothingness, and takes its place among the shadows
of history.

4. After he had arrived at the principle of the inclined plane, it was easy for Stevinus to apply that principle to the other machines and thereby to explain their action. He makes, for example, the following application.

We have, let us suppose, an inclined plane (Fig. 22) and on it a load Q. We pass a string over the pulley A at the summit and imagine the load Q held in equilibrium by the load P.

Stevinus, now, proceeds by a method similar to that later taken by Galileo. He remarks that it is not necessary that the load Q should lie directly on the inclined plane. Provided

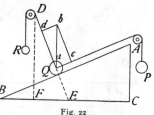

Fig. 22

only the form of the machine's motion be preserved, the proportion between force and load will in all cases remain the same. We may therefore equally well conceive the load Q to be attached to a properly weighted string passing over a pulley D: which string is normal to the inclined plane. If we carry out this alteration, we shall have a so-called funicular machine. We now perceive that we can ascertain very easily the portion of weight with which the body on the inclined plane tends downwards. We have only to draw a vertical line and to cut off on it a portion ab corresponding to the load Q. Then drawing on aA the perpendicular bc, we have $P/Q = AC/AB = ac/ab$. Therefore, ac represents the tension of the string aA. Nothing prevents us, now, from making the two strings change functions and from imagining the load Q to lie on the dotted inclined plane EDF. Similarly, here, we obtain *ad* for the tension of the second string. In this manner, accordingly, Stevinus indirectly arrives at a knowledge of the statical relations of the funicular machine and of the so-called parallelogram of forces; at first, of course, only for the particular case of strings (or forces) *ac, ad* at right angles to one another.

Subsequently, indeed, Stevinus employs the principle of the composition and resolution of forces in a more general form; yet the method by which he

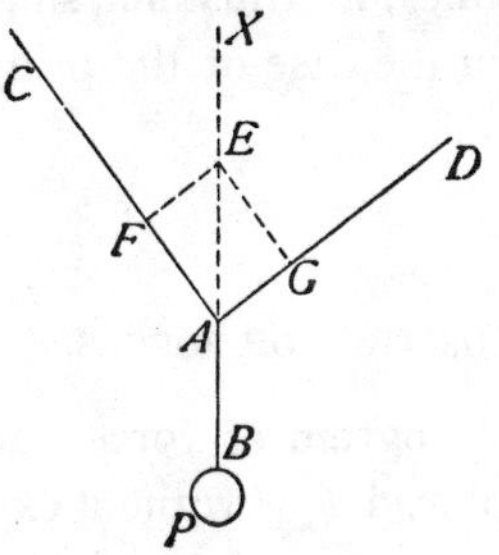

Fig. 23. 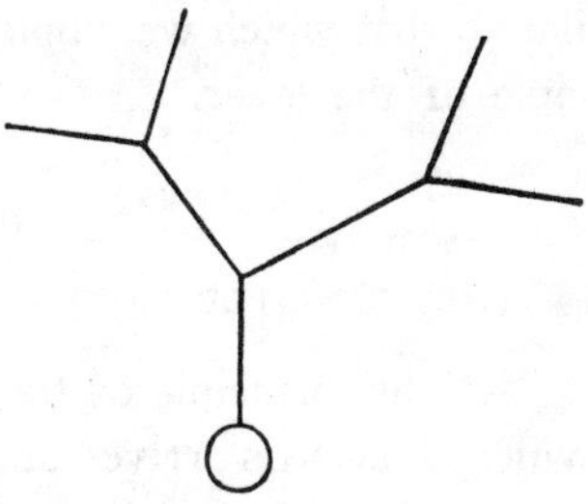   Fig. 24.

reached the principle, is not very clear, or at least is not obvious. He remarks, for example, that if we have three strings *AB, AC, AD*, stretched at any given angles, and the weight *P* is suspended from the first, the tensions may be determined in the following manner. We produce (Fig. 23) *AB* to *X* and cut off on it a portion *AE*. Drawing from the point *E, EF* parallel to *AD* and *EG* parallel to *AC,* the tensions of *AB, AC, AD* are respectively proportional to *AE, AF, AG*.

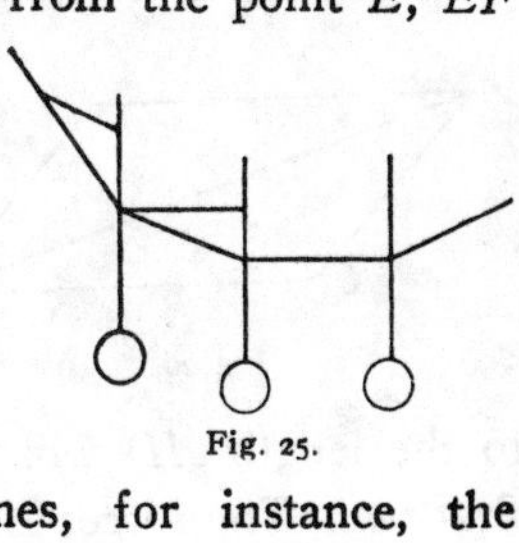

Fig. 25.

With the assistance of this principle of construction Stevinus solves highly complicated problems. He determines, for instance, the tensions of a system of ramifying strings like that illustrated in Fig. 24; in doing which of course he starts from the given tension of the vertical string.

The relations of the tensions of a funicular polygon are likewise ascertained by construction, in the manner indicated in Fig. 25.

We may therefore, by means of the principle of the inclined plane, seek to elucidate the conditions of operation of the other simple machines, in a manner similar to that which we employed in the case of the principle of the lever.

III.

THE PRINCIPLE OF THE COMPOSITION OF FORCES.

1. The principle of the parallelogram of forces, at which STEVINUS arrived and employed, (yet without expressly formulating it) consists, as we know, of the following truth. If a body A (Fig. 26) is acted upon by two forces whose directions coincide with the lines AB and AC, and whose magnitudes are proportional to the lengths AB and AC, these two forces produce the same effect as a single force, which acts in the direction of the diagonal AD of the parallelogram $ABCD$ and is proportional to that diagonal. For instance, if on the

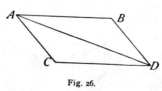

Fig. 26.

strings AB, AC weights exactly proportional to the lengths AB, AC be supposed to act, a single weight acting on the string AD exactly proportional to the length AD will produce the same effect as the first two. The forces AB and AC are called the components, the force *AD* the resultant. It is furthermore obvious, that conversely, a *single* force is replaceable by two or several other forces.

2. We shall now endeavor, in connection with the investigations of Stevinus, to give ourselves some idea of the manner in which the general proposition of the parallelogram of forces might have been reached. The relation, discovered by Stevinus, that exists between two mutually perpendicular forces and a third force that equilibrates them, we shall assume as (indirectly) given. We suppose now (Fig. 27) that there act on three strings *OX*, *OY*, *OZ*, pulls which balance each other. Let us endeavor to determine the nature of these pulls. Each pull holds the two remaining ones in equilibrium. The pull *OY* we will replace (following Stevinus's principle) by two new rectangular pulls, one in the direction *Ou* (the prolongation of *OX*), and one at right angles thereto in the direction *Ov*. And let us similarly resolve the pull *OZ* in the directions *Ou* and *Ow*. The sum of the pulls in the direction *Ou*, then, must balance the pull *OX*, and the two pulls in the directions *Ov* and *Ow* must mutually destroy each other. Taking the two latter as equal and opposite, and representing them by *Om* and *On*, we determine coincidently with the operation the com-

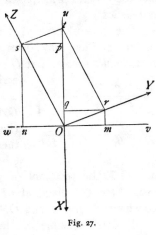

Fig. 27.

ponents Op and Oq parallel to Ou, as well also as the pulls Or, Os. Now the sum $Op + Oq$ is equal and opposite to the pull in the direction of OX; and if we draw st parallel to OY, or rt parallel to OZ, either line will cut off the portion $Ot = Op + Oq$: with which result the general principle of the parallelogram of forces is reached.

The general case of composition may be deduced in still another way from the special composition of rectangular forces. Let OA and OB be the two forces

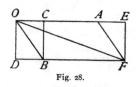

Fig. 28.

acting at O. For OB substitute a force OC acting parallel to OA and a force OD acting at right angles to OA. There then act for OA and OB the two forces $OE = OA + OC$ and OD, the resultant of which forces OF is at the same time the diagonal of the parallelogram $OAFB$ constructed on OA and OB as sides.

3. The principle of the parallelogram of forces, when reached by the method of Stevinus, presents itself as an *indirect* discovery. It is exhibited, as a consequence and as the condition of known facts. We perceive, however, merely that it *does* exist, not, as yet *why* it exists; that is, we cannot reduce it (as in dynamics) to still simpler propositions. In statics, indeed, the principle was not fully admitted until the time of Varignon, when dynamics, which leads directly to the principle, was already so far advanced that its adoption therefrom presented no difficulties. The principle of the parallelogram of forces was first clearly enunciated by NEWTON in his *Principles of Natural Philosophy*. In the same year, VARIGNON, independently of Newton, also enunciated the principle, in a work sub-

mitted to the Paris Academy (but not published until after its author's death), and, by the aid of a geometrical theorem, made extended practical application
of it.*

The geometrical theorem referred to is this. If we
consider (Fig. 29) a parallelogram the sides of which
are p and q, and the diagonal is r, and from any point m
in the plane of the par-
allelogram we draw per-
pendiculars on these
three straight lines,
which perpendiculars
we will designate as
u, v, w, then $p \cdot w +$
$q \cdot v = r \cdot w$. This is
easily proved by draw-
ing straight lines from

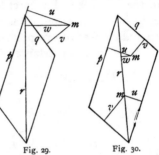

Fig. 29.   Fig. 30.

m to the extremities of the diagonal and of the sides of
the parallelogram, and considering the areas of the
triangles thus formed, which are equal to the halves
of the products specified. If the point m be taken
within the parallelogram and perpendiculars then be
drawn, the theorem passes into the form $p \cdot u - q \cdot v$
$= r \cdot w$. Finally, if m be taken on the diagonal and
perpendiculars again be drawn, we shall get, since the
perpendicular let fall on the diagonal is now zero,
$p \cdot u - q \cdot v = 0$ or $p \cdot u = q \cdot v$.

With the assistance of the observation that forces
are proportional to the motions produced by them in
equal intervals of time, Varignon easily advances from
the composition of motions to the composition of forces.
Forces, which acting at a point are represented in

* In the same year, 1687, Father Bernard Lami published a little appen-
dix to his *Traité de méchanique*, developing the same principle.—*Trans.*

magnitude and direction by the sides of a parallelogram, are replaceable by a single force, similarly represented by the diagonal of that parallelogram.

If now, in the parallelogram considered, p and q represent the concurrent forces (the components) and r the force competent to take their place (the resultant), then the products pu, qv, rw are called the moments of these forces with respect to the point m. If the point m lie in the direction of the resultant, the two moments pu and qv are with respect to it equal to each other.

4. With the assistance of this principle Varignon is now in a position to treat the machines in a much simpler manner than were his predecessors. Let us consider, for example, (Fig. 31) a rigid body capable of rotation about an axis passing through O. Perpendicular to the axis we conceive a plane, and select therein two

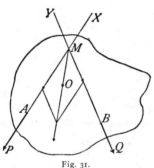

Fig. 31.

points A, B, on which two forces P and Q in the plane are supposed to act. We recognize with Varignon that the effect of the forces is not altered if their points of application be displaced along their line of action, since all points in the same direction are rigidly connected with one another and each one presses and pulls the other. We may, accordingly, suppose P applied at any point in the direction AX, and Q at any point in the direction BY, consequently also at their point of intersection M. With the forces as displaced to M, then, we construct a parallelogram, and replace the forces by their resultant. We have now to do only with the effect of the latter. If it acts only on movable points, equilibrium will not obtain. If, however, the direction of its action passes through the axis, through the point O, which is not movable, no motion can take place and equilibrium will obtain. In the latter case O is a point on the resultant, and if we drop the perpendiculars u and v from O on the directions of the forces p, q, we shall have, in conformity with the theorem before mentioned, $p \cdot u = q \cdot v$. With this we have deduced the law of the lever from the principle of the parallelogram of forces.

Varignon explains in like manner a number of other cases of equilibrium by the equilibration of the resultant force by some obstacle or restraint. On the inclined plane, for example, equilibrium exists if the resultant is found to be at right angles to the plane. In fact, Varignon rests statics in its entirety on a dynamic foundation; to his mind, it is but a special case of dynamics. The more general dynamical case constantly hovers before him, and he restricts himself voluntarily in his investigation to the case of equilibrium. We are confronted here with a dynamical statics, such as was possible only after the researches of Galileo. Incidentally, it may be remarked, that the majority of the theorems and methods of presentation which make up the statics of modern elementary text-books is derived from Varignon.

5. As we have already seen, purely statical considerations also lead to the proposition of the parallelogram of forces. In special cases, in fact, the principle admits of being very easily verified. We recognize at once, for instance, that any number whatsoever of equal forces acting (by pull or pressure) in the same plane at a point, around which their successive lines make equal

angles, are in equilibrium. If, for example, (Fig. 32)

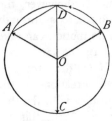

Fig. 32.

the three equal forces *OA*, *OB*, *OC* act on the point *O* at angles of 120°, each two of the forces holds the third in equilibrium. We see immediately that the resultant of *OA* and *OB* is equal and opposite *OC*. It is represented by *OD* and is at the same time the diagonal of the parallelogram *OADB*, which readily follows from the fact that the radius of a circle is also the side of the hexagon included by it.

6. If the concurrent forces act in the same or in opposite directions, the resultant is equal to the sum

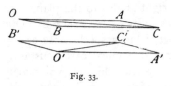

Fig. 33.

or the difference of the components. We recognize both cases without any difficulty as particular cases of the principle of the parallelogram of forces. If in the two drawings of Fig. 33 we imagine the angle *AOB* to be gradually reduced to the value 0°, and the angle *A' O' B'* increased to the value 180°, we shall perceive that *OC* passes into *OA* + *AC = OA + OB* and *O' C'* into *O' A' — A' C' = O' A' — O' B'*. The principle of the parallelogram of forces includes, accordingly, propositions which are generally made to precede it as independent theorems.

7. The principle of the parallelogram of forces, in the form in which it was set forth by Newton and Varignon, clearly discloses itself as a proposition derived from experience. A point acted on by two forces describes with accelerations proportional to the forces

two mutually independent motions. On this fact the parallelogram construction is based. DANIEL BER-NOULLI, however, was of opinion that the proposition of the parallelogram of forces was a *geometrical* truth, independent of physical experience. And he attempted to furnish for it a geometrical demonstration, the chief features of which we shall here take into consideration, as the Bernoullian view has not, even at the present day, entirely disappeared.

If two equal forces, at right angles to each other (Fig. 34), act on a point, there can be no doubt, according to Bernoulli, that the line of bisection of the angle (conformably to the principle of symmetry) is the direction of the resultant *r*. To determine geometrically also the magnitude of the resultant, each of the forces *p* is decomposed into two equal forces

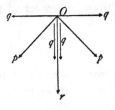

Fig 34.

q, parallel and perpendicular to *r*. The relation in respect of magnitude thus produced between *p* and *q* is consequently the same as that between *r* and *p*. We have, accordingly:

$$p = \mu\, q \text{ and } r = \mu\, p; \text{ whence } r = \mu^2 q.$$

Since, however, the forces *q* acting at right angles to *r* destroy each other, while those parallel to *r* constitute the resultant, it further follows that

$$r = 2q; \text{ hence } \mu = \sqrt{2}, \text{ and } r = \sqrt{2}\,.\,p.$$

The resultant, therefore, is represented also in respect of magnitude by the diagonal of the square constructed on *p* as side.

Similarly, the magnitude may be determined of the

resultant of unequal rectangular components. Here,
however, nothing is known before-
hand concerning the direction of
the resultant *r*. If we decompose
the components *p*, *q* (Fig. 35),
parallel and perpendicular to the
yet undetermined direction *r*, into
the forces *u*, *s* and *v*, *t*, the new
forces will form with the components *p*, *q* the same angles that *p*,
q form with *r*. From which fact the following relations in respect of magnitude are determined:

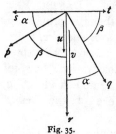
Fig. 35.

$$\frac{r}{p}=\frac{p}{u} \text{ and } \frac{r}{q}=\frac{q}{v}, \frac{r}{q}=\frac{p}{s} \text{ and } \frac{r}{p}=\frac{q}{t},$$

from which two latter equations follows $s = t = pq/r$.
On the other hand, however,

$$r = u + v = \frac{p^2}{r}+\frac{q^2}{r} \text{ or } r^2 = p^2+q^2.$$

The diagonal of the rectangle constructed on *p* and
q represents accordingly the magnitude of the resultant.

Therefore, for all rhombs, the *direction* of the resultant is determined; for all rectangles, the *magnitude;* and for squares both magnitude *and* direction.
Bernoulli then solves the problem of substituting for
two equal forces acting at one given angle, other equal,
equivalent forces acting at a different angle; and finally
by circumstantial considerations, not wholly exempt
from mathematical objections, but amended later by
Poisson, he arrives at the general principle.

8. Let us now examine the physical aspect of this
question. As a proposition derived from experience,
the principle of the parallelogram of forces was already

known to Bernoulli. What Bernoulli really does, therefore, is to simulate towards himself *a complete ignorance* of the proposition and then attempt to philosophize it abstractly out of the fewest possible assumptions. Such work is by no means devoid of meaning and purpose. On the contrary, we discover by such procedures, how few and how imperceptible the *experiences* are that suffice to supply a principle. Only we must not deceive ourselves, as Bernoulli did; we must keep before our minds *all* the assumptions, and should overlook no experience which we involuntarily employ. What are the assumptions, then, contained in Bernoulli's deduction?

9. Statics, primarily, is acquainted with force only as a pull or a pressure, that from whatever source it may come always admits of being replaced by the pull or the pressure of a weight. All forces thus may be regarded as quantities *of the same kind* and be measured by weights. Experience further instructs us, that the particular factor of a force, which is determinative of equilibrium or determinative of motion, is contained not only in the *magnitude* of the force but also in its *direction,* which is made known by the direction of the resulting motion, by the direction of a stretched cord, or in some like manner. We may ascribe magnitude indeed to other things given in physical experience, such as temperature, potential function, but not direction. The fact that both magnitude *and* direction are determinative in the efficiency of a force impressed on a point is an important though it may be an unobtrusive experience.

Granting, then, that the magnitude and direction of forces impressed on a point *alone* are decisive, it will be perceived that two equal and opposite forces, as they cannot *uniquely* and precisely determine *any* motion,

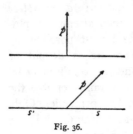

Fig. 36.

are in equilibrium. So, also, at
right angles to its direction, a
force *p* is unable uniquely to determine a motional effect. But
if a force *p* is inclined at an angle to another direction *s s'*
(Fig. 36), it *is* able to determine a motion in that direction.
Yet experience alone can in-
form us, that the motion is determined in the direction
of *s' s* and not in that of *s s';* that is to say, in the
direction of the side of the *acute* angle or in the direction of the *projection* of *p* on *s' s*.

Now this latter experience is made use of by Bernoulli at the very start. The *sense* of the resultant of
two equal forces acting at right angles to one another
is obtainable only on the ground of this experience.
From the principle of symmetry follows only, that the
resultant falls in the *plane* of the forces and coincides
with the line of bisection of the angle, but not that it
falls in the *acute* angle. But if we surrender this latter
determination, our whole proof is exploded before it
is begun.

10. If, now, we have reached the conviction that
our knowledge of the effect of the *direction* of a force is
obtainable solely from experience, still less then shall
we believe it in our power to ascertain by another way
the *form* of this effect. It is utterly beyond our power
to divine, that a force *p* acts in a direction *s* that makes
with its own direction the angle *α*, exactly as a force
p cos *α* in the direction *s*; a statement equivalent to the
proposition of the parallelogram of forces. Nor was
it in Bernoulli's power to do this. Nevertheless, he
makes use, scarcely perceptible it is true, of expe-

riences that involve by implication this very mathematical fact.

A person already *familiar* with the composition and resolution of forces is well aware that several forces acting at a point are, as regards their effect, replaceable, in *every* respect and in *every* direction, by a *single* force. This knowledge, in Bernoulli's mode of proof, is expressed in the fact that the forces p, q are regarded as absolutely qualified to replace in all respects the forces s, u and t, v, in the direction of r as well as in every other direction. Similarly r is regarded as the equivalent of p and q. It is further assumed as wholly indifferent, whether we estimate s, u, t, v first in the directions of p, q, and then p, q in the direction of r, or whether s, u, t, v be estimated directly and from the outset in the direction of r. But this is something that only a person can know who has antecedently acquired a very extensive experience concerning the composition and resolution of forces. We reach most simply the knowledge of the fact referred to, by starting from the knowledge of another fact, namely that a force p acts in a direction making with its own an angle α, with an effect equivalent to $p \cdot \cos \alpha$. As a fact, this is the way the perception of the truth *was* reached.

Let the coplanar forces $P, P', P''. . .$ be applied to one and the same point at the angles $\alpha, \alpha', \alpha'' . . .$ with a given direction X. Let us suppose these forces are replaceable by a single force Π, which makes with X an angle μ. By the familiar principle we have then

$$\Sigma P \cos\alpha = \Pi \cos\mu$$

If Π is still to remain the substitute of this system of forces, whatever direction X may take on the system being turned through any angle δ, we shall further have

$$\Sigma P \cos(\alpha + \delta) = \Pi \cos(\mu + \delta),$$

or

$$(\Sigma P \cos\alpha - \Pi \cos\mu) \cos\delta - (\Sigma P \sin\alpha - \Pi \sin\mu) \sin\delta = 0.$$

If we put

$$\Sigma P \cos\alpha - \Pi \cos\mu = A,$$
$$-(\Sigma P \sin\alpha - \Pi \sin\mu) = B,$$
$$\tan\tau = \frac{B}{A},$$

it follows that

$$A \cos\delta + B \sin\delta = \sqrt{A^2 + B^2} \sin(\delta + \tau) = 0,$$

which equation can subsist for *every* δ only on the condition that

$$A = \Sigma P \cos\alpha - \Pi \cos\mu = 0$$

and

$$B = (\Sigma P \sin\alpha - \Pi \sin\mu) = 0 ;$$

whence results

$$\Pi \cos\mu = \Sigma P \cos\alpha$$
$$\Pi \sin\mu = \Sigma P \sin\alpha.$$

From these equations follow for Π and μ the determinate values

$$\Pi = \sqrt{[(\Sigma P \sin\alpha)^2 + (\Sigma P \cos\alpha)^2]}$$

and

$$\tan\mu = \frac{\Sigma P \sin\alpha}{\Sigma P \cos\alpha}.$$

Granting, therefore, that the effect of a force in every direction can be measured by its projection on that direction, then truly every system of forces acting at a point is replaceable by a *single* force, determinate in magnitude and direction. This reasoning does not hold, however, if we put in the place of cos α any general function of an angle, $\varphi(\alpha)$. Yet if this be done, and we still regard the resultant as determinate, we shall obtain for

$\varphi(a)$, as may be seen, for example, from Poisson's deduction, the form $\cos a$. The experience that several forces acting at a point are always, in every respect, replaceable by a single force, is therefore *mathematically equivalent* to the principle of the parallelogram of forces or to the principle of projection. The principle of the parallelogram or of projection is, however, much more easily reached by observation than the more general experience mentioned above by statical observations. And as a fact, the principle of the parallelogram was reached earlier. It would require indeed an almost superhuman power of perception to deduce mathematically, without the guidance of any further knowledge of the actual conditions of the question, the principle of the parallelogram from the general principle of the equivalence of several forces to a single one. We criticize this, accordingly in the deduction of Bernoulli, that that which is easier to observe is reduced to that which is more difficult to observe. This is a violation of the economy of science. Bernoulli is also deceived in imagining that he does not proceed from any fact whatever of observation.

We must further remark that the fact that the forces are independent of one another, which is involved in the law of their composition, is another experience which Bernoulli throughout tacitly employs. As long as we have to do with uniform or symmetrical systems qf forces, all equal in magnitude, each can be affected by the others, even if they are not independent, only to the same extent and in the same way. Given but three forces, however, of which two are symmetrical to the third, even then the reasoning, provided we admit that the forces may not be independent, presents considerable difficulties.

11. Once we have been led, directly or indirectly,

to the principle of the parallelogram of forces, once we have *perceived* it, the principle is just as much an observation as any other. If the observation is recent, it of course is not accepted with the same confidence as old and frequently verified observations. We then seek to support the new observation by the old, to demonstrate their agreement. By and by the new observation acquires equal standing with the old. It is then no longer necessary constantly to reduce it to the latter. Deduction of this character is expedient only in cases in which observations that are difficult to obtain directly can be reduced to simpler ones more easily obtained, as is done with the principle of the parallelogram of forces in dynamics.

Fig. 37.

12. The proposition of the parallelogram of forces has also been illustrated by experiments especially instituted for the purpose. An apparatus very well adapted to this end was contrived by Varignon. The center of a horizontal divided circle (Fig. 37) is marked by a pin. Three threads f, f', f'', tied together at a

point, are passed over grooved wheels r, r', r'', which
can be fixed at any point in the circumference of the
circle, and are loaded by the weights p, p', p''. If three
equal weights be attached, for instance, and the wheels
placed at the marks of division 0, 120, 240, the point at
which the strings are knotted will assume a position
just above the center of the circle. Three equal forces
acting at angles of 120°, accordingly, are in equilibrium.

If we wish to represent another and different case,
we may proceed as follows. We imagine any two
forces p, q acting at any angle a, represent (Fig. 38)
them by lines, and construct on them as sides a parallelogram. We supply, further, a force
equal and opposite to the resultant r.
The three forces $p, q, -r$ hold each
other in equilibrium, at the angles visible from the construction. We now
place the wheels of the divided circle on
the points of division o, $a, a + \beta$, and
load the appropriate strings with the
weights p, q, r. The point at which the

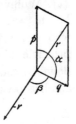

Fig. 38.

strings are knotted will come to a position exactly
above the middle point of the circle.

IV.

THE PRINCIPLE OF VIRTUAL VELOCITIES.

1. We now proceed to the discussion of the principle of virtual (possible) displacements.* In my ex-

* Termed in English the principle of "virtual velocities," this being the
original phrase (*vitesse virtuelle*) introduced by John Bernoulli. See the
text, page 68. The word *virtualis* seems to have been the fabrication of Duns
Scotus (see the *Century Dictionary*, under *virtual*); but *virtualiter* was used
by Aquinas, and *virtus* had been employed for centuries to translate δύναμις,
and therefore as a synonym for *potentia*. Along with many other scholastic
terms, *virtual* passed into the ordinary vocabulary of the English language.
Everybody remembers the passage in the third book of *Paradise Lost*,

> " Love not the heav'nly Spirits, and how thir Love
> Express they, by looks onely, or do they mix
> Irradiance, *virtual* or immediate touch? "—*Milton*.

position in preceding editions, E. Wohlwill finds that the achievements of Stevinus are overestimated as compared with those of del Monte and Galileo. As a matter of fact, del Monte, in his *Mechanicorum liber* (Pisauri, 1577), noticed the lengths of the paths which are described simultaneously by the weights in the cases of the lever, the pulleys, and the wheel and axle. To be sure his consideration is more geometrical than mechanical. Also, there is lacking in del Monte the apprehension of the principle by which the character of wonder is taken away from the operation of machines (*cf.* Wohlwill, *Galilei,* i, pp. 142 *et seq.*). Thus del Monte was out-distanced by other mediæval writers who concerned themselves with the heritage of the principle of virtual velocities which had been handed down by the ancients, and who are to be mentioned on another occasion. Now, at the end of the sixteenth

---

So, we all remember how it was claimed before our revolution that America had *"virtual* representation" in parliament. In these passages, as in Latin, *virtual* means: existing in effect, but not actually. In the same sense, the word passed into French; and was made pretty common among philosophers by Leibniz. Thus, he calls innate ideas in the mind of a child, not yet brought to consciousness, "des connoissances *virtuelles.*"

The principle in question was an extension to the case of more than two forces of the old rule that "what a machine gains in *power,* it loses in *velocity."* *Bernoulli's* modification reads that the sum of the products of the forces into their virtual velocities must vanish to give equilibrium. He says, in effect: give the system any possible and infinitesimal motion you please, and then the simultaneous displacements of the points of application of the forces, *resolved in the directions of those forces,* though they are not exactly velocities, since they are only displacements in one time, are, nevertheless, *virtually* velocities, for the purpose of applying the rule that what a machine gains in power, it loses in *velocity.*

Thomson and Tait say: "If the point of application of a force be displaced through a small space, the resolved part of the displacement in the direction of the force has been called its *Virtual Velocity.* This is positive or negative according as the virtual velocity is in the same, or in the opposite, direction to that of the force." This agrees with Bernoulli's definition which may be found in Varignon's *Nouvelle mécanique,* Vol. II, Chap. ix.—*Trans.*

century, Stevinus did *not* advance beyond his immediate predecessor, del Monte.

First of all, Stevinus treats combinations of pulleys in the same way that they are treated at the present day.

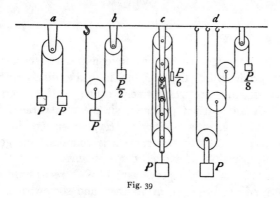

Fig. 39

In the case *a* (Fig. 39) equilibrium obtains, when an equal weight P is suspended at each side, for reasons already familiar. In *b*, the weight P is suspended by two parallel cords, each of which accordingly supports the weight $P/2$, with which weight in the case of equilibrium the free end of the cord must also be loaded. In *c*, P is suspended by six cords, and the weighting of the free extremity with $P/6$ will accordingly produce equilibrium. In *d*, the so-called Archimedean or potential pulley,* P in the first instance is suspended by two cords, each of which supports $P/2$; one of these two cords in turn is suspended by two others, and so on to the end, so that the free extremity will be held in equilibrium by the weight $P/8$. If we impart to these assemblages of pulleys displacements corresponding to a de-

---

* These terms are not in use in English.—*Trans.*

scent of the weight P through the distance h, we shall observe that as a result of the arrangement of the cords

the counterweight P				a distance h in a
"	"	$P/2$	will ascend	" " $2h$ " b
"	"	$P/6$		" " $6h$ " c
"	"	$P/8$		" " $8h$ " d

In a system of pulleys in equilibrium, therefore, the products of the weights into the displacements they sustain are respectively equal. ("Ut spatium agentis ad spatium patientis, sic potentia patientis ad potentiam agentis."—Stevini, *Hypomnemata*, T. IV, lib. 3, p. 172.) In this remark is contained the germ of the principle of virtual displacements.

2. GALILEO had previously (1594) recognized the truth of the principle in another, and somewhat more general case, in its application to the inclined plane. On

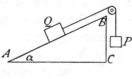

Fig. 40.

an inclined plane (Fig. 40), the length of which AB is double the height BC, a load Q placed on AB is held in equilibrium by the load P acting along the height BC, if $P = Q/2$. If the machine be set in motion, $P = Q/2$ will descend, say, the vertical distance h, and Q will ascend the same distance h along the incline AB. Galileo, now, allowing the phenomenon to exercise its full effect on his mind, perceives, that equilibrium is determined not by the weights alone but also by their *possible approach to and recession from the center of the earth.* Thus, while $Q/2$ descends along the vertical height the distance h, Q ascends h along the inclined length, vertically, however, only $h/2$; the result being that the products $Q(h/2)$ *and* $(Q/2)h$ come out equal on both sides. The elucidation that Galileo's observation affords and the light it diffuses, can hardly be emphasized strongly enough. The observation is so natural and unforced, moreover, that we admit it at once. What can appear simpler than that no motion takes place in a system of heavy bodies when on the whole no heavy mass can descend? Such a fact appears instinctively acceptable.

E. Wohlwill emphasizes that Galileo laid stress on the loss of velocity which corresponds to the economy of force in machines (*cf. Galilei, i*, pp. 141, 142). If we use the modern conception of work to the development of which Galileo contributed so much, we can say without equivocalness: in machines *work* is not economized.

Galileo's conception of the inclined plane strikes us as much less ingenious than that of Stevinus, but we recognize it as more natural and more profound. It is in this fact that Galileo discloses such scientific greatness: that he had the *intellectual audacity* to see, in a subject long before investigated, *more* than his predecessors had seen, and to trust to his own perceptions. With the frankness that was characteristic of him he unreservedly places before the reader his own view, together with the considerations that led him to it.

3. TORRICELLI, by the employment of the notion of "center of gravity," has put Galileo's principle in a form in which it appeals still more to our instincts, but in which it is also incidentally applied by Galileo himself. According to Torricelli, equilibrium exists in a machine when, if a displacement is imparted to it, the center of gravity of the weights attached thereto cannot

descend. On the supposition of a displacement in the inclined plane last dealt with, P, let us say, descends the distance h, in compensation wherefor Q vertically ascends h . sin α. Assuming that the center of gravity does not descend, we shall have

$$\frac{P \cdot h - Q \cdot h \sin \alpha}{P + Q} = 0, \text{ or } P \cdot h - Q \cdot h \sin \alpha = 0,$$

or

$$P = Q \sin \alpha = Q \frac{BC}{AB}.$$

If the weights bear to one another some different proportion, then the center of gravity *can* descend when a displacement is made, and equilibrium will not obtain. We expect the state of equilibrium *instinctively*, when the center of gravity of a system of heavy bodies cannot descend. The Torricellian form of expression, however, contains in no respect more than the Galilean.

4. As with systems of pulleys and with the inclined plane, so also the validity of the principle of virtual displacements is easily demonstrable for the other machines: for the lever, the wheel and axle, and the rest. In a wheel and axle, for instance, with the radii R, r and the respective weights P, Q, equilibrium exists, as we know, when $PR = Qr$. If we turn the wheel and axle through the angle α, P will descend $R\alpha$, and Q will ascend $r\alpha$. According to the conception of Stevinus and Galileo, when equilibrium exists, $P \cdot R\alpha = Q \cdot r\alpha$, which equation expresses the same thing as the preceding one.

5. When we compare a system of heavy bodies in which motion is taking place, with a similar system which is in equilibrium, the question forces itself upon us: What constitutes the difference between the two

cases? What is the factor operative here that determines motion, the factor that disturbs equilibrium—the factor that is present in the one case and absent in the other? Having put this question to himself, Galileo discovers that not only the weights, but also the distances of their vertical descents (the amounts of their vertical displacements) are the factors that determine motion. Let us call $P, P', P'' \ldots$ the weights of a system of heavy bodies, and $h, h', h'' \ldots$ their respective, simultaneously possible vertical displacements, where displacements downwards are reckoned as positive, and displacements upwards as negative. Galileo finds then, that the criterion or test of the state of equilibrium is contained in the fulfilment of the condition $Ph + P'h' + P'' h'' + \ldots = 0$. The sum $Ph + P' h' + P'' h'' + \ldots$ is the factor that destroys equilibrium, the factor that determines motion. Owing to its importance this sum has in recent times been characterized by the special designation *work*.

6. Whereas the earlier investigators, in the comparison of cases of equilibrium and cases of motion, directed their attention to the weights and their distances from the axis of rotation and recognized *the statical moments* as the decisive factors involved, Galileo fixes his attention on the weights and their distances of descent and discerns *work* as the decisive factor involved. It cannot of course be prescribed to the inquirer *what* mark or criterion of the condition of equilibrium he shall take account of, when several are present to choose from. The result alone can determine whether his choice is the right one. But if we cannot, for reasons already stated, regard the significance of the statical moments as given independently of experience, as something self-evident, no more can we entertain this

view with respect to the import of work. Pascal errs, and many modern inquirers share this error with him, when he says, on the occasion of applying the principle of virtual displacements to fluids: "Etant clair que c'est la même chose de faire faire un pouce de chemin à cent livres d'eau, que de faire faire cent pouces de chemin à une livre d'eau." This is correct only on the supposition that we have already come to recognize work as the decisive factor; and that it is so is a fact which experience alone can disclose.

If we have an equal-armed, equally-weighted lever before us, we recognize the equilibrium of the lever as the only effect that is uniquely determined, whether we regard the weights and the distances or the weights and the vertical displacements as the conditions that determine motion. Experimental knowledge of this or a similar character must, however, in the necessity of the case precede any judgment of ours with regard to the phenomenon in question. The particular way in which the disturbance of equilibrium depends on the conditions mentioned, that is to say, the significance of the statical moment (PL) or of the work (Ph), is even less capable of being philosophically excogitated than the general fact of the dependence.

7. When two equal weights with equal and opposite possible displacements are opposed to each other, we recognize at once the subsistence of equilibrium. We might now be tempted to reduce the more general case of the weights P, P' with the capacities of displacement h,h' where $Ph = P'h'$, to the simpler case. Suppose we have, for example, (Fig. 41) the weights $3\,P$ and $4\,P$ on a wheel and axle with the radii 4 and 3. We divide the weights into equal portions of the definite magnitude P, which we designate

by *a, b, c, d, e, f, g*. We
then transport *a, b, c* to
the level + 3, and *d, e, f*
to the level — 3. The
weights will, of them-
selves, neither enter on
this displacement nor
will they resist it. We
then take simultaneously

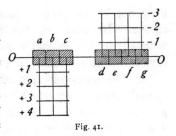

Fig. 41.

the weight *g* at the level 0 and the weight *a* at the level
+ 3, push the first upwards to —1 and the second
downwards to + 4, then again, and in the same way,
g to — 2 and *b* to + 4, *g* to — 3 and *c* to + 4. To all
these displacements the weights offer no resistance,
nor do they produce them of themselves. Ultimately,
however, *a, b, c* (or 3*P*) appear at the level + 4 and
d, e, f, g (or 4*P*) at the level — 3. Consequently,
with respect also to the last-mentioned total displacement, the weights neither produce it of themselves
nor do they resist it; that is to say, given the ratio of
displacement here specified, the weights will be in
equilibrium. The equation $4 \cdot 3P - 3 \cdot 4P = 0$ is, therefore, characteristic of equilibrium in the case assumed.
The generalization $(Ph - P'h' = 0)$ is obvious.

If we carefully examine the reasoning of this case,
we shall quite readily perceive that the inference involved cannot be drawn unless we take for granted
that the *order* of the operations performed and the *path*
by which the transferences are effected, are indifferent,
that is unless we have previously discerned that work
is determinative. We should commit, if we accepted
this inference, the same error that Archimedes committed in his deduction of the law of the lever; as has
been set forth at length in a preceding section and

need not in the present case be so exhaustively discussed. Nevertheless, the reasoning we have presented is useful, in that it makes perceptible the relationship of the simple and the complicated cases.

8. The *universal* applicability of the principle of
virtual displacements to all cases of equilibrium, was
perceived by JOHN BERNOULLI, who communicated his
discovery to Varignon in a letter written in 1717. We
will now enunciate the principle in its most general
form. At the points $A, B, C \ldots$ (Fig. 42) the forces
$P, P', P'' \ldots$ are applied. Impart to the points any
infinitely small displacements $v, v', v'' \ldots$ compatible
with the character of the connections of the points (so-
called virtual displacements), and construct the projections p, p', p'' of these displacements on the directions of the forces. These projections we consider

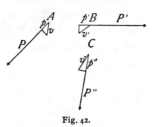

Fig. 42.

positive when they fall in
the direction of the force,
and negative when they fall
in the opposite direction.
The products $Pp, P'p',$
$P'' p'', \ldots$ are called virtual
moments, and in the two
cases just mentioned have
contrary signs. Now, the principle asserts, that for the
case of equilibrium $Pp + P' p' + P'' p'' + \ldots = 0$, or
more briefly $\Sigma Pp = 0$.

9. Let us now examine a few points more in detail.
Previous to Newton a force was almost universally
conceived simply as the pull or the pressure of a heavy
body. The mechanical researches of this period dealt
almost exclusively with heavy bodies. When, now,
in the Newtonian epoch, the generalization of the idea

of force was effected, all mechanical principles known
to be applicable to heavy bodies could be transferred
at once to any forces whatsoever. It was possible to
replace every force by the pull of a heavy body on a
string. In this sense we may also apply the principle
of virtual displacements, at first discovered only for
heavy bodies, to any forces whatsoever.

Virtual displacements are displacements consistent
with the character of the connections of a system and
with one another. If, for example, the two points of
a system, A and B, at which forces act, are connected
(Fig. 43, 1) by a rectangularly bent lever, free to revolve about C, then, if $CB = 2CA$, all virtual displacements of B and A are elements of the arcs of circles having C as center; the displacements of B are
always double the displacements of A, and both are in
every case at right angles to each other. If the points
A, B (Fig. 43, 2) be connected by a thread of the
length l, adjusted to slip
through stationary rings at
C and D, then all those displacements of A and B are
virtual in which the points

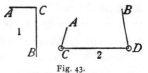

Fig. 43.

referred to move upon or within two spherical surfaces described with the radii r_1 and r_2 about C and D
as centers, where $r_1 + r_2 + CD = l$.

The use of *infinitely* small displacements instead of
finite displacements, such as Galileo assumed, is justified by the following consideration. If two weights
are in equilibrium on an inclined plane (Fig. 44), the
equilibrium will not be disturbed if the inclined plane,
at points at which it is not in immediate contact with
the bodies considered, passes into a surface of a different form. The essential condition is, therefore, the

momentary possibility of displacement in the momen-

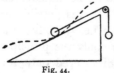

Fig. 44.

tary configuration of the system. To judge of equilibrium we must assume displacements vanishingly small and such only; as otherwise the system might be carried over into an entirely different adjacent configuration, for which perhaps equilibrium would not exist.

That the displacements themselves are not decisive but only the extent to which they occur in the directions of the forces, that is only their *projections* on the lines of the forces, was, in the case of the inclined plane, perceived clearly enough by Galileo himself.

With respect to the expression of the principle, it will be observed, that no problem whatever is presented if all the material points of the system on which forces act, are independent of each other. Each point thus conditioned can be in equilibrium only in the event that it is not movable in the direction in which the force acts. The virtual moment of each such point vanishes separately. If some of the points be independent of each other, while others in their displacements are dependent on each other, the remark just made holds good for the former; and for the latter the fundamental proposition discovered by Galileo holds, that the sum of their virtual moments is equal to zero. Hence, the sum-total of the virtual moments of all jointly is equal to zero.

10. Let us now endeavor to get some idea of the significance of the principle, by the consideration of a few simple examples that cannot be dealt with by the ordinary method of the lever, the inclined plane, and the like.

The differential pulley of Weston (Fig. 45) consists of two coaxial rigidly connected cylinders of slightly different radii r_1 and r_2 $< r_1$. A cord or chain is passed round the cylinders in the manner indicated in the figure. If we pull in the direction of the arrow with the force P, and rotation takes place through the angle φ, the weight Q attached below will be raised. In the case of equilibrium there will exist between the two virtual moments involved the equation

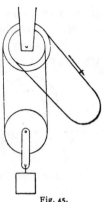

Fig. 45.

$$Q \frac{(r_1 - r_2)}{2} \varphi = P r_1 \varphi, \text{ or } P = Q \frac{r_1 - r_2}{2 r_1}.$$

A wheel and axle of weight Q (Fig. 46), which on the unrolling of a cord to which the weight P is attached rolls itself up on a second cord wound round the axle and rises, gives for the virtual moments in the case of equilibrium the equation

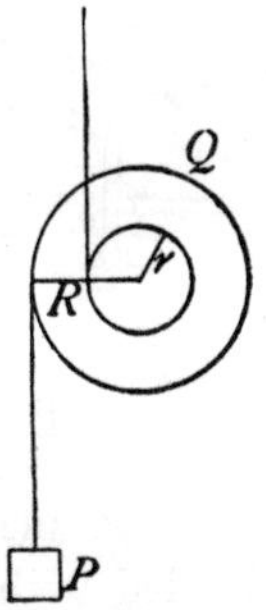

$$P (R - r) \varphi = Q r \varphi, \text{ or } P = \frac{Q r}{R - r}.$$

In the particular case $R - r = 0$, we must also put, for equilibrium, $Q r = 0$, or, for finite values of r, $Q = 0$. In reality the string behaves in this case like a loop in which the weight Q is placed. The latter can, if it be different from zero, continue to roll itself downwards on the string without moving the weight P. If, however, when $R = r$, we also put $Q = 0$, the result will be $P = \frac{0}{0}$, an indeterminate value. As a mat-

Fig. 46.

ter of fact, *every* weight P holds the apparatus in equilibrium, since when $R = r$ none can possibly descend.

A double cylinder (Fig. 47) of the radii r, R lies with friction on a horizontal surface, and a force Q is brought

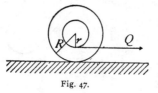

Fig. 47.

to bear on the string attached to it. Calling the resistance due to friction P, equilibrium exists when $P = (\overline{R - r}/R)\, Q$. If $P > (\overline{R - r}/R)Q$, the cylinder, on the application of the force, will roll itself up on the string.

Roberval's Balance (Fig. 48) consists of a parallelogram with variable angles, two opposite sides of which, the upper and lower, are capable of rotation about their middle points A, B. To the two remaining sides, which are always vertical, horizontal rods are fastened. If from these rods we suspend two equal weights P, equilibrium will subsist independently of

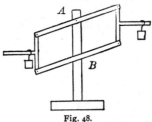

Fig. 48.

the position of the points of suspension, because on displacement the descent of the one weight is always equal to the ascent of the other.

At three fixed points A, B, C (Fig. 49) let pulleys be placed, over which three strings are passed loaded with equal weights and knotted at O. In what position of the strings will equilibrium exist? We will call the lengths of the three strings $AO = s_1$, $BO = s_2$, $CO = s_3$. To obtain the equation of equilibrium, let us displace the point O in the directions s_2 and s_3 the infinitely small distances δs_2 and δs_3, and note that by so doing every direction of displacement in the plane

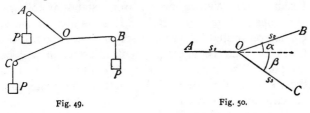

Fig. 49. Fig. 50.

ABC (Fig. 50) can be produced. The sum of the virtual moments is

$$
\begin{aligned}
P\delta s_2 - P\delta s_2 \cos\alpha + P\delta s_2 \cos(\alpha+\beta) \\
+ P\delta s_3 - P\delta s_3 \cos\beta + P\delta s_3 \cos(\alpha+\beta)
\end{aligned} \Bigg\} = 0,
$$

or

$$
[1 - \cos\alpha + \cos(\alpha+\beta)]\,\delta s_2 + [1 - \cos\beta + \cos(\alpha+\beta)]\,\delta s_3 = 0.
$$

But since each of the displacements δs_2, δs_3 is arbitrary, and each independent of the other, and may by itself be taken $= 0$, it follows that

$$
1 - \cos\alpha + \cos(\alpha+\beta) = 0
$$
$$
1 - \cos\beta + \cos(\alpha+\beta) = 0.
$$

Therefore

$$
\cos\alpha = \cos\beta,
$$

and each of the two equations may be replaced by

$$
1 - \cos\alpha + \cos 2\alpha = 0;
$$
$$
\text{or } \cos\alpha = \tfrac{1}{2},
$$
$$
\text{wherefore } \alpha + \beta = 120°.
$$

Accordingly, in the case of equilibrium, each of the strings makes with the others angles of 120°; which is, moreover, directly obvious, since three equal forces can be in equilibrium only when such an arrangement exists. This once known, we may find the position of the point O with respect to ABC in a number of different ways. We may proceed for instance as follows. We construct on AB, BC, CA, severally, as sides, equilateral triangles. If we describe circles about these triangles, their common point of intersection will be the point O sought; a result which easily follows from the well-known relation of the angles at the center and circumference of circles.

A bar OA (Fig. 51) is revolvable about O in the plane of the paper and makes with a fixed straight line

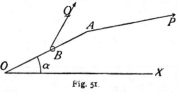

Fig. 51.

OX the variable angle α. At A there is applied a force P which makes with OX the angle γ, and at B, on a ring displaceable

along the length of the bar, a force Q, making with OX the angle β. We impart to the bar an infinitely small rotation, in consequence of which B and A move forward the distances δs and δs_1 at right angles to OA, and we also displace the ring the distance δr along the bar. The variable distance OB we will call r, and we will let $OA = a$. For the case of equilibrium we have then

$$Q\delta r \cos (\beta - \alpha) + Q\delta s \sin (\beta - \alpha) +$$
$$P\delta s_1 \sin (\alpha - \gamma) = 0.$$

As the displacement δr has no effect whatever on the other displacements, the virtual moment therein involved must, by itself, $= 0$, and since δr may be of any magnitude we please, the coefficient of this virtual moment must also $= 0$. We have, therefore,

$$Q \cos (\beta - \alpha) = 0,$$

or when Q is different from zero,

$$\beta - \alpha = 90°.$$

Further, in view of the fact that $\delta s_1 = (a/r)\delta s$, we also have

$$rQ \sin (\beta - \alpha) + a P \sin (\alpha - \gamma) \, 0,$$

or since $\sin (\beta - \alpha) = 1$,

$$rQ + aP \sin (\alpha - \gamma) = 0;$$

wherewith the relation of the two forces is obtained.

11. An advantage, not to be overlooked, which every general principle, and therefore also the principle of virtual displacements, furnishes, consists in the fact that it saves us to a great extent the necessity of considering every new particular case presented. In the possession of this principle we need

Fig. 52.

not, for example, trouble ourselves about the details of a machine. If by chance a new machine were so enclosed in a box (Fig. 52), that only two levers projected as points of application for the force P and the weight P', and we should find the simultaneous displacements of these levers to be h and h', we should know immediately that in the case of equilibrium $Ph = P' h'$, whatever the construction of the machine might be. Every principle of this character possesses therefore a distinct *economical* value.

12. We return to the general expression of the principle of virtual displacements, in order to add a few further remarks. If at the points A, B, C . . . . the forces P, P', P'' . . . . act, and p, p', p'' . . . . are the projections of infinitely small

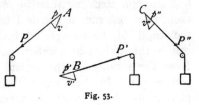

Fig. 53.

mutually compatible displacements, we shall have for the case of equilibrium

$$Pp + P'p' + P''p'' + \ldots = 0.$$

If we replace the forces by strings which pass over pulleys in the directions of the forces and attach thereto the appropriate weights, this expression simply asserts that the *center of gravity* of the system of weights as a whole cannot descend. If, however, in certain displacements it were possible for the center of gravity to *rise*, the system would still be in equilibrium, as the heavy bodies would not, of themselves, enter on any such motion. In this case the sum above given would be negative, or less than zero. The general expression of the condition of equilibrium is, therefore,

$$Pp + P'p' + P''p'' + \ldots \lesseqgtr 0.$$

When for every virtual displacement there exists another *equal* and *opposite* to it, as is the case, for example, in the simple machines, we may restrict ourselves to the upper sign, to the *equation*. For if it were possible for the center of gravity to ascend in certain displacements, it would also have to be possible, in consequence of the assumed reversibility of all the virtual displacements, for it to descend. Consequently, in the present case, a possible rise of the center of gravity is incompatible with equilibrium.

The question assumes a different aspect, however, when the displacements are not all reversible. Two bodies connected together by strings can approach each other but cannot recede from each other beyond the length of the strings. A body is able to slide or roll on the surface of another body; it can move away from the surface of the second body, but it cannot penetrate it. In these cases, therefore, there are displacements that cannot be reversed. Consequently,

for certain displacements a *rise* of the center of gravity
may take place, while the contrary displacements, to
which the *descent* of the center of gravity corresponds,
are impossible. We must therefore hold fast to the
more general condition of equilibrium, and say, the sum
of the virtual moments is *equal to or less* than zero.

13. LAGRANGE in his *Analytical Mechanics* attempted a deduction of the principle of virtual displacements, which we will now consider. At the points
$A, B, C \ldots$ (Fig. 54) the forces $P, P', P'' \ldots$ act.
We imagine rings placed at the points in question, and
other rings $A', B', C' \ldots$ fastened to points lying in
the directions of the forces. We seek some common
measure $Q/2$ of the forces $P, P', P'' \ldots$ that enables
us to put:

$$2n \cdot \frac{Q}{2} = P,$$

$$2n' \cdot \frac{Q}{2} = P',$$

$$2n'' \cdot \frac{Q}{2} = P'',$$

where $n, n', n'' \ldots$ are whole numbers. Further, we
make fast to the ring A' a string, carry this string *back*

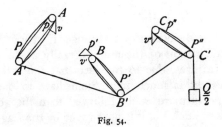

Fig. 54.

and *forth* n times between A' and A, then through B',
n' times back and forth between B' and B, then through

C', n'' times back and forth between C' and C, and, finally, let it drop at C', attaching to it there the weight $Q/2$. As the string has, now, in all its parts the tension $Q/2$, we replace by these ideal pulleys all the forces present in the system by the single *force* $Q/2$. If then the virtual (possible) displacements in any given configuration of the system are such that, these displacements, occurring, a descent of the weight $Q/2$ can take place, the weight will actually descend and produce those displacements, and equilibrium therefore will not obtain. But on the other hand, no motion will ensue, if the displacements leave the weight $Q/2$ in its original position, or raise it. The expression of this condition, reckoning the projections of the virtual displacements in the directions of the forces positive, and having regard for the number of the turns of the string in each single pulley, is

$$2np + 2n'p' + 2n''p'' + \ldots \gtreqless 0.$$

Equivalent to this condition, however, is the expression

$$2n\,\frac{Q}{2}\,p + 2n'\,\frac{Q}{2}\,p' + 2n''\,\frac{Q}{2}\,p'' + \ldots \gtreqless 0,$$

or

$$Pp + P'p' + P''p'' + \ldots \gtreqless 0.$$

14. The deduction of Lagrange, if stripped of the rather odd fiction of the pulleys, really possesses convincing features, due to the fact that the action of a single weight is much more immediate to our experience and is more easily followed than the action of several weights. Yet it is not proved by the Lagrangian deduction that work is the factor determinative of the disturbance of equilibrium, but is, by the employment of the pulleys, rather assumed by it. As a matter of

fact every pulley involves the fact enunciated and recognized by the principle of virtual displacements. The
replacement of all the forces by a single weight that
does the same work, presupposes a knowledge of the
import of work, and can be proceeded with on this assumption alone. The fact that some certain cases are
more familiar to us and more immediate to our experience has, as a necessary result, that we accept them
without analysis and make them the foundation of our
deductions without clearly instructing ourselves as to
their real character.

It often happens in the course of the development
of science that a new principle perceived by some inquirer in connection with a fact, is not immediately
recognized and rendered familiar in its entire generality.
Then, every expedient calculated to promote these
ends is, as is proper and natural, called into service.
All manner of facts, in which the principle, although
contained in them, has not yet been recognized by inquirers, but which from other points of view are more
familiar, are called in to furnish a support for the new
conception. It does not, however, beseem mature
science to allow itself to be deceived by procedures of
this sort. If, throughout all facts, we clearly *see* and *discern* a principle which, though not admitting of proof,
can yet be known to *prevail*, we have advanced much
farther in the consistent conception of nature than if
we suffered ourselves to be overawed by a specious
demonstration. If we have reached this point of view,
we shall, it is true, regard the Lagrangian deduction
with quite different eyes; yet it will engage nevertheless our attention and interest, and excite our satisfaction from the fact that it makes palpable the similarity of the simple and the complicated cases.

15. Maupertuis discovered an interesting proposition relating to equilibrium, which he communicated to the Paris Academy in 1740 under the name of the "Lois de repos." This principle was more fully discussed by Euler in 1751 in the Proceedings of the Berlin Academy. If we cause infinitely small displacements in any system, we produce a sum of virtual moments $Pp + P'p' + P''p'' + \ldots$, which only reduces to zero in the case of equilibrium. This sum is the work corresponding to the displacements, or since for infinitely small displacements it is itself infinitely small, the corresponding element of work. If the displacements are continuously increased till a finite displacement is produced, the elements of the work will, by summation, produce a finite amount of work. So, if we start from any given initial configuration of the system and pass to any given final configuration, a certain amount of work will have to be done. Now Maupertuis observed that the work done when a final configuration is reached which is a configuration of equilibrium, is generally a maximum or a minimum; that is, if we carry the system through the configuration of equilibrium the work done is previously and subsequently less or previously and subsequently greater than at the configuration of equilibrium itself. For the configuration of equilibrium,

$$Pp + P'p' + P''p'' + \ldots = 0,$$

that is, the element of the work or the differential (more correctly the variation) of the work is equal to zero. If the differential of a function can be put equal to zero, the function has generally a maximum or minimum value.

16. We can produce a very clear representation to

the eye of the import of Maupertuis's principle.

We imagine the forces of a system replaced by Lagrange's pulleys with the weight $Q/2$. We suppose that each point of the system is restricted to movement on a certain curve and that the motion is such that when one point occupies a definite position on its curve all the other points assume uniquely determined positions on their respective curves. The simple machines are as a rule systems of this kind. Now, while imparting displacements to the system, we may carry a vertical sheet of white paper horizontally over the weight $Q/2$, while this is ascending and descending on a vertical line, so that a pencil which it carries shall describe a curve upon the paper (Fig. 55). When the pencil stands at the points a, c, d of the curve, there are, we see, adjacent positions in the system of points at which the weight $Q/2$ will stand higher or lower than in the configuration given. The weight will then, if the system be left to itself, pass into this lower position and

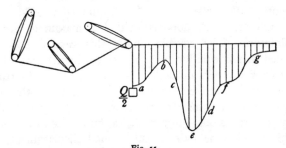

Fig. 55.

displace the system with it. Accordingly, under conditions of this kind, equilibrium does not subsist. If the pencil stands at e, then there exist only adjacent configurations for which the weight $Q/2$ stands higher. But of itself the system will not pass into the last-named configurations. On the contrary, every displacement in such a direction, will, by virtue of the tendency of the weight to move downwards, be reversed. *Stable equilibrium, therefore, is the condition that corresponds to the lowest position of the weight or to a maximum of work done in the system.* If the pencil stands at *b,* we see that every appreciable displacement brings the weight $Q/2$ lower, and that the weight therefore will continue the displacement begun. But, assuming infinitely small displacements, the pencil moves in the horizontal tangent at *b,* in which event the weight cannot descend. Therefore, *unstable equilibrium is the state that corresponds to the highest position of the weight $Q/2$, or to a minimum of work done in the system.* It will be noted, however, that conversely every case of equilibrium is not the correspondent of a maximum or a minimum of work performed. If the pencil is at *f,* at a point of horizontal contrary flexure, the weight in the case of infinitely small displacements neither rises nor falls. Equilibrium exists, although the work done is neither a maximum nor a minimum. The equilibrium of this case is the so-called *mixed* equilibrium*: for some disturbances it is stable, for others unstable. Nothing prevents us from regarding mixed equilibrium as belonging to the unstable class. When the pencil stands at *g,* where the curve runs along horizontally a finite distance, equilibrium likewise exists. Any small displacement, in the configuration in question, is neither continued nor reversed. This kind of equilibrium, to which likewise neither a maximum nor a minimum corresponds, is

---

* This term is not used in English, because our writers hold that no equilibrium is conceivable which is not stable or neutral for some possible displacements. Hence what is called *mixed* equilibrium in the text is called unstable equilibrium by English writers, who deny the existence of equilibrium unstable in every respect.—*Trans.*

termed [*neutral* or] *indifferent.* If the curve described by $Q/2$ has a cusp pointing upwards, this indicates a minimum of work done but no equilibrium (not even unstable equilibrium). To a cusp pointing downwards a maximum and stable equilibrium correspond. In the last named case of equilibrium the sum of the virtual moment is not equal to zero, but is negative.

17. In the reasoning just presented, we have assumed that the motion of a point of a system on one curve determines the motion of all the other points of the system on their respective curves. The movability of the system becomes multiplex, however, when each point is displaceable on a surface, in a manner such that the position of one point on its surface determines uniquely the position of all the other points on their surfaces. In this case, we are not permitted to consider the *curve* described by $Q/2$, but are obliged to picture to ourselves a surface described by $Q/2$. If, to go a step further, each point is movable throughout a space, we can no longer represent to ourselves in a purely geometrical manner the circumstances of the motion, by means of the locus of $Q/2$. In a correspondingly higher degree is this the case when the position of *one* of the points of the system does not determine conjointly all the other positions, but the character of the system's motion is more multiplex still. In all these cases, however, the curve described by $Q/2$ (Fig. 55) can serve us as a symbol of the phenomena to be considered. In these cases also we rediscover the Maupertuisian propositions.

We have also supposed, in our considerations up to this point, that constant forces, forces independent of the position of the points of the system, are the forces that act in the system. If we assume that the forces do depend on the position of the points of the system (but not on the time), we are no longer able to conduct our operations with simple pulleys, but must devise apparatus the force active in which, still exerted by $Q/2$, varies with the displacement: the ideas we have reached, however, still obtain. The depth of the descent of the weight $Q/2$ is in every case the measure of the work performed, which is always the same in the same configuration of the system and is independent of the path of transference. A contrivance which would by means of a constant weight develop a force varying with the displacement, would be, for example, a wheel and axle (Fig. 56) with a non-circular wheel. It would not repay the trouble, however, to enter into the details of the reasoning indicated in this case, since we perceive at a glance its feasibility.

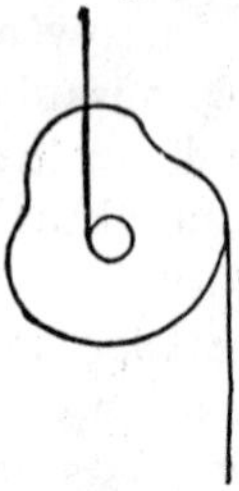

Fig. 56.

18. If we know the relation that subsists between the work done and the so-called *vis viva* of a system, a relation established in dynamics, we arrive easily at the principle communicated by COURTIVRON in 1749 to the Paris Academy, which is this: For the configuration of $\frac{\text{stable}}{\text{unstable}}$ equilibrium, at which the work done is a $\frac{\text{maximum}}{\text{minimum}}$, the vis viva of the system, in motion, is also a $\frac{\text{maximum}}{\text{minimum}}$ in its transit through these configurations.

19. A heavy, homogeneous triaxial ellipsoid resting on a horizontal plane is admirably adapted to illustrate the various classes of equilibrium. When the ellipsoid rests on the extremity of its smallest axis, it is in stable equilibrium, for any displacement it may suffer elevates its center of gravity. If it rests on its longest

axis, it is in unstable equilibrium. If the ellipsoid stands
on its mean axis, its equilibrium
is mixed. A homogeneous sphere
or a homogeneous right cylinder on a horizontal plane illustrates the case of indifferent
equilibrium. In Fig. 57 we have

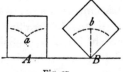

Fig. 57.

represented the paths of the center of gravity of a
cube rolling on a horizontal plane about one of its
edges. The position *a* of the center of gravity is the
position of stable equilibrium, the position *b*, the position of unstable equilibrium.

20. We will now consider an example which at
first sight appears very complicated but is elucidated
at once by the principle of virtual displacements. John
and James Bernoulli, on the occasion of a conversa-
tion on mathematical topics during a walk in Basel,
lighted on the question of what form a chain would
take that was freely suspended and fastened at both
ends. They soon and easily agreed in the view that
the chain would assume that form of equilibrium at
which its center of gravity lay in the lowest possible
position. As a matter of fact we really do perceive
that equilibrium subsists when all the links of the chain
have sunk as low as possible, when none can sink lower
without raising an equivalent mass in consequence of
the connections of the system equally high or higher.
When the center of gravity has sunk as low as it possibly can sink, when all has happened that can happen,
stable equilibrium exists. The *physical* part of the
problem is disposed of by this consideration. The determination of the curve that has the lowest center of
gravity for a given length between the two points *A*,
B, is simply a *mathematical* problem. (See Fig. 58.)

21. Collecting all that has been presented, we see that there is contained in the principle of virtual displacements simply the recognition of a fact that was instinctively familiar to us long previously, only that we had not apprehended it so precisely and clearly. This fact consists in the circumstance that heavy bodies, of themselves, move only downwards. If several such bodies are joined together so that they can suffer no displacement independently of each other, they will then move only in the event that some heavy mass is *on the whole* able to descend, or as the principle, with a more perfect adaptation of our ideas to the facts, more exactly expresses it, only in the event that *work* can be performed. If, extending the notion of force, we transfer the principle to forces other than those due to gravity, the recognition is again contained therein of the fact that the natural occurrences in question take place, of themselves, *only in a definite sense* and not in the opposite sense. Just as heavy bodies descend downwards, so differences of temperature and electrical potential cannot increase of their own accord but only diminish, and so on. If occurrences of this kind are so connected that they can take place only in the contrary sense, the principle then establishes, more precisely than our instinctive apprehension could do this, the factor *work* as determinative and decisive of the direction of the occurrences. The equilibrium equation of the principle may be reduced in every case to the trivial statement, that *when nothing can happen nothing does happen.*

22. It is important to obtain clearly the perception, that we have to deal, in the case of all principles, merely with the ascertainment and establishment of a *fact*. If we neglect this, we shall always be sensible of some deficiency and will seek a verification of the

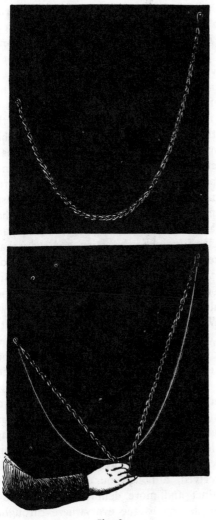

Fig. 58.

principle, that is not to be found. Jacobi states in his *Lectures on Dynamics* that Gauss once remarked that Lagrange's equations of motion had not been proved, but only historically enunciated. And this view really seems to us to be the correct one in regard to the principle of virtual displacements.

The task of the early inquirers, who lay the foundations of any department of investigation, is entirely different from that of those who follow. It is the business of the former to seek out and to establish the facts of most cardinal importance only; and, as history teaches, more brains are required for this than is generally supposed. When the most important facts are once furnished, we are then placed in a position to work them out deductively and logically by the methods of mathematical physics; we can then organize the department of inquiry in question, and show that in the acceptance of some *one* fact a whole series of others is included which were not to be immediately discerned in the first. The one task is as important as the other. We should not however confound the one with the other. We cannot prove by mathematics that nature *must* be exactly what it is. But we can prove, that one set of observed properties determines conjointly another set which often are not directly manifest.

Let it be remarked in conclusion, that the principle of virtual displacements, like every general principle, brings with it, by the insight which it furnishes, *disillusionment* as well as elucidation. It brings with it disillusionment to the extent that we recognize in it facts which were long before known and even instinctively perceived, although our present recognition is more distinct and more definite; and elucidation, in that it enables us to see everywhere throughout the most complicated relations the same simple facts.

RETROSPECT OF THE DEVELOPMENT OF STATICS.

1. Having passed successively in review the prin-'ciples of statics, we are now in a position to take a brief supplementary survey of the development of the principles of the science as a whole. This development, falling as it does in the earliest period of mechanics —the period which begins in Grecian antiquity and reaches its close at the time when Galileo and his younger contemporaries were inaugurating modern mechanics—illustrates in an excellent manner the process of the formation of science generally. All conceptions, all methods are here found in their simplest form, and as it were in their infancy. These beginnings point unmistakably to their origin in the experiences of the manual arts. To the necessity of putting these experiences into *communicable* form and of disseminating them beyond the confines of class and craft, science owes its origin. The collector of experiences of this kind, who seeks to preserve them in written form, finds before him many different, or at least supposedly different, experiences. His position is one that enables him to review these experiences more frequently, more variously, and more impartially than the individual workingman, who is always limited to a narrow province. The facts and their dependent rules are brought into closer temporal and spatial proximity in his mind and writings, and thus acquire the opportunity of revealing their relationship, their connection, and their gradual transition the one into the other. The desire to simplify and abridge the labor of communication supplies a further impulse in the same direction. Thus, from

economical reasons, in such circumstances, great numbers of facts and the rules that spring from them are condensed into a system and comprehended in a *single* expression.

2.   A collector of this character has, moreover, opportunity to take note of some *new* aspect of the facts before him—of some aspect which former observers had not considered.   A rule, reached by the observation of facts, cannot possibly embrace the *entire* fact, in all its infinite wealth, in all its inexhaustible manifoldness; on the contrary, it can furnish only a rough *outline* of the fact, one-sidedly emphasizing the feature that is of importance for the given technical (or scientific) aim in view.   *What* aspects of a fact are taken notice of, will consequently depend upon circumstances, or even on the caprice of the observer.   Hence there is always opportunity for the discovery of new aspects of the fact, which will lead to the establishment of new rules of equal validity with, or superior to, the old.   So, for instance, the weights and the lengths of the lever-arms were regarded at first, by Archimedes, as the conditions that determined equilibrium.   Afterwards, by Da Vinci and Ubaldi the weights and the perpendicular distances from the axis of the lines of force were recognized as the determinative conditions.   Still later, by Galileo, the weights and the amounts of their displacements, and finally by Varignon the weights and the directions of the pulls with respect to the axis were taken as the elements of equilibrium, and the enunciation of the rules modified accordingly.

3.   Whoever makes a new observation of this kind, and establishes such a new rule, knows, of course, our liability to error in attempting to represent the fact mentally, whether by concrete images or in abstract conceptions, which we must do in order to have the mental

model we have constructed always at hand as a substitute for the fact when the latter is partly or wholly inaccessible. The circumstances, indeed, to which we have to attend, are accompanied by so many other, collateral circumstances, that it is frequently difficult to single out and consider those that are essential to the purpose in view. Just think how the facts of friction, the rigidity of ropes and cords, and like conditions in machines, obscure and obliterate the pure outlines of the main facts. No wonder, therefore, that the discoverer or verifier of a new rule, urged by mistrust of himself, seeks after a *proof* of the rule whose validity he believes he has discerned. The discoverer or verifier does not at the outset fully trust in the rule; or, it may be, he is confident only of a part of it. So, Archimedes, for example, doubted whether the effect of the action of weights on a lever was *proportional* to the lengths of the lever-arms, but he accepted without hesitation the fact of their influence in some way. Daniel Bernoulli does not question the influence of the direction of a force generally, but only the form of its influence. As a matter of fact, it is far easier to observe that a circumstance *has* influence in a given case, than to determine *what* influence it has. In the latter inquiry we are in much greater degree liable to error. The attitude of the investigators is therefore perfectly natural and defensible.

The proof of the correctness of a new rule can be attained by the repeated application of it, the frequent comparison of it with experience, the putting of it to the *test* under the most diverse circumstances. This process would, in the natural course of events, be carried out in time. The discoverer, however, hastens to reach his goal more quickly. He compares the results that flow from his rule with all the experiences with which he is familiar, with all older rules, repeatedly tested in times gone by, and watches to see if he does not light on contradictions. In this procedure, the greatest credit is, as it should be, conceded to the oldest and most familiar experiences, the most thoroughly tested rules. Our instinctive experiences, those generalizations that are made involuntarily, by the irresistible force of the innumerable facts that press in upon us, enjoy a peculiar authority; and this is perfectly warranted by the consideration that it is precisely the elimination of subjective caprice and of individual error that is the object aimed at.

In this manner Archimedes *proves* his law of the lever, Stevinus his law of inclined pressure, Daniel Bernoulli the parallelogram of forces, Lagrange the principle of virtual displacements. Galileo alone is perfectly aware, with respect to the last-mentioned principle, that his new observation and perception are of equal rank with *every former* one—that it is derived from *the same* source of experience. He attempts no demonstration. Archimedes, in his proof of the principle of the lever, uses facts concerning the center of gravity, which he had probably proved by means of the very principle now in question; yet we may suppose that these facts were otherwise so familiar, as to be unquestioned—so familiar indeed, that it may be doubted whether he remarked that he had employed them in demonstrating the principle of the lever. The instinctive elements embraced in the views of Archimedes and Stevinus have been discussed at length in the proper place.

4. It is quite in order, on the making of a new discovery, to resort to all proper means to bring the new rule to the test. When, however, after the lapse of a reasonable period of time, it has been sufficiently often

subjected to direct testing, it becomes science to recognize that any other proof than that has become quite needless; that there is no sense in considering a rule as the better established for being founded on others that have been reached by the very same method of observation, only earlier; that one well-considered and tested observation is as good as another. Today, we should regard the principles of the lever, of statical moments, of the inclined plane, of virtual displacements, and of the parallelogram of forces as discovered by *equivalent* observations. It is of no importance *now*, that some of these discoveries were made directly, while others were reached by roundabout ways and as dependent upon other observations. It is more in keeping, furthermore, with the economy of thought and with the æsthetics of science, directly to *recognize* a principle (say that of the statical moments) as the key to the understanding of *all* the facts of a department, and *really see* how it *pervades* all those facts, rather than to hold ourselves obliged first to make a clumsy and lame deduction of it from unobvious propositions that involve the same principle but that happen to have become earlier familiar to us. This process science and the individual (in historical study) may go through once for all. But having done so both are free to adopt a more convenient point of view.

5. In fact, this mania for demonstration in science results in a rigor that is *false* and *mistaken*. Some propositions are held to be possessed of more certainty than others and even regarded as their necessary and incontestable foundation; whereas actually no higher, or perhaps not even so high, a degree of certainty attaches to them. Even the rendering clear of the degree of certainty which exact science aims at, is not at-

tained here. Examples of such mistaken rigor are to
be found in almost every text-book. The deductions
of Archimedes, not considering their historical value,
are infected with this erroneous rigor. But the most
conspicuous example of all is furnished by Daniel Bernoulli's deduction of the parallelogram of forces (*Comment. Acad. Petrop.* T. I.).

6. As already seen, instinctive knowledge enjoys
our exceptional confidence. No longer knowing *how*
we have acquired it, we cannot criticize the logic by
which it was inferred. We have personally contributed
nothing to its production. It confronts us with a force
and irresistibleness foreign to the products of voluntary reflective experience. It appears to us as something free from subjectivity, and extraneous to us, although we have it constantly at hand so that it is more
ours than are the individual facts of nature.

All this has often led men to attribute knowledge of
this kind to an entirely different source, namely, to view
it as existing *a priori* in us (previous to all experience).
That this opinion is untenable was fully explained in
our discussion of the achievements of Stevinus. Yet
even the authority of instinctive knowledge, however
important it may be for actual processes of development, must ultimately give place to that of a clearly and
deliberately observed principle. Instinctive knowledge
is, after all, only experimental knowledge, and as such
is liable, we have seen, to prove itself utterly insufficient and powerless, when some new region of experience is suddenly opened up.

7. The *true* relation and connection of the different
principles is the *historical* one. The one extends farther
in this domain, the other farther in that. Notwithstanding that some one principle, say the principle of

virtual displacements, may control with facility a greater number of cases than other principles, still no assurance can be given that it will always maintain its supremacy and will not be outstripped by some new principle. All principles single out, more or less arbitrarily, now this aspect now that aspect of the same facts, and contain an abstract summarized rule for the refigurement of the facts in thought. We can never assert that this process has been definitely completed. Whosoever holds to this opinion, will not stand in the way of the advancement of science.

8. Let us, in conclusion, direct our attention for a moment to the conception of force in statics. Force is any circumstance of which the consequence is motion. Several circumstances of this kind, however, each single one of which determines motion, may be so conjoined that in the result there shall be no motion. Now statics investigates what this mode of conjunction, in general terms, is. Statics does not further concern itself about the particular character of the motion conditioned by the forces. The circumstances determinative of motion that are best known to us, are our own volitional acts—our innervations. In the motions which we ourselves determine, as well as in those to which we are forced by external circumstances, we are always sensible of a pressure. Thence arises our habit of representing all circumstances determinative of motion as something akin to volitional acts—as *pressures*. The attempts we make to set aside this conception, as subjective, animistic, and unscientific, fail invariably. It cannot profit us, surely, to do violence to our own natural-born thoughts and to doom ourselves, in that regard, to voluntary mental penury. We shall subsequently have occasion to observe, that the conception

referred to also plays a part in the foundation of dynamics.

We are able, in a great many cases, to replace the circumstances determinative of motion, which occur in nature, by our innervations, and thus to reach the idea of a gradation of the intensity of forces. But in the estimation of this intensity we are thrown entirely on the resources of our memory, and are also unable to communicate our sensations. Since it is possible, however, to represent *every* condition that determines motion by a weight, we arrive at the perception that all circumstances determinative of motion (all forces) are alike in character and may be replaced and measured by quantities that stand for weight. The measurable weight serves us, as a certain, convenient, and communicable index, in mechanical researches, just as the thermometer in thermal researches is an exacter substitute for our perceptions of heat. As has previously been remarked, statics cannot wholly rid itself of all knowledge of phenomena of motion. This particularly appears in the determination of the direction of a force by the direction of the motion which it would produce if it acted alone. By the point of application of a force we mean that point of a body whose motion is still determined by the force when the point is freed from its connections with the other parts of the body.

Force accordingly is any circumstance that determines motion; and its attributes may be stated as follows. The direction of the force is the direction of motion which is determined by that force, alone. The point of application is that point whose motion is determined independently of its connections with the system. The magnitude of the force is that weight which, acting (say, on a string) in the direction determined, and applied at the point in question, determines the same motion or maintains the same equilibrium. The other circumstances that modify the determination of a motion, but by themselves alone are unable to produce it, such as virtual displacements, the arms of levers, and so forth, may be termed collateral conditions determinative of motion and equilibrium.

9. The knowledge of the development of a science rests on the study of writings in their historical sequence and in their historical connection. For ancient times many sources are, of course, lacking, and for other times the author is unknown or doubtful. In later centuries, especially before the discovery of printing, the bad habit was general of the author seldom referring to his predecessors where he uses their works, and usually only doing so where he thinks he has to contradict those predecessors. By these circumstances, the above study is made very difficult and makes the highest demands on criticism.

P. DUHEM develops in his book, *Les origines de la statique* (Paris, 1905, vol. I), the view that E. Wohlwill had already taken, that modern scientific civilization is much more intimately connected with ancient scientific civilization than people usually suppose. The scientific thoughts of the Renaissance developed very slowly and gradually from those of ancient Greece, particularly from those of the peripatetic and Alexandrian school. I will here emphasize that Duhem's book contains a mine of stimulating, instructive and enlightening details condensed in a small space. To the knowledge of these details we could only otherwise attain by a wearisome study of old books and manuscripts. For that alone reading Duhem's work excites much admiration and is very fruitful.

In especial, Duhem ascribes to Jordanus Nemorarius, a writer of the thirteenth century who was an interpreter and developer of ancient thoughts, and to a later elaborator of the *Liber Jordani de ratione ponderis,* whom he calls the "forerunner of Leonardo da Vinci," a great influence on Leonardo, Cardano, and Benedetti. The most important corrections to *Jordani opusculum de ponderositate,* which Tartaglia published as his own and used in *Questi et inventioni diverse* without naming Jordanus or his later elaborator, are contained in a manuscript under the title *Liber Jordani de ratione ponderis,* which Duhem found in the national library at Paris (*fond latin,* No. 7378 A). This leads to the supposition of the anonymous "forerunner." Also, Leonardo's manuscripts, which were not carefully preserved and were unprotected from unauthorized use, have had, according to Duhem, in spite of their delayed publication, an effect on Cardano and Benedetti. The authors named above influenced, above all, Galileo in Italy, Stevinus in Holland, and their works reached France by both channels. There they found, in the first place, fruitful soil in Roberval and Descartes. Consequently, the continuity between ancient and modern statics was never broken.

Let us now consider some details. The author of the *Mechanical Problems* mentioned on p. 12 remarks about the lever that the weights which are in equilibrium are inversely proportional to the arms of the lever or to the arcs described by the end-points of the arms when a motion is imparted to them*. With great

---

* According to the view of E. Wohlwill, it may be considered to be decided that the *Mechanical Problems* cannot be due to Aristotle. *Cf.* Zeller, *Philosophie der Griechen,* 3rd ed., pt. ii, § ii, note on p. 90. But then a thorough investigation as to whether the lately found Arabic translation (published in 1893) of Hero's *Mechanics,* if not the older text, is necessary. *Cf. Heron's Werke,* edited by L. Nix and W. Schmidt (Leipzig 1900), vol. II.

freedom of interpretation we can regard this remark as the incomplete expression of the principle of virtual displacements. But, with Jordanus Nemorarius (Duhem, *op. cit.,* pp. 121, 122), the equilibrium of the lever is characterized by the inverse proportionality of the height to which the weights are raised (or the depths to which they fall) to the weights which are in equilibrium. The essential point is brought into prominence by this. Jordanus also knows that a weight does not always act in the same way, and introduces—though only qualitatively—the conception of weight according to position: "secundum situm gravius, quando in eodem situ minus obliquus est descensus" (*op. cit.,* p. 118). The "forerunner" of Leonardo improves and completes the exposition of Jordanus. He recognizes the equilibrium of an angular lever whose axis lies above the weights, by the consideration of the possible depths of falling and heights of rising, as *stable* (*op. cit.,* p. 142). He knows also that such a lever directs itself in such a manner that the weights are proportional to their distances from the vertical through the axis (*op. cit.,* pp. 142, 143), and thus arrives in essentials at the use of the conception of *moment.* Thus the "gravitas secundum situm" here attains a *quantitative* form and is used in a brilliant way for the solution of the problem of the inclined plane (*opt. cit.,* p. 145). If two weights on inclined planes of equal heights but different lengths are so connected by a rope and pulley that the one must rise when the other sinks, the weights are, in the case of equilibrium, inversely as the *vertical* displacements, that is to say, vary directly as the lengths of the inclined planes. Consequently in this the "forerunner" anticipated the essential elements of modern statics.

The study of the manuscripts of Leonardo, which have only been published in part, is extremely profitable. The comparison of his various occasional notes shows clearly his knowledge of the principle of virtual displacements, or rather of the concept of work, though he does not use any special nomenclature. "When a force carries (raises?) a body (a weight?) in a certain time through a definite path, the same force can carry (raise?) half of the body (the weight?) in the same time through a path double in length." This theorem is applied to machines, levers, pulleys, and so on, and by this the rather doubtful meaning of the above words is more closely determined. If we have a definite quantity of water which can sink to a definite depth, we can, according to Leonardo, drive one or even two equal mills with it, but in the second case we can only accomplish as much as in the first case. The perception of the "potential lever," to which Leonardo attained by a stroke of genius, put him in the position to gain all the insight which was reached later by the conception of "moment." His figures make us suspect that the consideration of the pulley and the wheel and axle showed him the way to his conception (*cf. Mechanics*, p. 28). Leonardo's constructions concerning the pulls on combinations of cords visibly rest, too, on the thought of the potential lever. Leonardo was less happy in the treatment of the problem of the inclined plane. By the side of sketches in which sometimes a correct view is expressed, we find many incorrect constructions. However, we must consider Leonardo's scribblings as leaves of a diary, which fix the most various sudden ideas and points of view and beginnings of investigations, and do not attempt to carry out these investigations according to a unitary principle. To explain the fact that Leonardo was not master of all the problems which had been completely solved in the thirteenth century, we must remember that it by no means suffices, as we must recognize with Duhem, that an insight should be once attained and made known, but years and centuries are often necessary for this insight to be *generally* recognized and understood (Duhem, *op. cit.*, p. 182).

The idea of the impossibility of perpetual motion is developed with Leonardo to great clearness. His consideration about the mill shows this: "No impetus without life can press or draw a body without accompanying the body moved; these impetuses can be nothing else than forces or gravity. When gravity presses or draws, it effects motion only because it strives for rest; no body can, by its motion of falling, rise to the height from which it fell; its motion reaches an end" (*op. cit.*, p. 53). "Force is a spiritual and invisible power which is impregnated in bodies by motion (here we certainly have to think of what at the present time is called *vis viva*); the greater it is the more quickly does it expend itself" (*op. cit.*, p. 54). Cardano has a similar view in which we may judge an influence of Leonardo to be probable if we have grounds for doubting Cardano's independence (*op. cit.*, pp. 40, 57, 58). Also Aristotle's idea that only the circular motion of the heavens is eternal appears again with Cardano. Duhem considers that Cardano is not a common plagiarist. He used without acknowledgment the works of his predecessors, especially those of Leonardo, but brought these works into a better connection and, by that, improved the position of the sixteenth century (*op. cit.*, pp. 42, 43). Cardano does not overcome the problem of the inclined plane; his opinion is that the weight of the body on the inclined plane is to the whole weight as the angle of elevation of the plane is to a right angle. Benedetti put himself in opposition to all his predecessors, and this opposition had a good effect, especially in criticism of the dynamical doctrines of Aristotle. But Benedetti often opposed much that was right. In his writings occur again thoughts of Leonardo's, and errors of Leonardo's as well.

If we regard the discoveries we have just spoken of as sufficiently known and accessible to the successors of the above men, there remains for these successors —especially for Stevinus and Galileo—not very much more to do in statics. Stevinus's solution of the problem of the inclined plane (*cf. Mechanics,* pp. 32-41) is indeed quite original, but the "forerunner" of Leonardo already knew the *result* of the considerations of Stevinus and Galileo, and Galileo's considerations join on to those of Cardano. From the consideration of the inclined plane Stevinus attained to the composition and resolution of *rectangular* components according to the principle of the parallelogram, and considered this principle to be generally valid without being able to prove it. Roberval filled up this gap. He imagined a weight R supported by pulleys and held in equilibrium by a cord of any direction loaded with counter-weights P and Q. If, first, we consider one cord as a rod which can rotate about the pulley and apply Leonardo's principle of the potential lever, and then proceed in a similar way with respect to the other cord, we find the relations of R to P and Q and all the theorems which hold for the triangle of forces or the parallelogram of *forces* (*op. cit.,* esp. p. 319). Descartes finds in the principle of virtual displacements the foundation for the understanding of *all* machines. He sees in work, the product of weight and distance of falling (in his nomenclature, "force"), the determining circumstance or cause of the behavior of machines, the *Why* and not merely the *How* of the event. It is not a question of the velocity, but of the height of raising and the depth of falling. "For it is the same thing to raise a hundred pounds two feet or two hundred pounds one foot" (*op. cit.,* p. 328; *cf.* 66 of *Mechanics* on Pascal's statement). Descartes denies the unmistakable influence on his thoughts of all his predecessors from Jordanus to Roberval; and yet his developments show everywhere important progress, and throughout he emphasizes essential points (*op. cit.,* pp. 327-352).

With respect to details we must refer to Duhem's brilliant book. Here I will only give expression to my somewhat different opinion on the relation of ancient to modern natural science. Natural science grows in two ways. In the first place, it grows by our retaining in memory the observed facts or processes, reproducing them in our presentation, and trying to reconstruct them in our thoughts. But, as the observations are continued, these attempts at construction, which are successively or simultaneously taken in hand, always show certain defects by which the agreement of these constructions both with the facts and with one another is disturbed. Thus there results a need for material correction and logical harmonization of the constructions. This is the *second* process which builds up natural science. If everyone had only himself to rely on, he would have to begin anew with his observations and thoughts alone, and consequently could not get far. This holds both for single human beings and for single nations. Thus we cannot treasure highly enough the heritage which our immediate predecessors in civilization—the Greek students of nature, astronomers and mathematicians—have bequeathed to us. We enter on investigation under favorable conditions, since we are in possession of an image of the world—although this image be insufficient—and are, above all, equipped with the logical and critical education of the Greek mathematicians. This possession makes the continuance of the work easier for us. But we must consider not only our scientific heritage but also *material* civilization—in our special case the machines and tools which have been handed down to us as well as the tradition of their use. We can easily set up observations on this material heritage, or repeat and extend those which led the investigators of ancient times to their science, and thus for the first time really learn to understand this science. It appears to me that this material heritage — continually waking up anew, as it does, our independent activity—is too little esteemed in comparison with the literary heritage. For can we suppose that the paltry remarks of the author of the *Mechanical Problems* about the lever, and even the far more exact remarks of the Alexandrian mathematicians, would not have continually obtruded themselves upon the observing men who were busied with machines, even if these remarks were not preserved in writing? Does not this hold good, say, about the knowledge of the impossibility of perpetual motion, which must present itself to everybody who does not seek the wonderful in mechanics, as a dreamer after the fashion of the alchemists, but is busied, as a calm investigator, in practice with machines? Even when such finds are transferred to those who come after, they must be gained independently by these followers.

The sole advantage a follower has consists in the start that he has gained by a quicker passage over the same course, by which he outstrips his predecessors. An incomplete knowledge put into words forms a relatively firm prop for fleeting thoughts, from which the thoughts, seeking among facts, set out, and to which, modifying it by criticism and comparison, they continually return. Now, whether these props are made stronger by newer experience or are gradually shifted, or are even at last recognized as invalid, they have helped us on. But if the predecessor becomes a great authority, and if even his errors are prized as marks of deep insight, we get a state of things which can only act in a hurtful way on the followers of this man. Thus, by many passages in the writings of E. Wohlwill and P. Duhem, it seems that even Galileo was sometimes hindered, even in his later years, by the traditional peripatetic burden from perceiving undisturbed his own far stronger light. In our estimation of the importance of an investigator, then, it is only a question of what *new* use he has made of old views and under what *opposition* of his contemporaries and followers *his own* views have come to be held. From this point of view, Duhem seems to me to go rather too far in his feeling of reverence towards the memory of Aristotle. With Aristotle (*De coelo,* book iii, 2) there are, for example, among unclear and unpromising utterances, the passages: "Whatever the moving force may be, the less and the lighter receive more motion from the same force. . . . The velocity of the less heavy body will be to that of the heavier body as the heavier to the lighter body." If we disregard the fact that Aristotle cannot be credited with a clear distinction of path, velocity, and acceleration, we can recognize in this the expression of a primitive but correct experience which led at length to the conception of mass. But, after what has been said in the whole of the second chapter, it seems hardly thinkable to refer this passage to the raising of weights by machines, to combine it with what Aristotle has said about the lever, and then to see in it the germ of the conception of work (Duhem, *op. cit.*, pp. 6, 7; *cf.* Vailati, *Bolletino di bibliografia e storia di scienze matematiche*, Feb. and March, 1906, p. 3). Further, Duhem blames Stevinus for his peripatetic tendencies. But Stevinus seems to me to be in the right when he puts himself in opposition to the "wonderful" circles of Aristotle, which are not described in the case of equilibrium. This is just as justifiable as the protest of Gilbert and Galileo against the hypothesis of the effectiveness of a mere position or a point (see *Mechanics,* p. 230). Only from a broader point of view, when work is recognized as that which determines motion, does the dynamical derivation of equilibrium attain the merit of greater rationality and generality. Before that, hardly anything could be urged against Stevinus's inspired deductions on the grounds of instinctive experience and after the manner of Archimedes.

VI.

THE PRINCIPLES OF STATICS IN THEIR APPLICATION TO FLUIDS.

1. The consideration of fluids has not supplied statics with many essentially new points of view, yet numerous applications and confirmations of the principles already known have resulted therefrom, and physical experience has been greatly enriched by the investiga-

tions of this domain. We shall devote, therefore, a few pages to this subject.

2. To ARCHIMEDES also belongs the honor of founding the domain of the statics of liquids. To him we owe the well-known proposition concerning the buoyancy, or loss of weight, of bodies immersed in liquids, of the discovery of which Vitruvius, *De Architectura*, Lib. IX, gives the following account:

"Though Archimedes discovered many curious matters that evince great intelligence, that which I am about to mention is the most extraordinary. Hiero, when he obtained the regal power in Syracuse, having, on the fortunate turn of his affairs, decreed a votive crown of gold to be placed in a certain temple to the immortal gods, commanded it to be made of great value, and assigned for this purpose an appropriate weight of the metal to the manufacturer. The latter, in due time, presented the work to the king, beautifully wrought; and the weight appeared to correspond with that of the gold which had been assigned for it.

"But a report having been circulated, that some of the gold had been abstracted, and that the deficiency thus caused had been supplied by silver, Hiero was indignant at the fraud, and, unacquainted with the method by which the theft might be detected, requested Archimedes would undertake to give it his attention. Charged with this commission, he by chance went to a bath, and on jumping into the tub, perceived that, just in the proportion that his body became immersed, in the same proportion the water ran out of the vessel. Whence, catching at the method to be adopted for the solution of the proposition, he immediately followed it up, leapt out of the vessel in joy, and returning home naked, cried out with a loud voice that he had found

that of which he was in search, for he continued ex-
claiming, in Greek, εὕρηκα, εὕρηκα, (I have found it,
I have found it!)"

3. The observation which led Archimedes to his
proposition, was accordingly this, that a body immersed in water must *raise* an equivalent quantity of
water; exactly as if the body lay on one pan of a balance and the water on the other. This conception,
which at the present day is still the most natural and
the most direct, also appears in Archimedes's treatises
On Floating Bodies, which unfortunately have not been
completely preserved but have in part been restored
by F. Commandinus.

The assumption from which Archimedes starts
reads thus:

"It is assumed as the essential property of a liquid
that in all uniform and continuous positions of its parts
the portion that suffers the lesser pressure is forced
upwards by that which suffers the greater pressure.
But each part of the liquid suffers pressure from the
portion perpendicularly above it if the latter be sinking
or suffer pressure from another portion."

Archimedes now, to present the matter briefly, conceives the entire spherical earth as fluid in constitution,
and cuts out of it pyramids the vertices of which lie at
the center (Fig. 59). All these pyramids must, in the
case of equilibrium, have the same weight, and the
similarly situated parts of the same must all suffer the
same pressure. If we plunge a body *a* of the same
specific gravity as water into one of the pyramids, the
body will completely submerge, and in the case of
equilibrium, will supply by its weight the pressure of
the displaced water. The body *b,* of less specific gravity, can sink, without disturbance of equilibrium, only

to the point at which the water beneath it suffers the
same pressure from the weight of the body as it would
if the body were taken out
and the submerged portion
replaced by water. The body
c, of a greater specific gravity, sinks as deep as it possibly can. That its weight is
lessened in the water by an
amount equal to the weight
of the water displaced, will
be manifest if we imagine the

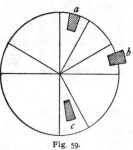

Fig. 59.

body joined to another of less specific gravity so that
a third body is formed having the same specific gravity as water, which just completely submerges.

4. When in the sixteenth century the study of the
works of Archimedes was again taken up, the principles
of his researches were scarcely understood. The complete comprehension of his deductions was at that time
impossible.

STEVINUS rediscovered by a method of his own the
most important principles of hydrostatics and the deductions therefrom. It was principally two ideas from
which Stevinus derived his fruitful conclusions. The
one is quite similar to that relating to the endless chain.
The other consists in the assumption that the solidifi-
cation of a fluid in equilibrium does not disturb its
equilibrium.

Stevinus first lays down this principle. Any given
mass of water A (Fig. 60), immersed in water, is in
equilibrium in all its parts. If A was not supported
by the surrounding water but should, let us say, descend,
then the portion of water taking the place of A and

placed thus in the same circumstances, would, on the
same assumption, also have to de-
scend. This assumption leads, therefore, to the establishment of a
perpetual motion, which is contrary
to our experience and to our instinctive knowledge of things.

Fig. 60.

Water immersed in water loses accordingly its
whole weight. If, now, we imagine the surface of the
submerged water solidified, the vessel formed by this
surface, the *vas superficiarium* as Stevinus calls it, will
still be subjected to the same circumstances of pressure. If *empty*, the vessel so formed will suffer an
upward pressure in the liquid equal to the weight of
the water displaced. If we fill the solidified surface
with some other substance of any specific gravity we
may choose, it will be plain that the diminution of the
weight of the body will be equal to the weight of the
fluid displaced on immersion.

In a rectangular, vertically placed parallelepipedal
vessel filled with a liquid, the pressure on the horizontal
base is equal, also, for all parts of the bottom of the
same area. When now Stevinus imagines portions of
the liquid to be cut out and replaced by rigid immersed
bodies of the same specific gravity, or, what is the
same thing, imagines parts of the liquid to become solidified, the relations of pressure in the vessel will not
be altered by the procedure. But we easily obtain in
this way a clear view of the law that the pressure on
the base of a vessel is independent of its form, as well
as of the laws of pressure in communicating vessels,
and so forth.

5. GALILEO treats the equilibrium of liquids in communicating vessels and the problems connected therewith by the help of the principle of virtual displacements. *NN* (Fig. 61) being the common level of a liquid in equilibrium in two communicating vessels, Galileo explains the equilibrium here presented by observing that in the case of any disturbance the displacements of the columns are to each other in the inverse proportion of the areas of the transverse sections and of the weights of the columns

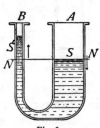

Fig. 61.

—that is, as with machines in equilibrium. But this is not quite correct. The case does not exactly correspond to the cases of equilibrium investigated by Galileo in machines, which present indifferent equilibrium. With liquids in communicating tubes every disturbance of the common level of the liquids produces an elevation of the center of gravity. In the case represented in Fig. 61, the center of gravity *S* of the liquid displaced from the shaded space in *A* is elevated to *S'*, and we may regard the rest of the liquid as not having been moved. Accordingly, in the case of equilibrium, the center of gravity of the liquid lies at its lowest possible point.

6. PASCAL likewise employs the principle of virtual displacements, but in a more correct manner, leaving the weight of the liquid out of account and considering only the pressure at the surface. If we imagine two communicating vessels to be closed by pistons (Fig. 62), and these pistons loaded with weights proportional to their surface-areas, equilibrium will obtain, because

in consequence of the invariability of the volume of
the liquid the displacements in every disturbance are

Fig. 62.

inversely proportional to the weights.
For Pascal, accordingly, it *follows,* as
a necessary consequence, from the
principle of virtual displacements, that
in the case of equilibrium every pressure on a superficial portion of a
liquid is propagated with undiminished
effect to every other superficial portion, however and
in whatever position it be placed. No objection is to
be made to *discovering* the principle in this way. Yet
we shall see later on that the more natural and satisfactory conception is to regard the principle as immediately given.

7. We shall now, after this historical sketch, again
examine the most important cases of liquid equilibrium,
and from such different points of view as may be convenient.

The fundamental property of liquids given us by
experience consists in the flexure of their parts on the
slightest application of pressure. Let us picture to ourselves an element of volume of a liquid, the gravity of
which we disregard—say a tiny cube. If the slightest
excess of pressure is exerted on one of the surfaces of
this cube, (which we now conceive, for the moment,
as a fixed geometrical locus, containing the fluid but
not of its substance) the liquid (supposed to have previously been in equilibrium and at rest) will yield and
pass out in all directions through the other five surfaces
of the cube. A solid cube can stand a pressure on its
upper and lower surfaces different in magnitude from
that on its lateral surfaces; or *vice versa.* A fluid cube,

on the other hand, can retain its shape only if the same perpendicular pressure be exerted on all its sides. A similar train of reasoning is applicable to all polyhedrons. In this conception, as thus geometrically elucidated, is contained nothing but the crude experience that the particles of a liquid yield to the slightest pressure, and that they retain this property also in the interior of the liquid when under a high pressure; it being observable, for example, that under the conditions cited minute heavy bodies sink in fluids, and so on.

With the mobility of their parts liquids combine still another property, which we will now consider. Liquids suffer through pressure a diminution of volume which is proportional to the pressure exerted on unit of surface. Every alteration of pressure carries along with it a proportional alteration of volume and density. If the pressure diminish, the volume becomes greater, the density less. The volume of a liquid continues to diminish therefore when the pressure is increased, till the point is reached at which the elasticity generated within it equilibrates the increase of the pressure.

8. The earlier inquirers, as for instance those of the Florentine Academy, were of the opinion that liquids were incompressible. In 1761, however, JOHN CANTON performed an experiment by which the compressibility of water was demonstrated. A thermometer glass is filled with water, boiled, and then sealed. (Fig. 63.) The liquid reaches to *a*. But since the space above *a* is airless, the liquid supports no atmospheric pressure. If the sealed end be broken off, the liquid will sink to *b*. Only a portion, however, of this displacement is to be placed to the credit of the compression of the liquid by atmospheric pressure. For if we place the

Fig. 63.

glass before breaking off the top under an air-pump and exhaust the chamber, the liquid will sink to *c*. This last phenomenon is due to the fact that the pressure that bears down on the exterior of the glass and diminishes its capacity, is now removed. On breaking off the top, this exterior pressure of the atmosphere is compensated for by the interior pressure then introduced, and an enlargement of the capacity of the glass again sets in. The portion *c b,* therefore, answers to the actual compression of the liquid by the pressure of the atmosphere.

The first to institute exact experiments on the compressibility of water, was OERSTED, who employed to

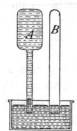

Fig. 64.

this end a very ingenious method. A thermometer glass *A* (Fig. 64) is filled with boiled water and is inverted, with open mouth, into a vessel of mercury. Near it stands a manometer tube *B* filled with air and likewise inverted with open mouth in the mercury. The whole apparatus is then placed in a vessel filled with water, which is compressed by the aid of a pump. By this means the water in *A* is also compressed, and the filament of quicksilver which rises in the capillary tube of the thermometer-glass indicates this compression. The alteration of capacity which the glass *A* suffers in the present instance, is merely that arising from the pressing together of its walls by forces which are equal on all sides.

The most delicate experiments on this subject have been conducted by GRASSI with an apparatus constructed by Regnault, and computed with the assistance of Lamé's correction-formulae. To give a tangible idea of the compressibility of water, we will remark that Grassi observed for boiled water at 0° under an increase of one atmospheric pressure a diminution of the original volume amounting to 5 in 100,000 parts. If we imagine, accordingly, the vessel A to have the capacity of one liter (1000 ccm.), and affix to it a capillary tube of 1 sq. mm. cross-section, the quicksilver filament will ascend in it 5 cm. under a pressure of one atmosphere.

9 Surface-pressure, accordingly, induces a physical alteration in a liquid (an alteration in density), which can be detected by sufficiently delicate means—even optical. We are always at liberty to think that portions of a liquid under a higher pressure are more dense, though it may be very slightly so, than parts under a less pressure.

Let us imagine now, we have in a liquid (in the interior of which no forces act and the gravity of which we accordingly neglect) two portions subjected to unequal pressures and contiguous to one another. The portion under the greater pressure, being denser, will expand, and press against the portion under the less pressure, until the forces of elasticity as lessened on the one side and increased on the other establish equilibrium at the bounding surface and both portions are equally compressed.

If we endeavor, now, quantitatively to elucidate our mental conception of these two facts, the easy mobility and the compressibility of the parts of a liquid, so that they will fit the most diverse classes of experience, we shall arrive at the following proposition: When equilibrium subsists in a liquid, in the interior of which no forces act and the gravity of which we neglect, the same equal pressure is exerted on each and every equal surface-element of that liquid, however and wherever situated. The pressure, therefore, is the same at all points and is independent of direction.

Special experiments in demonstration of this principle have, perhaps, never been instituted with the requisite degree of exactitude. But the proposition has by our experience of liquids been made very familiar, and readily explains it.

10. If a liquid is enclosed in a vessel (Fig. 65) which is supplied with a piston A, the cross-section of which is unit in area, and with a piston B which

Fig. 65.

for the time being is made stationary, and on the piston A a load p be placed, then the same pressure p, gravity neglected, will prevail throughout all the parts of the vessel. The piston will penetrate inward and the walls of the vessel will continue to be deformed till the point is reached at which the elastic forces of the rigid and fluid bodies perfectly equilibrate one another. If then we imagine the piston B, which has the cross-section f, to be movable, a force $f \cdot p$ alone will keep it in equilibrium.

Concerning Pascal's deduction of the proposition before discussed from the principle of virtual displacements, it is to be remarked that the conditions of displacement which he perceived hinge wholly upon the fact of the ready mobility of the parts and on the equality of the pressure throughout every portion of the liquid. If it were possible for a greater compression to take place in one part of a liquid than in another, the ratio of the displacements would be disturbed and Pascal's deduction would no longer be admissible. That the property of the equality of the pressure is a

property given in experience, is a fact that cannot be
escaped; as we shall readily admit if we recall to mind
that the same law that Pascal deduced for liquids also
holds good for gases, where even approximately there
can be no question of a constant volume. This latter
fact does not afford any difficulty to our view; but to
that of Pascal it does. In the case of the lever also, be
it incidentally remarked, the ratios of the virtual displacements are assured by the elastic forces of the
lever-body, which do not permit of any great deviation from these relations.

11. We shall now consider the action of liquids under the influence of gravity. The upper surface of a
liquid in equilibrium is horizontal,
NN (Fig. 66). This fact is at once
rendered intelligible when we reflect that every alteration of the surface in question elevates the center
of gravity of the liquid, and pushes
the liquid mass resting in the shaded

Fig. 66.

space beneath NN and having the center of gravity S
into the shaded space above NN having the center of
gravity S'. Which alteration, of course, is at once reversed by gravity.

Let there be in equilibrium in a vessel a heavy
liquid with a horizontal upper surface. We consider
(Fig. 67) a small rectangular
parallelepipedon in the interior.
The area of its horizontal base,
we will say, is a, and the length
of its vertical edges dh. The
weight of this parallelepipedon is
therefore $a\,d\,h\,s$, where s is its

Fig. 67.

specific gravity. If the parallelepipedon does not sink,

this is possible only on the condition that a greater pressure is exerted on the lower surface by the fluid than on the upper. The pressures on the upper and lower surfaces we will respectively designate as ap and $a(p + dp)$. Equilibrium obtains when $adh \cdot s = adp$ or $dp/dh = s$, where h in the downward direction is reckoned as positive. We see from this that for equal increments of h vertically downwards the pressure p must, correspondingly, also receive equal increments. So that $p = hs + q$; and if q, the pressure at the upper surface, which is usually the pressure of the atmosphere, becomes $= 0$, we have, more simply, $p = hs$, that is, the pressure is proportional to the depth beneath the surface. If we imagine the liquid to be pouring into a vessel, and this condition of affairs not yet attained, every liquid particle will then sink until the compressed particle beneath balances by the elasticity developed in it the weight of the particle above.

From the view we have here presented it will be further apparent, that the increase of pressure in a liquid takes place solely in the direction in which gravity acts. Only at the lower surface, at the base, of the parallelepipedon, is an excess of elastic pressure on the part of the liquid beneath required to balance the weight of the parallelepipedon. Along the two sides of the vertical containing surfaces of the parallelepipedon, the liquid is in a state of equal compression, since no force acts in the vertical containing surfaces that would determine a greater compression on the one side than on the other.

If we picture to ourselves the totality of all the points of the liquid at which the same pressure p acts, we shall obtain a surface—a so-called *level surface*. If we displace a particle in the direction of the action of gravity, it undergoes a change of pressure. If we displace it at right angles to the direction of the action of gravity, no alteration of pressure takes place. In the latter case it remains on the same level surface, and the element of the level surface, accordingly, stands at right angles to the direction of the force of gravity.

Imagining the earth to be fluid and spherical, the level surfaces are concentric spheres, and the directions of the forces of gravity (the radii) stand at right angles to the elements of the spherical surfaces. Similar observations are admissible if the liquid particles be acted on by other forces than gravity, magnetic forces, for example.

The level surfaces afford, in a certain sense, a diagram of the force-relations to which a fluid is subjected; a view further elaborated by analytical hydrostatics.

12. The increase of the pressure with the depth below the surface of a heavy liquid may be illustrated by a series of experiments which we chiefly owe to Pascal. These experiments also well illustrate the fact, that the pressure is independent of the direction. In Fig. 68, 1, is an empty glass tube *g* ground off at the bottom and closed by a metal disc *p p,* to which a string is attached, and the whole plunged into a vessel of water. When immersed to a sufficient depth we may let the string go, without the metal disc, which is supported by the pressure of the liquid, falling. In 2, the metal disc is replaced by a tiny column of mercury. In (3) we dip an open siphon tube filled with mercury into the water, we shall see the mercury, in consequence of the pressure at *a*, rise into the longer arm. In 4, we see a tube, at the lower extremity of which a leather bag filled with mercury is tied: continued immersion forces the mercury higher and higher into the tube. In 5, a piece of wood *h* is driven by the

pressure of the water into the small arm of an empty
siphon tube. In 6, a piece of
wood H immersed in mercury
adheres to the bottom of the
vessel, and is pressed firmly
against it for as long a time as
the mercury is kept from
working its way underneath it.

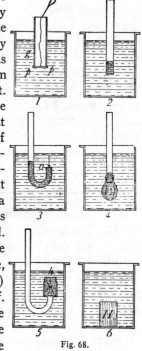

13. Once we have made
quite clear to ourselves that
the pressure in the interior of
a heavy liquid increases proportionally to the depth below the surface, the law that
the pressure at the base of a
vessel is independent of its
form will be readily perceived.
The pressure increases as we
descend at an equal rate,
whether the vessel (Fig. 69)
has the form $a\,b\,c\,d$ or $e\,b\,c\,f$.
In both cases the walls of the
vessel where they meet the
liquid, go on deforming till the

Fig. 68.

point is reached at which they equilibrate by the elasticity developed in them the pressure exerted by the
fluid, that is, take the place as regards pressure of the
fluid adjoining. This fact is a direct justification of
Stevinus's fiction of the solidified fluid supplying the
place of the walls of the vessel. The pressure on the
base always remains $P = A\,h\,s$, where A denotes the
area of the base, h the depth of the horizontal plane
base below the level, and s the specific gravity of the
liquid.

The fact that, the walls of the vessel being neglected, the vessels 1, 2, 3 of Fig. 70 of equal base-area and equal pressure-height

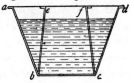

Fig. 69.

weigh differently in the balance, of course in no wise contradicts the laws of pressure mentioned. If we take into account the lateral pressure, we shall see that in the case of 1 we have left an extra component downwards, and in the case of 3 an extra component upwards, so that on the whole the resul-

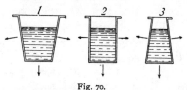

Fig. 70.

tant superficial pressure is always equal to the weight.

14. The principle of virtual displacements is admirably adapted to the acquisition of clearness and comprehensiveness in cases of this character, and we shall accordingly make use of it. To begin with, however, let the following be noted. If the weight q (Fig. 71) descend from position 1 to position 2, and a weight of exactly the same size move at the same time from 2 to 3, the

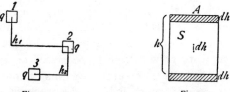

Fig. 71. Fig. 72.

work performed in this operation is $q h_1 + q h_2 = q (h_1 + h_2)$, the same, that is, as if the weight q passed directly from 1 to 3 and the weight at 2 remained in its original position. The observation is easily generalized.

Let us consider a heavy homogeneous rectangular parallelepipedon, with vertical edges of the length h, base A, and the specific gravity s (Fig. 72). Let this parallelepipedon (or, what is the same thing, its center of gravity) descend a distance $d\,h$. The work done is then $A\,h\,s{\cdot}d\,h$, or, also, $A\,d\,h\,s{\cdot}h$. In the first expression we conceive the whole weight $A\,h\,s$ displaced the vertical distance $d\,h;$ in the second we conceive the weight $A\,d\,h\,s$ as having descended from the upper shaded space to the lower shaded space the distance h,

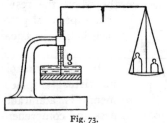

and leave out of account the rest of the body. Both methods of conception are admissible and equivalent.

15. With the aid of this observation we shall obtain a clear insight into

Fig. 73.

the paradox of Pascal, which consists of the following. The vessel g (Fig. 73), fixed to a separate support and consisting of a narrow upper and a very broad lower cylinder, is closed at the bottom by a movable piston, which, by means of a string passing through the axis of the cylinders, is independently suspended from the extremity of one arm of a balance. If g be filled with water, then, despite the smallness of the quantity of water used, there will have to be placed on the other scale-pan, to balance it, several weights of considerable size, the sum of which will be $A\,h\,s,$ where A is the piston-area, h the height of the liquid, and S its specific gravity. But if the liquid be frozen and the mass loosened from the walls of the vessel, a very small weight will be sufficient to preserve equilibrium.

Let us look to the virtual displacements of the two

cases (Fig. 74). In the first case, supposing the piston to be lifted a distance $d\,h$, the virtual moment is $A\,d\,h\,s{\cdot}h$ or $A\,h\,s{\cdot}d\,h$. It thus comes to the same thing,

whether we consider the mass that the motion of the piston displaces to be lifted to the upper surface of the fluid through the entire pressure-height, or consider the entire weight $A\,h\,s$ lifted the dis-

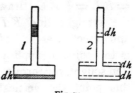

Fig. 74.

tance of the piston-displacement $d\,h$. In the second case, the mass that the piston displaces is not lifted to the upper surface of the fluid, but suffers a displacement which is much smaller—the displacement, namely, of the piston. If A, a are the sectional areas respectively of the greater and the less cylinder, and k and l their respective heights, then the virtual moment of the present case is $A\,d\,h\,s{\cdot}k + a\,d\,h\,s{\cdot}l = (A\,k + a\,l)\,s{\cdot}d\,h;$ which is equivalent to the lifting of a much smaller weight $(A\,k + a\,l)\,s$, the distance $d\,h$.

16. The laws relating to the lateral pressure of liquids are but slight modifications of the laws of basal pressure. If we have, for example, a cubical vessel of 1 decimeter on the side, which is a vessel of liter capacity, the pressure on any one of the vertical lateral walls $ABCD$, when the vessel is filled with water, is easily determinable. The deeper the migratory element considered descends beneath the surface, the greater the pressure will be to which it is subjected. We easily perceive, thus, that the pressure on a lateral wall is represented by a wedge of water $A\,B\,C\,D\,H\,I$ resting upon the wall horizontally placed, where $I\,D$ is at right angles to $B\,D$ and $I\,D = H\,C = A\,C$. The lateral pressure accordingly is equal to half a kilogram.

To determine the point of application of the resultant pressure, conceive $A B C D$ again horizontal with the water-wedge resting upon it. We cut off $A K = B L = \frac{2}{3} A C$, draw the straight line $K L$ and bisect it at M; M is the point of application sought, for through this point the vertical line cutting the center of gravity of the wedge passes.

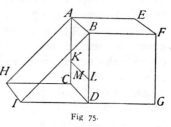

Fig 75.

A plane inclined figure forming the base of a vessel filled with a liquid, is divided into the elements a, a', a'' . . . with the depths h, h', h'' . . . below the level of the liquid. The pressure on the base is

$$(a h + a' h' + a'' h'' + \ldots)\, s.$$

If we call the total base-area A, and the depth of its center of gravity below the surface H, then

$$\frac{\alpha h + \alpha' h' + \alpha'' h'' + \ldots}{\alpha + \alpha' + \alpha'' + \ldots} = \frac{\alpha h + \alpha' h' + \ldots}{A} = H,$$

whence the pressure on the base is $A H s$.

17. The principle of Archimedes can be deduced in various ways. After the manner of Stevinus, let us conceive in the interior of the liquid a portion of it solidified. This portion now, as before, will be supported by the circumnatant liquid. The resultant of the forces of pressure acting on the surfaces is accordingly applied at the center of gravity of the liquid displaced by the solidified body, and is equal and opposite to its weight. If now we put in the place of the solidified liquid another different body of the same form, but

of a different specific gravity, the forces of pressure at
the surfaces will remain the same. Accordingly, there
now act on the body two forces, the weight of the body,
applied at the center of gravity of the body, and the upward buoyancy, the resultant of the surface-pressures,
applied at the center of gravity of the displaced liquid.
The two centers of gravity in question coincide only in
the case of homogeneous solid bodies.

If we immerse a rectangular parallelepipedon of altitude h and base a, with edges vertically placed, in a
liquid of specific gravity s, then the pressure on the
upper basal surface, when at a depth k below the level
of the liquid is $a\,k\,s$, while the pressure on the lower
surface is $a\,(k+h)\,s$. As the lateral pressures destroy
each other, an excess of pressure $a\,h\,s$ upwards remains; or, where v denotes the volume of the parallelepipedon, an excess $v \cdot s$.

We shall approach nearest the fundamental conception from which Archimedes started, by recourse to
the principle of virtual displacements. Let a parallelepipedon (Fig. 76) of the specific gravity σ, base a,
and height h sink the distance $d\,h$. The virtual moment of the transference from the upper into the lower
shaded space of the figure will be $a\,d\,h \cdot \sigma\,h$. But while

Fig. 76.

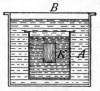

Fig. 77.

this is done, the liquid rises from the lower into the upper space, and its moment is $a\,d\,h\,s\,h$. The total virtual moment is therefore $a\,h\,(\sigma - s)\,d\,h = (p - q)$

d h, where *p* denotes the weight of the body and *q* the
weight of the displaced liquid.

18. The question might occur to us, whether the
upward pressure of a body in a liquid is affected by the
immersion of the latter in another liquid. As a fact,
this very question has been proposed. Let therefore
(Fig. 77) a body *K* be submerged in a liquid *A* and the
liquid with the containing vessel in turn submerged in
another liquid *B*. If in the determination of the loss
of weight in *A* it were proper to take account of the
loss of weight of *A* in *B*, then *K's* loss of weight would
necessarily vanish when the fluid *B* became identical
with *A*. Therefore, *K* immersed in *A* would suffer a
loss of weight and it would suffer none. Such a rule
would be nonsensical.

With the aid of the principle of virtual displacements, we easily comprehend the more complicated
cases of this character. If a body be first gradually
immersed in *B*, then partly in *B* and partly in *A*,
finally in *A* wholly; then, in the second case, considering the virtual moments, both liquids are to be taken
into account in the proportion of the volume of the
body immersed in them. But as soon as the body is
wholly immersed in *A*, the level of *A* on further displacement no longer rises, and therefore *B* is no longer
of consequence.

19. Archimedes's principle may be illustrated by a
pretty experiment. From the one extremity of a scale-
beam (Fig. 78) we hang a hollow cube *H*, and beneath
it a solid cube *M*, which exactly fits into the first cube.
We put weights into the opposite pan, until the scales
are in equilibrium. If now *M* be submerged in water
by lifting a vessel which stands beneath it, the equilibrium will be disturbed; but it will be immediately
restored if *H*, the hollow cube, be filled with water.

A counter-experiment is the following. *H* is left suspended alone at the one extremity of the balance, and into the opposite pan is placed a vessel of water, above which on an independent support *M* hangs by a thin wire. The scales are brought to equilibrium. If now *M* be lowered until it is immersed in the water, the equilibrium of the scales will be disturbed; but on filling *H* with water, it will be restored.

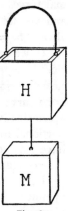

Fig. 78.

At first glance this experiment appears a little paradoxical. We feel, however, instinctively, that *M* cannot be immersed in the water without exerting a pressure that affects the scales. When we reflect, that the level of the water in the vessel rises, and that the solid body *M* equilibrates the surface-pressure of the water surrounding it, that is to say represents and takes the place of an equal volume of water, it will be found that the paradoxical character of the experiment vanishes.

20. The most important statical principles have been reached in the investigation of solid bodies. This course is accidentally the *historical* one, but it is by no means the only possible and *necessary* one. The different methods that Archimedes, Stevinus, Galileo, and the rest, pursued, place this idea clearly enough before the mind. As a matter of fact, general statical principles, might, with the assistance of some very simple propositions from the statics of rigid bodies, have been reached in the investigation of liquids. Stevinus certainly came very near such a discovery. We shall stop a moment to discuss the question.

Let us imagine a liquid, the weight of which we neglect. Let this liquid be enclosed in a vessel and subjected to a definite pressure. A portion of the liquid, let us suppose, solidifies. On the closed surface normal forces act proportional to the elements of the area, and we see without difficulty that their resultant will always be $= 0$.

If we mark off by a closed curve a portion of the closed surface, we obtain, on either side of it, a non-closed surface. All surfaces which are bounded by the same curve (of double curvature) and on which forces act normally (in the same sense) proportional to the elements of the area, have lines coincident in position for the resultants of these forces.

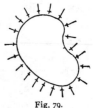

Fig. 79.

Let us suppose, now, that a fluid cylinder, determined by any closed plane curve as the perimeter of its base, solidifies. We may neglect the two basal surfaces, perpendicular to the axis. And instead of the cylindrical surface the closed curve simply may be considered. From this method follow quite analogous propositions for normal forces proportional to the elements of a plane curve.

If the closed curve pass into a triangle, the consideration will shape itself thus. The resultant normal

Fig. 80.

forces applied at the middle points of the sides of the triangle, we represent in direction, sense, and magnitude by straight lines (Fig. 80). The lines mentioned intersect at a point — the center of the circle described about the triangle. It will further be noted, that by the simple parallel displacement of the lines representing the forces a triangle is constructible which is similar and congruent to the original triangle.

Thence follows this proposition:

Any three forces, which, acting at a point, are proportional and parallel in direction to the sides of a triangle, and which on meeting by parallel displacement form a congruent triangle, are in equilibrium. We see at once that this proposition is simply a different form of the principle of the parallelogram of forces.

If instead of a triangle we imagine a polygon, we shall arrive at the familiar proposition of the polygon of forces.

We conceive now in heavy liquid of specific gravity $\varkappa$ a portion solidified. On the element a of the closed encompassing surface there acts a normal force $a\varkappa z$, where z is the distance of the element from the level of the liquid. We know from the outset the result.

If normal forces which are determined by $a\varkappa z$, where a denotes an element of area and z its perpendicular distance from a given plane E, act on a closed surface inwards, the resultant will be $V\cdot\varkappa$, in which expression V represents the enclosed volume. The resultant acts at the center of gravity of the volume, is perpendicular to the plane mentioned, and is directed towards this plane.

Under the same conditions let a rigid curved surface be bounded by a plane curve, which encloses on the plane the area A. The resultant of the forces acting on the curved surface is R, where

$$R^2 = (AZ\varkappa)^2 + (V\varkappa)^2 - AZV\varkappa^2 \cos v,$$

in which expression Z denotes the distance of the center of gravity of the surface A from E, and v the normal angle of E and A.

In the proposition of the last paragraph mathematically practised readers will have recognized a particular case of Green's Theorem, which consists in the reduction of surface-integrations to volume-integrations or *vice versa*.

We may, accordingly, *see into* the force-system of a fluid in equilibrium, or, if you please, *see out* of it, systems of forces of greater or less complexity, and thus reach by a short path propositions *a posteriori*. It is a mere accident that Stevinus did not light on these propositions. The method here pursued corresponds exactly to his. In this manner new discoveries can still be made.

21. The paradoxical results that were reached in the investigation of liquids, supplied a stimulus to further reflection and research. It should also not be left unnoticed, that the conception of a *physico-mechanical continuum* was first formed on the occasion of the investigation of liquids. A much freer and much more fruitful mathematical mode of view was developed thereby, than was possible through the study even of systems of several solid bodies. The origin, in fact, of important modern mechanical ideas, as for instance that of the potential, is traceable to this source.

VII.

THE PRINCIPLES OF STATICS IN THEIR APPLICATION TO GASEOUS BODIES.

1. The same views that subserve the ends of science in the investigation of liquids are applicable with but slight modifications to the investigation of gaseous bodies. To this extent, therefore, the investigation of gases does not afford mechanics any very rich returns. Nevertheless, the first steps that were taken in this

province possess considerable significance from the point of view of the progress of civilization and so have a high import for science generally.

Although the ordinary man has abundant opportunity, by his experience of the resistance of the air, by the action of the wind, and the confinement of air in bladders, to perceive that air is of the nature of a body, yet this fact manifests itself infrequently, and never in the obvious and unmistakable way that it does in the case of solid bodies and fluids. It is known, to be sure, but is not sufficiently familiar to be prominent in popular thought. In ordinary life the presence of the air is scarcely ever thought of.

Our modern notions with regard to the nature of air are a direct continuation of the ancient ideas. Anaxagoras proves the corporeality of air from its resistance to compression in closed bags of skin, and from the gathering up of the expelled air (in the form of bubbles?) by water (Aristotle, *Physics,* IV., 9). According to Empedocles, the air prevents the water from penetrating into the interior of a vessel immersed with its aperture downwards (Gomperz, *Griechische Denker,* I., p. 191). Philo of Byzantium employs for the same purpose an inverted vessel having in its bottom an orifice closed with wax. The water will not penetrate into the submerged vessel until the wax cork is removed, whereupon the air escapes in bubbles. An entire series of experiments of this kind is performed, in almost the precise form customary in the schools today (*Philonis lib. de ingeniis spiritualibus,* in V. Rose's *Anecdota græca et latina*). Hero describes in his *Pneumatics* many of the experiments of his predecessors, with additions of his own; in theory he is an adherent of Strato, who occupied an intermediate position between Aristotle and Democritus. An

absolute and continuous vacuum, he says, can be
produced only artificially, although numberless tiny
vacua exist between the particles of bodies, including
air, just as air does among grains of sand. This is
proved, in quite the same ingenious fashion as in our
present elementary books, from the possibility of rare-
fying and compressing bodies, including air (inrush-
ing and outrushing of the air in Hero's ball). An argument of Hero's for the existence of vacua (pores)
between corporeal particles rests on the fact that rays
of light penetrate water. The result of artificially increasing a vacuum, according to Hero and his predecessors, is always the attraction and solicitation of
adjacent bodies. A light vessel with a narrow aperture remains hanging to the lips after the air has been
exhausted. The orifice may be closed with the finger
and the vessel submerged in water. "If the finger
be released, the water will rise in the vacuum created,
although the movement of the liquid upward is not
according to nature. The phenomenon of the cup-
ping-glass is the same; these glasses, when placed
on the body, not only do not fall off, although they
are heavy enough, but they also draw out adjacent
particles through the pores of the body." The bent
siphon is also treated at length. "The filling of the
siphon on exhaustion of the air is accomplished by
the liquid's closely following the exhausted air, for
the reason that a continuous vacuum is inconceivable." If the two arms of the siphon are of the same
length, nothing flows out. "The water is held in equilibrium as in a balance." Hero accordingly conceives
of the flow of water as analogous to the movement
of a chain hanging with unequal lengths over a pulley.
The union of the two columns, which for us is pre-

served by the pressure of the atmosphere, is cared for in his case by the "inconceivability of a continuous vacuum." It is shown at length, not that the smaller mass of water is attracted and drawn along by the greater mass, and that conformably to this principle water cannot flow upwards, but rather that the phenomenon is in harmony with the principle of communicating vessels. The many pretty and ingenious tricks which Hero describes in his *Pneumatics* and in his *Automata,* and which were designed partly to entertain and partly to excite wonder, offer a charming picture of the material civilization of the day rather than excite our scientific interest. The automatic sounding of trumpets and the opening of temple doors, with the thunder simultaneously produced, are not matters which interest science properly so called. Yet Hero's writings and notions contributed much toward the diffusion of physical knowledge (compare W. Schmidt, *Hero's Werke,* Leipzig, 1899, and Diels, *System des Strato, Sitzungsberichte der Berliner Akademie,* 1893).

Although the ancients, as we may learn from the accounts of Vitruvius, possessed instruments like the so-called hydraulic organs, which were based on the condensation of air, although the invention of the airgun is traced back to Ctesibius, and this instrument was also known to Guericke, the notions which people held with regard to the nature of the air even as late as the seventeenth century were exceedingly curious and loose. We must not be surprised, therefore, at the intellectual commotion which the first more important experiments in this direction evoked. The enthusiastic description which Pascal gives of Boyle's air-pump experiments is readily comprehended, if we transport our-

OTTO De GUERICKE

Sereniſſ. as Potentiſſ. Elector: Brandeb:

Conſiliarius et Civitat: Magdeb. Conſul:

selves back into the epoch of these discoveries. What indeed could be more wonderful than the sudden discovery that a thing which we do not see, hardly feel, and take scarcely any notice of, constantly envelopes us on all sides, penetrates all things; that it is the most important condition of life, of combustion, and of gigantic mechanical phenomena. It was first made manifest on this occasion, perhaps, by a great and striking disclosure, that physical science is not restricted to the investigation of palpable and grossly sensible processes.

To form some idea of the slowness with which the new notions about air became more familiar to men, it is enough to read the article on air which Voltaire,[1] one of the most enlightened men of his time, wrote in his *Dictionnaire Philosophique* from the *Encyclopédie*, in 1764—a century after Guericke, Boyle, and Pascal, and not long before the discoveries of Cavendish, Priestley, Volta, and Lavoisier—that air is not visible and, quite generally, is not perceptible; all the functions that we ascribe to the air can be discharged by the perceptible exhalations whose existence we have no grounds for doubting. How can the air enable us to hear the different notes of a melody simultaneously? Air and ether are, with respect to the certainty of their existence, put on the same level.

[1] [Voltaire's article "Air" in the first volume of his *Questions sur l' Encyclopédie par des Amateurs* was republished in the *Collection complette des Œuvres de Mr de* . . . (vol. xxi, Geneva, 1774, pp. 73-81; the part noticed in the text above, which contains Voltaire's own opinions, is on pp. 77-79). The *Questions* were first published in 1770-72 in seven volumes, and the article "Air" is in the first part (1770). The *Dictionnaire Philosophique* was first published in 1764, and was greatly augmented in various subsequent editions from 1767 to 1776. The editor, de Kehl, in 1785-89, included various works under the single title of *Dictionnaire Philosophique*, viz., the *Dictionnaire Philosophique*, the *Questions*. a manuscript dictionary entitled *L'Opinion par l' Alphabet*, Voltaire's articles in the great *Encyclopédie*, and several articles destined for the *Dictionnaire de l' Académie Française*. The article "Air" is contained in vol. xxvi of M. Beuchot's *Œuvres de Voltaire* (72 volumes, Paris, 1829), pp. 136-147.]

2. In Galileo's time philosophers explained the phenomenon of suction, the action of syringes and pumps by the so-called *horror vacui*—nature's abhorrence of a vacuum. Nature was thought to possess the power of preventing the formation of a vacuum by laying hold of the first adjacent thing, whatsoever it was, and immediately filling up with it any empty space that arose. Apart from the ungrounded speculative element which this view contains, it must be conceded, that to a certain extent it really represents the phenomenon. The person competent to enunciate it must actually have discerned some principle in the phenomenon. This principle, however, does not fit all cases. Galileo is said to have been greatly surprised at hearing of a newly constructed pump accidentally supplied with a very long suction-pipe which was not able to raise water to a height of more than eighteen Italian ells. His first thought was that the *horror vacui* (or the *resistenza del vacuo*) possessed a measurable power. The greatest height to which water could be raised by suction he called *altezza limitatissima*. He sought, moreover, to determine directly the weight able to draw out of a closed pump-barrel a tightly fitting piston resting on the bottom.

3. TORRICELLI hit upon the idea of measuring the resistance to a vacuum by a column of mercury instead of a column of water, and he expected to obtain a column of about 1/14 of the length of the water column. His expectation was confirmed by the experiment performed in 1643 by Viviani in the well-known manner, and which bears today the name of the Torricellian experiment. A glass tube somewhat over a meter in length, sealed at one end and filled with mercury, is stopped at the open end with the finger, inverted in a

dish of mercury, and placed in a vertical position. Removing the finger, the column of mercury falls and remains stationary at a height of about 76 cm. By this experiment it was rendered quite probable, that some very definite pressure forced the fluids into the vacuum. What pressure this was, Torricelli very soon divined.

Galileo had endeavored, some time before this, to determine the weight of the air, by first weighing a glass bottle containing nothing but air and then again weighing the bottle after the air had been partly expelled by heat. It was known, accordingly, that the air was heavy. But to the majority of men the *horror vacui* and the weight of the air were very distantly connected notions. It is possible that in Torricelli's case the two ideas came into sufficient proximity to lead him to the conviction that all phenomena ascribed to the *horror vacui* were explicable in a simple and logical manner by the pressure exerted by the weight of a fluid column—a column of air. Torricelli discovered, therefore, the pressure of the atmosphere; he also first observed by means of his column of mercury the variations of the pressure of the atmosphere.

4. The news of Torricelli's experiment was circulated in France by Mersenne, and came to the knowledge of Pascal in the year 1644. The accounts of the theory of the experiment were presumably so imperfect that PASCAL found it necessary to reflect independently thereon. (*Pesanteur de l' air.* Paris, 1663.)

He repeated the experiment with mercury and with a tube of water, or rather of red wine, 40 feet in length. He soon convinced himself by inclining the tube that the space above the column of fluid was really empty; and he found himself obliged to defend this view against the violent attacks of his countrymen. Pascal pointed out an easy way of producing the vacuum which they regarded as impossible, by the use of a glass syringe, the nozzle of which was closed with the finger under water and the piston then drawn back without much difficulty. Pascal showed, in addition, that a curved siphon 40 feet high filled with water does not flow, but can be made to do so by a sufficient inclination to the perpendicular. The same experiment was made on a smaller scale with mercury. The same siphon flows or does not flow according as it is placed in an inclined or a vertical position.

In a later performance, Pascal refers expressly to the fact of the weight of the atmosphere and to the pressure due to this weight. He shows, that minute animals, like flies, are able, without injury to themselves, to stand a high pressure in fluids, provided only the pressure is equal on all sides; and he applies this at once to the case of fishes and of animals that live in the air. Pascal's chief merit, indeed, is to have established a complete analogy between the phenomena conditioned by liquid pressure (water-pressure) and those conditioned by atmospheric pressure.

5. By a series of experiments Pascal shows that mercury in consequence of atmospheric pressure rises into a space containing no air in the same way that, in consequence of water-pressure, it rises into a space containing no water. If into a deep vessel filled with water (Fig. 81) a tube be sunk at the lower end of which a bag of mercury is tied, but so inserted that the upper end of the tube projects out of the water and thus contains only air, then the deeper the tube is sunk into the water the higher will the mercury, subjected to the constantly increasing pressure of the

water, ascend into the tube. The experiment can also
be made, with a siphon-tube, or with a
tube open at its lower end.

Fig. 81.

Undoubtedly it was the attentive consideration of this very phenomenon that
led Pascal to the idea that the barometer-
column must necessarily stand lower at
the summit of a mountain than at its
base, and that it could accordingly be employed to determine the height of mountains. He communicated this idea to his brother-in-law, Perier, who
forthwith successfully performed the experiment on
the summit of the Puy de Dôme. (Sept. 19, 1648.)

Pascal referred the phenomena connected with ad-
hesion-plates to the pressure of the atmosphere, and
gave as an illustration of the principle involved the resistance experienced when a large hat lying flat on a
table is suddenly lifted. The cleaving of wood to the
bottom of a vessel of quicksilver is
a phenomenon of the same kind.

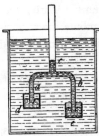

Fig. 82.

Pascal imitated the flow produced in a siphon by atmospheric
pressure, by the use of water-pressure. The two open unequal arms
a and b of a three-armed tube
$a\,b\,c$ (Fig. 82) are dipped into the
vessels of mercury e and d. If the
whole arrangement then be im-
mersed in a deep vessel of water, yet so that the long
open branch shall always project above the upper surface, the mercury will gradually rise in the branches
a and b, the columns finally unite, and a stream begin
to flow from the vessel d to the vessel e through the
siphon-tube open above to the air.

The Torricellian experiment was modified by Pascal in a very ingenious manner. A tube of the form

Fig. 83.

$a b c d$ (Fig. 83), of double the length of an ordinary barometer-tube, is filled with mercury. The openings a and b are closed with the fingers and the tube placed in a dish of mercury with the end a downwards. If now a be opened, the mercury in $c d$ will all fall into the expanded portion at c, and the mercury in $a b$ will sink to the height of the ordinary barometer-column. A vacuum is produced at b which presses the finger closing the hole painfully inwards. If b also be opened the column in $a b$ will sink completely, while the mercury in the expanded portion c, being now exposed to the pressure of the atmosphere, will rise in $c d$ to the height of the barometer-column. Without an air-pump it was hardly possible to combine the experiment and the counter-experiment in a simpler and more ingenious manner than Pascal thus did.

6. With regard to Pascal's mountain-experiment, we shall add the following brief supplementary remarks. Let b_0 be the height of the barometer at the level of the sea, and let it fall, say, at an elevation of m meters, to $k b_0$, where k is a proper fraction. At a further elevation of m meters, we must expect to obtain the barometer-height $k \cdot k \, b_0$, since we here pass through a stratum of air the density of which bears to that of the first the proportion of $k : 1$. If we pass upwards to the altitude $h = n \cdot m$ meters, the barometer-height corresponding thereto will be

$$b_h = k^n \cdot b_o \text{ or } n = \frac{\log b_h - \log b_o}{\log k} \text{ or}$$

$$h = \frac{m}{\log k} (\log b_h - \log b_o).$$

The principle of the method is, we see, a very simple one; its difficulty arises solely from the multifarious collateral conditions and corrections that have to be looked to.

7. The most original and fruitful achievements in the domain of aërostatics we owe to OTTO VON GUERICKE. His experiments appear to have been suggested in the main by philosophical speculations. He proceeded entirely in his own way; for he first heard of the Torricellian experiment from Valerianus Magnus at the Imperial Diet of Ratisbon in 1654, where he demonstrated the experimental discoveries made by him about 1650. This statement is confirmed by his method of constructing a water-barometer which was entirely different from that of Torricelli.

Guericke's book (*Experimenta nova, ut vocantur, Magdeburgica.* Amsterdam. 1672) makes us realize the narrow views men took in his time. The fact that he was able gradually to abandon these views and to acquire broader ones by his individual endeavor speaks favorably for his intellectual powers. We perceive with astonishment how short a space of time separates us from the era of scientific barbarism, and can no longer marvel that the barbarism of the social order still so oppresses us.

In the introduction to this book and in various other places, Guericke, in the midst of his experimental investigations, speaks of the various objections to the Copernican system which had been drawn from the

Guericke   First Experiments.   (*Experim. Magdeb.*)

Bible, (objections which he seeks to invalidate,) and discusses such subjects as the locality of heaven, the locality of hell, and the day of judgment. Disquisitions on empty space occupy a considerable portion of the work.

Guericke regards the air as the exhalation or odor of bodies, which we do not perceive because we have been accustomed to it from childhood. Air, to him, is not an element. He knows that through the effects of heat and cold it changes its volume, and that it is compressible in Hero's Ball, or *Pila Heronis;* on the basis of his own experiments he gives its pressure at 20 ells of water, and expressly speaks of its weight, by which flames are forced upwards.

8. To produce a vacuum, Guericke first employed a wooden cask filled with water. The pump of a fire-engine was fastened to its lower end. The water, it was thought, in following the piston and the action of gravity, would fall and be pumped out. Guericke expected that empty space would remain. The fastenings of the pump repeatedly proved to be too weak, since in consequence of the atmospheric pressure that weighed on the piston considerable force had to be applied to move it. On strengthening the fastenings three powerful men finally accomplished the exhaustion. But, meantime the air poured in through the joints of the cask with a loud blast, and no vacuum was obtained. In a subsequent experiment the small cask from which the water was to be exhausted was immersed in a larger one, likewise filled with water. But in this case, too, the water gradually forced its way into the smaller cask.

Wood having proved in this way to be an unsuitable material for the purpose, and Guericke having remarked in the last experiment indications of success, the philosopher now took a large hollow sphere of copper and ventured to exhaust the air directly. At

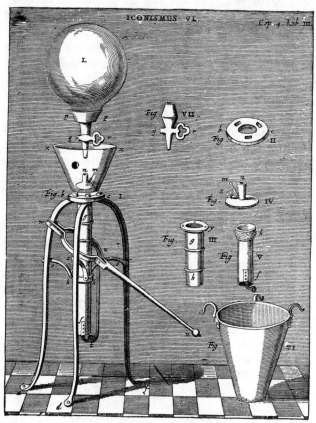

Guericke's Air-pump. (*Experim. Magdeb.*)

the start the exhaustion was successfully and easily conducted. But after a few strokes of the piston, the pumping became so difficult that four stalwart men (*viri quadrati*), putting forth their utmost efforts, could hardly budge the piston. And when the exhaustion had gone still further, the sphere suddenly collapsed, with a violent report. Finally by the aid of a copper vessel of perfect spherical form, the production of the vacuum was successfully accomplished. Guericke describes the great force with which the air rushed in on the opening of the cock.

9. After these experiments Guericke constructed an independent air-pump. A great glass globular receiver was mounted and closed by a large detachable tap in which was a stop-cock. Through this opening the objects to be subjected to experiment were placed in the receiver. To secure more perfect closure the receiver was made to stand, with its stop-cock under water, on a tripod, beneath which the pump proper was placed. Subsequently, separate receivers, connected with the exhausted sphere, were also employed in the experiments.

The phenomena which Guericke observed with this apparatus are manifold and various. The noise which water in a vacuum makes on striking the sides of the glass receiver, the violent rush of air and water into exhausted vessels suddenly opened, the escape on exhaustion of gases absorbed in liquids, the liberation of their fragrance, as Guericke expresses it, were immediately remarked. A lighted candle is extinguished on exhaustion, because, as Guericke conjectures, it derives its nourishment from the air. Combustion, as his striking remark is, is not an annihilation, but a transformation of the air.

A bell does not ring in a vacuum. Birds die in it. Many fishes swell up, and finally burst. A grape is kept fresh *in vacuo* for over half a year.

By connecting with an exhausted cylinder a long tube dipped in water, a water-barometer is constructed.

The column raised is 19-20 ells high; and Guericke explained all the effects that had been ascribed to the *horror vacui* by the principle of atmospheric pressure.

An important experiment consisted in the weighing of a receiver, first when filled with air and then when exhausted. The weight of the air was found to vary with the circumstances; namely, with the temperature and the height of the barometer. According to Guericke a definite ratio of weight between air and water does not exist.

But the deepest impression on the contemporary world was made by the experiments relating to atmospheric pressure. An exhausted sphere formed of two hemispheres tightly adjusted to one another was rent asunder with a violent report only by the traction of sixteen horses. The same sphere was suspended from a beam, and a heavily laden scale-pan was attached to the lower half.

The cylinder of a large pump is closed by a piston. To the piston a rope is tied which leads over a pulley and is divided into numerous branches on which a great number of men pull. The moment the cylinder is connected with an exhausted receiver, the men at the ropes are thrown to the ground. In a similar manner a huge weight is lifted.

Guericke mentions the compressed-air gun as something already known, and constructs independently an instrument that might appropriately be called a rarified-air gun. A bullet is driven by the external atmospheric pressure through a suddenly exhausted tube, forces aside at the end of the tube a leather valve which closes it, and then continues its flight with a considerable velocity.

Closed vessels carried to the summit of a mountain and opened, blow out air; carried down again in the same manner, they suck in air. From these and other experiments Guericke discovers that the air is elastic.

10. The investigations of Guericke which were, in part, demonstrated as early as 1654, were continued by an Englishman, ROBERT BOYLE, in 1660. The new experiments which Boyle had to supply were few. He observes the propagation of light in a vacuum and the action of a magnet through it; lights tinder by means of a burning glass; brings the barometer under the receiver of the air-pump, and was the first to construct a balance-manometer ["the statical manometer"]. The ebullition of heated fluids and the freezing of water on exhaustion were first observed by him.

Of the air-pump experiments common at the present day may be mentioned the experiment with falling bodies, which confirms simply the view of Galileo that when the resistance of the air has been eliminated light and heavy bodies both fall with the same velocity. In an exhausted glass tube a leaden bullet and a piece of paper are placed. Putting the tube in a vertical position and quickly turning it about a horizontal axis through an angle of 180°, both bodies will be seen to arrive simultaneously at the bottom of the tube.

Of the quantitative data we will mention the following. The atmospheric pressure that supports a column of mercury of 76 cm. is easily calculated from the specific gravity 13.60 of mercury to be 1.0336 kg. to 1 sq.cm. The weight of 1000 cu.cm. of pure, dry air at 0°C. and 760 mm. of pressure at Paris at an elevation of 6 meters will be found to be 1.293 grams, and the corresponding specific gravity, referred to water, to be 0.001293.

11. Guericke knew of only *one* kind of air. We may imagine therefore the excitement it created when

in 1755 BLACK discovered carbonic acid gas (fixed air)
and CAVENDISH in 1766 hydrogen (inflammable air),
discoveries which were soon followed by other similar
ones. The dissimilar
physical properties of
gases are very strik-
ing. Faraday has il-
lustrated their great
inequality of weight
by a beautiful lecture-
experiment. If from
a balance in equilib-

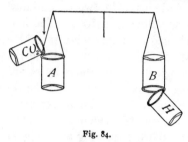

Fig. 84.

rium, we suspended (Fig. 84) two beakers A, B, the one
in an upright position and the other with its opening
downwards, we may pour heavy carbonic acid gas from
above into the one and light hydrogen from beneath
into the other. In both instances the balance turns in
the direction of the arrow. Today, as we know, the
decanting of gases can be made directly visible by the
optical method of Foucault and Toeppler.

12. Soon after Torricelli's discovery, attempts were
made to employ practically the vacuum thus produced.
The so-called mercurial air-pumps were tried. But no
such instrument was successful until the present century. The mercurial air-pumps now in common use
are really barometers of which the extremities are supplied with large expansions and so connected that their
difference of level may be easily varied. The mercury
takes the place of the piston of the ordinary air-pump.

13. The expansive force of the air, a property observed by Guericke, was more accurately investigated
by BOYLE, and, later, by MARIOTTE. The law which
both found is as follows: If V be called the volume of
a given quantity of air and P its pressure on unit area

of the containing vessel, then the product $V \cdot P$ is always = a constant quantity. If the volume of the enclosed air be reduced one-half, the air will exert double the pressure on unit of area; if the volume of the enclosed quantity be doubled, the pressure will sink to one-half; and so on. It is quite correct—as a number of English writers have maintained in recent times—that Boyle and not Mariotte is to be regarded as the discoverer of the law that usually goes by Mariotte's name. Not only is this true, but it must also be added that Boyle knew that the law did not hold exactly, whereas this fact appears to have escaped Mariotte.

The method pursued by Mariotte in the ascertainment of the law was very simple. He partially filled Torricellian tubes with mercury, measured the volume of the air remaining, and then performed the Torricellian experiment. The new volume of air was thus obtained, and by subtracting the height of the column of mercury from the barometer-height, also the new pressure to which the same quantity of air was now subjected.

Fig. 85.

To condense the air Mariotte employed a siphon-tube with vertical arms. The smaller arm in which the air was contained was sealed at the upper end; the longer, into which the mercury was poured, was open at the upper end. The volume of the air was read off on the graduated tube, and to the difference of level of the mercury in the two arms the barometer-height was added. At the present day both sets of experiments are performed in the simplest manner by fastening a cylindrical glass tube (Fig. 86) *r r*, closed at the top, to a vertical scale and connecting it by a caoutchouc

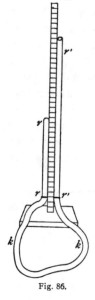

Fig. 86.

tube $k\,k$ with a second open glass tube $r'\,r'$, which is movable up and down the scale. If the tubes be partly filled with mercury, any difference of level whatsoever of the two surfaces of mercury may be produced by displacing $r'\,r'$, and the corresponding variations of volume of the air enclosed in $r\,r$ observed.

It struck Mariotte on the occasion of his investigations that any small quantity of air cut off completely from the rest of the atmosphere and therefore not directly affected by the latter's weight, also supported the barometer-column; as where, to give an instance, the open arm of a barometer-tube is closed. The simple explanation of this phenomenon, which, of course, Mariotte immediately found, is this, that the air before enclosure must have been compressed to a point at which its tension balanced the gravitational pressure of the atmosphere; that is to say, to a point at which it exerted an equivalent elastic pressure.

We shall not enter here into the details of the arrangement and use of air-pumps, which are readily understood from the law of Boyle and Mariotte.

14. It simply remains for us to remark, that the discoveries of aërostatics furnished so much that was new and wonderful that a valuable intellectual stimulus proceeded from the science.

CHAPTER II.

THE DEVELOPMENT OF THE PRINCIPLES OF DYNAMICS.

I.

GALILEO'S ACHIEVEMENTS.

1. We now pass to the discussion of the fundamental principles of dynamics. This is entirely a modern science. The mechanical speculations of the ancients, particularly of the Greeks, related wholly to statics. Only in mostly unsuccessful paths, does their thinking extend into dynamics. We shall readily recognize the correctness of this assertion if we but consider a moment a few propositions held by the Aristotelians of Galileo's time. To explain the descent of heavy bodies and the rising of light bodies, (in liquids for instance,) it was assumed that every object sought its *place*: the place of heavy bodies was below, the place of light bodies was above. Motions were divided into natural motions, as that of descent, and violent motions, as, for example, that of a projectile. From some few superficial experiences and observations, philosophers had concluded that heavy bodies fall more quickly and lighter bodies more slowly, or, more precisely, that bodies of greater weight fall more quickly and those of less weight more slowly. It is sufficiently obvious from this that the dynamical knowledge of the Greeks was very insignificant. Besides, the views of Aristotle found opponents even in antiquity. Especially the perverse opinion of Aristotle that the continued motion of a body which is projected is brought about by means of the *air* which has been set in motion at the same time plainly offered critics

an obvious point of attack. According to Wohlwill's researches, Philoponos, a writer of the sixth century

of the Christian era, expressly contested this view—a view contrary to every sound instinct. Why must the moving hand touch the stone at all if the air manages everything? This natural question asked by Philoponos did not fail to exercise an influence on Leonardo, Cardano, Benedetti, Giordano Bruno, and Galileo. Philoponos also contradicts the assertion that bodies of greater weight fall more quickly, and refers to observation. Finally, Philoponos shows a modern trait in that he denies any force to the *position in itself*, but attributes to bodies the effort to preserve their order (*cf.* Wohlwill, "Ein Vorgänger Galilei's im 6. Jahrhundert," *Physik. Zeitschrift von Riecke und Simon,* 7. Jahrg, No. 1, pp. 23-32).

One of the most important predecessors of Galileo, to whom we have already referred in another place, was Leonardo da Vinci (1452-1519). It was impossible for Leonardo's achievements to have influenced the development of science at the time, for the reason that they were first made known by the publication of Venturi in 1797. Leonardo knew the ratio of the times of descent down the slope and the height of an inclined plane. Frequently also knowledge of the law of inertia is attributed to him. Indeed, some sort of instinctive knowledge of the persistence of motion once begun will not be gainsaid to any normal man. But Leonardo seems to have gone much farther than this. He knows that from a column of checkers one of the pieces may be knocked out without disturbing the others; he knows that a body in motion will move longer according as the resistance is less, but he believes that the body will move a distance proportional to the impulse, and nowhere expressly speaks of the persistence of the motion when the resistance is altogether removed. (Compare Wohlwill, *Bibliotheca Mathematica,* Stockholm, 1888, p. 19).

Benedetti, an immediate predecessor of Galileo,

(1530-1590) knows that falling bodies are accelerated, and explains the acceleration as due to the summation of the impulses of gravity, just as the increase of the projectile-force of a stone by means of a sling is reduced to an aggregation of impulses. Such an impulse has, according to Benedetti, the tendency to force the body forward in a straight line. A body projected horizontally approaches the earth more slowly; consequently, the gravity of the earth appears to be partly taken away. A spinning top does not fall, but stands on the end of its axis, because its parts have the tendency to fly away tangentially and perpendicularly to the axis, and by no means to approach the earth. Benedetti ascribes the continued motion of a projected body not to the influence of the air but to a "virtus impressa," without attaining full clearness in relation to the problems (G. Benedetti, *Sulle proporzioni dei motu locali,* Venice, 1553; *Divers. speculat. math. et physic. liber,* Turin, 1585).

2. Galileo, in the works of his youth (in Pisa), as has become known by the recent critical edition, appears as an opponent of Aristotle, as doing honor to the "divine" Archimedes, and as the immediate successor of Benedetti, whom he follows both in the manner in which he puts questions to himself and also often in his way of expression, without, however, citing him. Like Benedetti, he supposes a gradually decreasing 'vis impressa" in cases of projection. If the projection is upwards, the impressed force is a transferred "lightness"; as this lightness decreases, the gravity receives an increasing preponderance directed down, and the motion of falling is accelerated. In this idea Galileo agrees with the ancient astronomer Hipparchus of the second century B. C., but does not do justice to Benedetti's view of the acceleration of

falling. For, according to Hipparchus and Galileo, the motion of falling would have to be *uniform* when the impressed force is wholly overcome.

In the former editions of this book, the exposition of Galileo's researches was based on his final work, *Discorsi e dimostrazioni matematiche* of 1638.[1] However, his original notes, which have become known later, lead to different views on the course of his development. With respect to these I adopt, in essentials, the conclusions of E. Wohlwill (*Galilei und sein Kampf für die Kopernikanische Lehre,* Hamburg and Leipzig, 1909). In the riper and more fruitful time of his residence in Padua, Galileo dropped the question as to the "why" and inquired the "how" of the many motions which can be observed. The consideration of the line of projection and its conception as a combination of a uniform horizontal motion and an accelerated motion of falling enabled him to recognize this line as a parabola, and consequently the space fallen through as proportional to the square of the time of falling. The statical investigations on the inclined plane led to the consideration of falling down such a plane, and also to the observation of the vibrating pendulum. From the foregoing comprehensive observations and experiments on the pendulum it appeared that a body which falls down a series of inclined planes can, by means of the velocity thus obtained, rise on any series of other planes only to the original height. In other words, the velocity obtained by the falling depends only on the distance fallen. Finally, Galileo reached a definition of uniformly accelerated motion which has the properties of the motion of falling, and from which, inversely, all

[1] [There is a convenient German annotated translation of the *Discorsi e dimostrazioni matematiche* by A. J. von Oettingen in *Ostwald's Klassiker der exakten Wissenschaften,* Nos. 11, 24, 25; and an English translation by Henry Crew and Alfenso de Salvio under the title *Dialogues concerning Two New Sciences,* New York, 1914.]

those provisional lemmas which led him to his view can be deductively derived.

G. Vailati, who has devoted much attention to Benedetti's investigations (*Atti della R. Acad. di Torino,* XXXIII., 1898), finds the chief merit of Benedetti to be that he subjected the Aristotelian views to mathematical and critical scrutiny and correction, and endeavored to lay bare their inherent contradictions, thus preparing the way for further progress. He knows that the assumption of the Aristotelians, that the velocity of falling bodies is inversely proportional to the density of the surrounding medium, is untenable and possible only in special cases. Let the velocity of descent be proportional to $p - q$, where p is the weight of the body and q the upward impulsion due to the medium. If only half the velocity of descent is set up in a medium of double the density, the equation $p - q = 2(p - 2q)$ must exist—a relation which is possible only in case $p = 3q$. Light bodies *per se* do not exist for Benedetti; he ascribes weight and upward impulsion even to air. Different-sized bodies of the same material fall, in his opinion, with the same velocity. Benedetti reaches this result by conceiving equal bodies falling alongside each other first disconnected and then connected, where the connection cannot alter the motion. In this he approaches the conception of Galileo, with the exception that the latter takes a profounder view of the matter. Nevertheless, Benedetti also falls into many errors. He believes, for example, that the velocity of descent of bodies of the same size and of the same shape is proportional to their weight, that is, to their density. His reflections on the oscillation of a body about the center of the earth in a canal bored through the earth, are interesting, and contain little to be criticized. He thus does not solve the riddle fully, but prepares the way for the solution, especially for the discovery of the law of inertia.

3. With respect to the definition of uniformly accelerated motion, Galileo hesitated for a long time. He first called that motion uniformly accelerated in which the increments of velocity are proportional to the lengths of path described; he held, according to a fragment dating from 1604 (*Edizione Nazionale,* VIII, pp. 373-374), and a letter to Sarpi written at the same time, that this conception corresponded to all facts, in which, however, he was mistaken. According to Wohlwill, it was probably about 1609 that he overcame the error and defined uniformly accelerated motion by the proportionality of the velocity to the *time* of motion. He then turned away from his first view on grounds just as insufficient as those on which he had accepted it earlier. The natural explanation of all this will, as in the older editions of this book, be spoken of later. We will now consider what heritage Galileo left to modern thinkers. Here it will appear clearly that he allowed himself to be led by suppositions which today can be conceived as more or less immediate corollaries from his law of falling bodies; this perhaps speaks most eloquently for his talent as an investigator and for his discoverer's instinct. Now, whether Galileo attained to knowledge of the uniformly accelerated motion of falling bodies by consideration of the parabola of projection or in another way, we cannot doubt that he tested the law experimentally *as well.* Salviati, chief advocate of Galileo's doctrines in the *Discorsi,* assures us of his repeatedly taking part in experiments, and describes the experiments very accurately (*Le opere di Galilei, Edizione Nazionale,* VIII, pp. 212-213).

It was difficult to prove by any direct means that the velocity acquired was proportional to the time of descent. It was easier, however, to investigate by what law the distance increased with the time; and he consequently deduced from his assumption the relation that obtained between the distance and the time, and tested this by experiment. The deduction is simple,

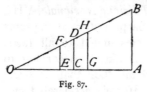

Fig. 87.

distinct, and perfectly correct. He draws (Fig. 87) a straight line, and on it cuts off successive portions that represent to him the times elapsed. At the extremities of these portions he erects perpendiculars (ordinates), and these represent the velocities acquired. Any portion *OG* of the line *OA* denotes, *therefore, the time of* descent elapsed, and the corresponding perpendicular *GH* the velocity acquired in such time.

If, now, we fix our attention on the progress of the velocities, we shall observe with Galileo the following fact: namely, that at the instant *C*, at which one-half *OC* of the time of descent *OA* has elapsed, the velocity *CD* is also one-half of the final velocity *AB*.

If now we examine two instants of time, *E* and *G*, equally distant in opposite directions from the instant *C*, we shall observe that the velocity *HG* exceeds the mean velocity *CD* by the same amount that *EF* falls short of it. For every instant antecedent to *C* there exists a corresponding one equally distant from it subsequent to *C*. Whatever loss is suffered in the first half of the motion, therefore, as compared with *uni-form* motion with half the final velocity, such loss is made up in the second half. We may consequently regard the distance fallen through as having been *uni-formly* described with half the final velocity. If, ac-

cordingly, we make the final velocity v proportional to
the time of descent t, we shall obtain $v = g\,t$, where
g denotes the final velocity acquired in unit of time—
the so-called acceleration. The space s descended
through is therefore given by the equation $s = (gt/2)$
t or $s = gt^2/2$. Motion of this sort, in which, according to the assumption, equal velocities constantly ac-
crue in equal intervals of time, we call *uniformly
accelerated motion.*

If we collect the times of descent, the final velocities, and the distances traversed, we shall obtain the
following table:

t.	*v.*	*s.*
1.	$1g.$	$1 \times 1 . \dfrac{g}{2}$
2.	$2g.$	$2 \times 2 . \dfrac{g}{2}$
3.	$3g.$	$3 \times 3 . \dfrac{g}{2}$
4.	$4g.$	$4 \times 4 . \dfrac{g}{2}$
$\vdots$	$\vdots$	$\vdots$
	$t\,g.$	$t \times t . \dfrac{g}{2}$

4. The relation obtaining between t and s admits
of experimental proof; and this *Galileo* accomplished
in the manner which we shall now describe.

We must first remark that no part of the knowledge
and ideas on this subject with which we are now so
familiar existed in Galileo's time, but that Galileo had
to create these ideas and means for us. Accordingly,
it was impossible for him to proceed as we should do
today, and he was obliged, therefore, to pursue a different method. He first sought to retard the motion

of descent, that it might be more accurately observed. He made observations on balls, which he caused to roll down inclined planes (grooves); assuming that only the velocity of the motion would be lessened here, but that the form of the law of descent would remain unmodified. If, beginning from the upper extremity, the distances 1, 4, 9, 16 . . . be notched off on the groove, the respective times of descent will be representable, it was assumed, by the numbers 1, 2, 3, 4 . . . , a result which was, be it added, confirmed. The observation of the times involved, Galileo accomplished in a very ingenious manner. There were no clocks of the modern kind in his day: such were first rendered possible by the dynamical knowledge of which Galileo laid the foundations. The mechanical clocks which were used were very inaccurate, and were available only for the measurement of great spaces of time. Moreover, it was chiefly water-clocks and sand-glasses that were in use—in the form in which they had been handed down from the ancients. Galileo, now, constructed a very simple clock of this kind, which he especially adjusted to the measurement of small spaces of time; a thing not customary in those days. It consisted of a vessel of water of very large transverse dimensions, having in the bottom a minute orifice which was closed with the finger. As soon as the ball began to roll down the inclined plane Galileo removed his finger and allowed the water to flow out on a balance; when the ball had arrived at the terminus of its path he closed the orifice. As the pressure-height of the fluid did not, owing to the great transverse dimensions of the vessel, perceptibly change, the weights of the water discharged from the orifice were proportional to the times. It was in this way actually shown that the times increased simply, while the spaces fallen through increased quadratically.

The inference from Galileo's assumption was thus confirmed by experiment, and with it the assumption itself.

If we are to understand Galileo's train of thought, we must bear in mind that he was already in possession of instinctive experiences prior to his resorting to experiment. Freely falling bodies are followed with more difficulty by the eye the longer and the farther they have fallen; their impact on the hand receiving them is in like measure sharper; the sound of their striking louder. The velocity accordingly increases with the time elapsed and the space traversed. But for scientific purposes our mental representations of the facts of sensual experience must be submitted to *conceptual* formulation. Only thus may they be used for discovering by abstract mathematical rules unknown properties conceived to be dependent on certain initial properties having definite and assignable arithmetic values; or, for completing what has been only partly given. This formulation is effected by isolating and emphasizing what is deemed of importance, by neglecting what is subsidiary, by *abstracting,* by idealizing. The experiment determines whether the form chosen is adequate to the facts. Without some preconceived opinion the experiment is impossible, because its form is determined by the opinion. For how and on what could we experiment if we did not previously have some suspicion of what we were about? The *complemental* function which the experiment is to fulfil is determined entirely by our prior experience. The experiment confirms, modifies, or overthrows our suspicion. The modern inquirer would ask in a similar predicament: Of what is v a function? What function of t is v? Galileo asks, in his ingenious and primitive way: is v proportional to s, is v proportional to t? Galileo, thus, *gropes* his way along synthetically, but reaches his goal nevertheless. Systematic, routine methods are the final outcome of research, and do not stand perfectly developed at the disposal of genius in the first steps it takes. (Compare the article "Ueber Gedankenexperimente," *Zeitschrift für den phys. und chem. Unterricht*, 1897, I.; *Erkenntnis und Irrtum*, 2nd Ed. Leipzig 1906.)

5. To form some notion of the relation which subsists between motion on an inclined plane and that of free descent, Galileo made the assumption, that a body which falls through the height of an inclined plane attains the same final velocity as a body which falls through its length. This is an assumption that will strike us as rather a bold one; but in the manner in which it was enunciated and employed by Galileo, it is quite natural. We shall endeavor to explain the way by which he was led to it. He says: If a body fall freely downwards, its velocity increases proportionally to the time. When, then, the body has arrived at a point below, let us imagine its velocity reversed and directed upwards; the body then, it is clear, will rise. We make the observation that its motion in this case is a reflection, so to speak, of its motion in the first case. As then its velocity increased proportionally to the time of descent, it will now, conversely, diminish in that proportion. When the body has continued to rise for as long a time as it descended, and has reached the height from which it originally fell, its velocity will be reduced to zero. We perceive, therefore, that a body will rise, in virtue of the velocity acquired in its descent, just as *high* as it has fallen. If, accordingly, a body falling down an inclined plane could acquire a velocity which would enable it, when placed on a differently inclined plane, to rise higher than the point from which it had fallen, we should be able to effect the elevation of

bodies by gravity alone. There is contained, accordingly, in this assumption, that the velocity acquired by a body in descent depends solely on the *vertical* height fallen through and is independent of the inclination of the path, nothing more than the uncontradictory apprehension and recognition of the *fact* that heavy bodies do not possess the tendency to rise, but only the tendency to fall. If we should assume that a body falling down the length of an inclined plane in some way or other attained a greater velocity than a body that fell through its height, we should only have to let the body pass with the acquired velocity to another inclined or vertical plane to make it rise to a greater vertical height than it had fallen from. And if the velocity attained on the inclined plane were less, we should only have to reverse the process to obtain the same result. In both instances a heavy body could, by an appropriate arrangement of inclined planes, be forced continually upwards solely by its own weight—a state of things which wholly contradicts our instinctive knowledge of the nature of heavy bodies.

6. Galileo, in this case, again, did not stop with the mere philosophical and logical discussion of his assumption, but tested it by comparison with experience.

He took a simple filar pendulum (Fig. 88) with a heavy ball attached. Lifting the pendulum, while elongated its full length, to the level of a given altitude, and then letting it fall, it ascended to the same level on the opposite side. If it does not do so *exactly,* Galileo said, the resistance of the air must be the cause of the deficit. This is inferrible from the fact that the deficiency is greater in the case of a cork ball than it is in the case of a heavy metal one. However, apart from this, the body ascends to the same altitude on the

opposite side. Now it is permissible to regard the motion of a pendulum in the arc of a circle as a motion
of descent along a series of inclined planes of different

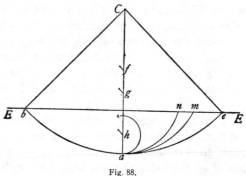

Fig. 88.

inclinations. Now, we can easily, with Galileo, cause
the body to rise on a different arc—on a different
series of inclined planes. This we accomplish by driving in at one side of the thread, as it hangs vertically,
a nail f or g, which will prevent any given portion of
the thread from taking part in the second half of the
motion. The moment the thread arrives at the line of
equilibrium and strikes the nail, the ball, which has
fallen through $b\,a$, will begin to ascend by a different
series of inclined planes, and describe the arc $a\,m$ or $a\,n$.
Now if the inclination of the planes had any influence
on the velocity of descent, the body could not rise to
the same horizontal level from which it had fallen.
But it does. By driving the nail sufficiently low down,
we may shorten the pendulum for half of an oscillation
as much as we please; the phenomenon, however, always remains the same. If the nail h is driven so low
down that the remainder of the string cannot reach to
the plane E, the ball will turn completely over and

wind the thread round the nail; because when it has attained the greatest height it can reach it still has a residual velocity left.

7. If we assume thus, that the same final velocity is attained on an inclined plane whether the body fall through the height or the length of the plane—in which assumption nothing more is contained than that a body rises by virtue of the velocity it has acquired in falling just as high as it has fallen—we shall easily arrive, with Galileo, at the perception that the times of the descent along the height and the length of an inclined plane are in the simple proportion of the height and the length; or, what is the same, that the accelerations are inversely proportional to the times of descent. The acceleration along the height will consequently bear to the acceleration along the length the proportion of the length to the height. Let AB (Fig. 89) be the height and AC the length of the inclined plane. Both will be descended through in uniformly accelerated motion in the times t and t_1 with the final velocity v. Therefore,

$$AB = \frac{v}{2}t \text{ and } AC = \frac{v}{2}t_1, \frac{AB}{AC} = \frac{t}{t_1}.$$

If the accelerations along the height and the length be called respectively g and g_1, we also have

$$v = gt \text{ and } v = g_1 t_1, \text{ whence } \frac{g_1}{g} = \frac{t}{t_1} = \frac{AB}{AC} = \sin \alpha.$$

In this way we are able to deduce from the acceleration on an inclined plane the acceleration of free descent.

From this proposition Galileo deduces several corollaries, some of which have passed into our elementary text-books. The accelerations along the height and length are in the inverse proportion of the height and length. If now we cause one body to fall along the length of an inclined plane and simultaneously another to fall freely along its height, and ask what the distances are that are traversed by the two in equal intervals of time, the solution of the problem will be readily found (Fig. 90) by simply letting fall from *B* a perpendicular on the length. The part *A D,* thus cut off, will be the distance traversed by the one body on the inclined plane, while the second body is freely falling through the height of the plane.

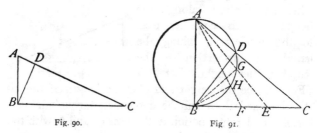

Fig. 90.　　　　　　　Fig 91.

If we describe (Fig. 91) a circle on *A B* as diameter, the circle will pass through *D*, because *D* is a right angle. It will be seen thus, that we can imagine any number of inclined planes, *A E, A F,* of any degree of inclination, passing through *A*, and that in every case the chords *A G, A H* drawn in this circle from the upper extremity of the diameter will be traversed in the same time by a falling body as the vertical diameter itself. Since, obviously, only the lengths and inclinations are essential here, we may also draw the chords in question from the lower extremity of the diameter, and say generally: The vertical diameter of a circle is described by a falling particle in the same time that any chord through either extremity is so described.

We shall present another corollary, which, in the pretty form in which Galileo gave it, is usually no longer incorporated in elementary expositions. We imagine gutters radiating in a vertical plane from a common point A at a number of different degrees of inclination to the horizon (Fig. 92). We place at their common extremity A a like number of heavy bodies and cause them to begin simultaneously their motion of descent. The bodies will always at any one instant of time form a circle. After the lapse of a longer time they will be found in a circle of larger radius, and the radii increase proportionally to the

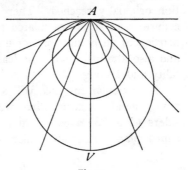

Fig. 92.

squares of the times. If we imagine the gutters to radiate in a space instead of a plane, the falling bodies will always form a sphere, and the radii of the spheres will increase proportionally to the squares of the times. This will be perceived by imagining the figure revolved about the vertical $A V$.

8. We see thus—as deserves again to be briefly noticed—that Galileo did not supply us with a *theory* of the falling of bodies, but investigated and established, wholly without preformed opinions, the *actual facts* of falling.

Gradually *adapting,* on this occasion, his thoughts to the facts, and everywhere logically abiding by the ideas he had reached, he hit on a conception, which to himself, perhaps less than to his successors, appeared in the light of a new law. In all his reasonings, Galileo followed, to the greatest advantage of science, a principle which might appropriately be called the *principle of continuity*. Once we have reached a theory that applies to a particular case, we proceed gradually to modify in thought the conditions of that case, as far as it is at all possible, and endeavor in so doing to adhere throughout as closely as we can to the conception originally reached. There is no method of procedure more surely calculated to lead to that comprehension of all natural phenomena which is the *simplest* and also attainable with the least expenditure of mentality and feeling.

A particular instance will show more clearly than any general remarks what we mean. Galileo considers (Fig. 93) a body which is falling down the inclined plane $A B$, and which, being placed with the

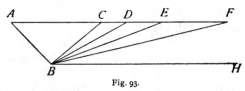

Fig. 93.

velocity thus acquired on a second plane BC, for example, ascends this second plane. On all planes $B C,$ $B D,$ and so forth, it ascends to the horizontal plane that passes through A. But, just as it falls on $B D$ with less *acceleration* than it does on $B C,$ so similarly it will ascend on $B D$ with less *retardation* than it will on $B C$. The nearer the planes $B C, B D, B E, B F$ approach to the horizontal plane $B H$, the less will the retardation of the body on those planes be, and the longer and further will it move on them. On the horizontal plane *B H* the retardation vanishes *entirely* (that is, of course, neglecting friction and the resistance of the air), and the body will continue to move infinitely long and infinitely far with *constant* velocity. Thus advancing to the limiting case of the problem presented, Galileo discovers the so-called law of inertia, according to which a body not under the influence of forces, i. e. of special circumstances that change motion, will retain forever its velocity (and direction). We shall presently revert to this subject.

In an exhaustive study in the *Zeitschrift für Völkerpsychologie,* 1884, XIV., pp. 365-410, and XV., pp. 70-135, 337-387, entitled "Die Entdeckung des Beharrunggesetzes," E. Wohlwill has shown that the predecessors and contemporaries of Galileo, nay, even Galileo himself, only *very gradually* abandoned the Aristotelian conceptions for the acceptance of the law of inertia. Even in Galileo's mind uniform circular motion and uniform horizontal motion occupy distinct places. Wohlwill's researches are very acceptable and show that Galileo had not attained perfect clearness in his own new ideas and was liable to frequent reversion to the old views, as might have been expected.

Indeed, from my own exposition the reader will have inferred that the law of inertia did not possess in Galileo's mind the degree of clearness and universality that it subsequently acquired. (Compare *"Erhaltung der Arbeit,"* p. 47). With regard to my exposition just presented, however, I still believe, in spite of the opinions of Wohlwill and Poske, that I have indicated the point which both for Galileo and his successors must have placed in the most favorable light the *transition* from the old conception to the new. How much was wanting to absolute comprehension, may be gathered from the fact that Baliani was able without difficulty to infer from Galileo's statement that acquired velocity could not be destroyed—a fact which Wohl-will himself points out (p. 112). It is not at all surprising that in treating of the motion of *heavy* bodies, Galileo applies his law of inertia almost exclusively to horizontal movements. Yet he knows that a musket-ball *possessing no weight* would continue rectilinearly on its path in the direction of the barrel. (*Dialogues on the two World-Systems*, German translation, Leipzig, 1891, p. 184.) His hesitation in enunciating in its most general terms a law that at first blush appears so startling, is not surprising.

9. The motion of falling that Galileo found actually to exist, is, accordingly, a motion of which the velocity increases proportionally to the time—a so-called uniformly accelerated motion.

It would be an anachronism and utterly unhistorical to attempt, as is sometimes done, to derive the uniformly accelerated motion of falling bodies from the constant action of the force of gravity. "Gravity is a constant force; *consequently* it generates in equal elements of time equal increments of velocity; thus, the motion produced is uniformly accelerated." Any exposition such as this would be unhistorical, and would put the whole discovery in a false light, for the reason that the notion of force as we hold it today was first created by Galileo. Before Galileo *force* was known solely as *pressure*. Now, no one can know, who has not learned it from experience, that generally pressure produces motion, much less *in what manner* pressure passes into motion; that not position, nor velocity, but acceleration, is determined by it. This cannot be philosophically deduced from the conception, itself. Conjectures may be set up concerning it. But experience alone can definitely inform us with regard to it.

10. It is not by any means self-evident, therefore, that the circumstances which determine motion, that is, forces, immediately produce accelerations. A glance at other departments of physics will at once make this clear. The differences of temperature of bodies also determine alterations. However, by differences of temperature not compensatory *accelerations* are determined, but compensatory *velocities*.

That it is *accelerations* which are the immediate effects of the circumstances that determine motion, that is, of the forces, is a fact which Galileo *perceived* in the natural phenomena. Others before him had also perceived many things. The assertion that everything seeks its place also involves a correct observation. The observation, however, does not hold good in all cases, and it is not exhaustive. If we cast a stone into the air, for example, it no longer seeks its place which is below. But the acceleration towards the earth, the retardation of the upward motion, the fact that Galileo perceived, is still present. His observation always remains correct; it holds true more generally; it embraces in *one* mental effort *much more*.

11. We have already remarked that Galileo discovered the so-called law of inertia quite incidentally. A body on which, as we are wont to say, no force acts, preserves its direction and velocity unaltered. The fortunes of this law of inertia have been strange. It appears never to have played a prominent part in Galileo's thought. But Galileo's successors, particularly Huygens and Newton, formulated it as an independent law. Nay, some have even made of inertia a general property of matter. We shall readily perceive, however, that the law of inertia is not at all an indepen-

dent law, but is contained implicitly in Galileo's perception that all circumstances determinative of motion, or forces, produce *accelerations*.

In fact, if a force determine, not position, not velocity, but acceleration, *change* of velocity, it stands to reason that where there is no force there will be no change of velocity. It is not necessary to enunciate this in independent form. The embarrassment of the neophyte, which also overcame the great investigators in the face of the great mass of new material presented, alone could have led them to conceive the *same* fact as two different facts and to formulate it twice.

In any event, to represent inertia as self-evident, or to derive it from the general proposition that "the effect of a cause persists," is totally wrong. Only a mistaken straining after rigid logic can lead us so out of the way. Nothing is to be accomplished in the present domain with scholastic propositions like the one just cited. We may easily convince ourselves that the contrary proposition, "cessante causa cessat effectus," is as well supported by reason. If we call the required velocity "the effect," then the first proposition is correct; if we call the acceleration "effect," then the second proposition holds.

12. We shall now examine Galileo's researches from another side. He began his investigations with the notions familiar to his time—notions developed mainly in the practical arts. One notion of this kind was that of velocity, which is very readily obtained from the consideration of a uniform motion. If a body traverse in every second of time the same distance c, the distance traversed at the end of t seconds will be $s = c\,t$. The distance c traversed in a second of time we call the velocity, and obtain it from the examination of any portion of the distance and the corresponding time by the

help of the equation $c = s/t$, that is, by dividing the number which is the measure of the distance traversed by the number which is the measure of the time elapsed.

Now, Galileo could not complete his investigations without tacitly modifying and extending the traditional idea of velocity. Let us represent for distinctness sake

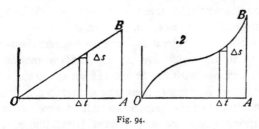

Fig. 94.

in 1 (Fig. 94) a uniform motion, in 2 a variable motion, by laying off as abscissæ in the direction $O A$ the elapsed times, and erecting as ordinates in the direction $A B$ the distances traversed. Now, in 1, whatever increment of the distance we may divide by the corresponding increment of the time, in all cases we obtain for the velocity c the *same* value. But if we were thus to proceed in 2, we should obtain widely differing values, and therefore the word "velocity" as ordinarily understood, ceases in this case to be unequivocal. If, however, we consider the increase of the distance in a sufficiently small element of time, where the element of the curve in 2 approaches to a straight line, we may regard the increase as uniform. The velocity in this element of the motion we may then define as the quotient, $\Delta s/\Delta t$, of the element of the time into the corresponding element of the distance. Still more precisely, the velocity at any instant is defined as the limiting value which the ratio $\Delta s/\Delta t$ assumes as the elements become infinitely small—a value designated by $d\,s/d\,t$. This new notion includes the old one as a particular case, and is, moreover, immediately applicable to uniform motion. Although the express formulation of this idea, as thus extended, did not take place till long after Galileo, we see none the less that he made use of it in his reasonings.

13. An entirely new notion to which Galileo was led is the idea of *acceleration*. In uniformly accelerated motion the velocities increase with the time according to the same law as in uniform motion the spaces increase with the times. If we call v the velocity acquired in time t, then $v = g\,t$. Here g denotes the increment of the velocity in unit of time or the acceleration, which we also obtain from the equation $g = v/t$. When the investigation of variably accelerated motions was begun, this notion of acceleration had to experience an extension similar to that of the notion of velocity. If in 1 and 2 the times be again drawn as abscissæ, but now the *velocities* as ordinates, we may go through anew the whole train of the preceding reasoning and define the acceleration as $d\,v/d\,t$, where $d\,v$ denotes an infinitely small increment of the velocity and $d\,t$ the corresponding increment of the time. In the notation of the differential calculus we have for the acceleration of a *rectilinear* motion, $\varphi = dv/dt = d^2s/dt^2$.

The ideas here developed are susceptible, moreover, of graphic representation. If we lay off the times as abscissæ and the distances as ordinates, we shall perceive, that the velocity at each instant is measured by the slope of the curve of the distance. If in a similar manner we put times and velocities together, we shall see that the acceleration of the instant is measured by the slope of the curve of the velocity. The course of the latter slope is, indeed, also capable of being traced in the curve of distances, as will be perceived from the following considerations. Let us imagine, in the

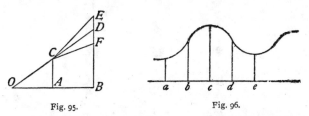

Fig. 95.   Fig. 96.

usual manner (Fig. 95), a uniform motion represented by a straight line *OCD*. Let us compare with this a motion *OCE* the velocity of which in the second half of the time is greater, and another motion *OCF* of which the velocity is in the same proportion smaller. In the first case, accordingly, we shall have to erect for the time $OB = 2\,OA$, an ordinate greater than $BD = 2\,AC$; in the second case, an ordinate less than *BD*. We see thus, without difficulty, that a curve of distance convex to the axis of the time-abscissæ corresponds to accelerated motion, and a curve concave thereto to retarded motion. If we imagine a lead-pencil to perform a vertical motion of any kind and in front of it during its motion a piece of paper to be uniformly drawn along from right to left and the pencil to thus execute the drawing in Fig. 96, we shall be able to read off from the drawing the peculiarities of the motion. At *a* the velocity of the pencil was directed upwards, at *b* it was greater, at *c* it was = 0, at *d* it was directed downwards, at *e* it was again = 0. At *a, b, d, e,* the acceleration was directed upwards, at *c* downwards; at *c* and *e* it was greatest.

14. The summary representation of what Galileo

discovered is best made by a table of times, required velocities, and traversed distances. But the numbers

t.	v.	s.
1	g	$1\dfrac{g}{2}$
2	$2g$	$4\dfrac{g}{2}$
3	$3g$	$9\dfrac{g}{2}$
. . . .		
t	tg	$t^2\dfrac{g}{2}$

follow so simple a law—one immediately recognizable —that there is nothing to prevent our replacing the table by a *rule for its construction.* If we examine the relation that connects the first and second columns, we shall find that it is expressed by the equation $v = gt$, which, in its last analysis, is nothing but an abbreviated direction for constructing the first two columns of the table. The relation connecting the first and third columns is given by the equation $s = g\,t^2/2$. The connection of the second and third columns is represented by $s = v^2/2g$.

Of the three relations

$$v = gt$$
$$s = \frac{gt^2}{2}$$
$$s = \frac{v^2}{2g},$$

strictly, the first two only were employed by Galileo. Huygens was the first who evinced a higher appreciation of the third, and laid, in thus doing, the foundations of important advances.

15. We may add a remark in connection with this table that is very valuable. It has been stated previously that a body, by virtue of the velocity it has acquired in its fall, is able to rise again to its original height, in doing which its velocity diminishes in the same way (with respect to time and space) as it increased in falling. Now a freely falling body acquires in double time of descent double velocity, but falls in this double time through four times the simple distance. A body, therefore, to which we impart a vertically upward double velocity will ascend twice as long a time, but *four times* as high as a body to which the simple velocity has been imparted.

It was remarked, very soon after Galileo, that there is inherent in the velocity of a body a something that corresponds to a force—a something, that is, by which a force can be overcome, a certain "efficacy," as it has been aptly termed. The only point that was debated was, whether this efficacy was to be reckoned proportional to the *velocity* or to the *square of the velocity*. The Cartesians held the former, the Leibnizians the latter. But it will be perceived that the question involves no dispute whatever. The body with the double velocity overcomes a given force through double the time, but through *four times* the distance. With respect to time, therefore, its efficacy is proportional to the velocity; with respect to distance, to the square of the velocity. D'Alembert drew attention to this misunderstanding, although in not very distinct terms. It is to be especially remarked, however, that Huygens's thoughts on this question were perfectly clear.

16. The experimental procedure by which, at the present day, the laws of falling bodies are verified, is somewhat different from that of Galileo. Two methods

may be employed.   Either the motion of falling, which
from its rapidity is difficult to observe directly, is so
retarded, without altering the law, as to be easily observed; or the motion of falling is not altered at all,
but our means of observation are improved in deli-
cacy.   On the first principle rest Galileo's inclined
gutter and Atwood's machine.   Atwood's machine
consists (Fig. 97) of an easily running pulley, over
which is thrown a thread, to whose extremities two
equal weights P are attached.   If upon one of the
weights P we lay a third small weight p, a
uniformly accelerated motion will be set up
by the added weight, having the acceleration
$(p/\overline{2\,P + p})g$—a result that will be readily
obtained when we shall have discussed the
notion of "mass."   Now by means of a graduated vertical standard connected with the
pulley it may easily be shown that in the times
1, 2, 3, 4 . . . . the distances 1, 4, 9, 16 . . . . . are traversed.   The final velocity corresponding to any given
time of descent is investigated by catching the small
additional weight, p, which is shaped so as to project
beyond the outline of P, in a ring through which the
falling body passes, after which the motion continues
without acceleration.

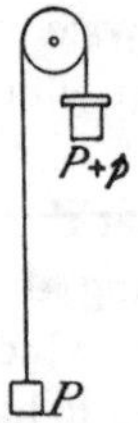

The apparatus of Morin is based on a different principle.   A body to which a writing pencil is attached
describes on a vertical sheet of paper, which is drawn
uniformly across it by a clock-work, a horizontal straight
line.   If the body fall while the paper is not in motion,
it will describe a vertical straight line.   If the two
motions are combined, a parabola will be produced,
of which the horizontal abscissæ correspond to the
elapsed times and the vertical ordinates to the dis-

Fig. 97.

tances of descent described. For the abscissæ 1, 2, 3, 4 .... we obtain the ordinates 1, 4, 9, 16 .... By an unessential modification, Morin employed instead of a plane sheet of paper, a rapidly rotating cylindrical drum with vertical axis, by the side of which the body

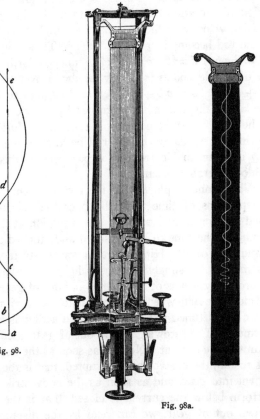

Fig. 98.

Fig. 98a.

fell down a guiding wire. A different apparatus, based on the same principle, was invented, independently, by

Laborde, Lippich, and Von Babo. A lampblacked sheet of glass (Fig. 98a) falls freely, while a horizontally vibrating vertical rod, which in its first transit through the position of equilibrium starts the motion of descent, traces, by means of a quill, a curve on the lampblacked surface. Owing to the constancy of the period of vibration of the rod combined with the increasing velocity of the descent, the undulations traced by the rod become longer and longer. Thus (Fig. 98) $bc = 3\,ab, cd = 5\,ab, de = 7\,b$, and so forth. The law of falling bodies is clearly exhibited by this, since $ab + cb = 4\,ab, ab + bc + cd = 9\,ab$, and so forth. The law of the velocity is confirmed by the inclinations of the tangents at the points a, b, c, d, and so forth. If the time of oscillation of the rod be known, the value of g is determinable from an experiment of this kind with considerable exactness.

Wheatstone employed for the measurement of minute portions of time a rapidly operating clock-work called a chronoscope, which is set in motion at the beginning of the time to be measured and stopped at the termination of it. Hipp has advantageously modified this method by simply causing a light index-hand to be thrown by means of a clutch in and out of gear with a rapidly moving wheel-work regulated by a vibrating reed of steel tuned to a high note, and acting as an escapement. The throwing in and out of gear is effected by an electric current. Now if, as soon as the body begins to fall, the current be interrupted, that is the hand thrown into gear, and as soon as the body strikes the platform below the current is closed, that is the hand thrown out of gear, we can read by the distance the index-hand has travelled the time of descent.

17. Among the further achievements of Galileo we

have yet to mention his ideas concerning the motion of the pendulum, and his refutation of the view that bodies of greater weight fall faster than bodies of less weight. We shall revert to both of these points on another occasion. It may be stated here, however, that Galileo, on discovering the constancy of the period of pendulum-oscillations, at once applied the pendulum to pulse-measurements at the sick-bed, as well as proposed its use in astronomical observations and to a certain extent employed it therein himself.

18. Of still greater importance are his investigations concerning the motion of projectiles. A free body, according to Galileo's view, constantly experiences a vertical acceleration g towards the earth. If at the beginning of its motion it is affected with a vertical velocity c, its velocity at the end of the time t will be $v = c + g t$. An initial velocity upwards would have to be reckoned negative here. The

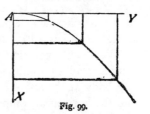

Fig. 99.

distance described at the end of time t is represented by the equations $s = a + c t + \frac{1}{2} g t^2$, where $c t$ and $\frac{1}{2} g t^2$ are the portions of the traversed distance that correspond respectively to the uniform and the uniformly accelerated motion. The constant a is to be put $= 0$ when we reckon the distance from the point that the body passes at time $t = 0$. When Galileo had once reached his fundamental conception of dynamics, he easily recognized the case of horizontal projection as a combination of two *independent* motions, a horizontal uniform motion, and a vertical uniformly accelerated motion. He thus introduced into use the principle of the *parallelogram of motions.* Even oblique projection no longer presented the slightest difficulty.

If a body receives a horizontal velocity c, it describes in the horizontal direction in time t the distance $y = c t$, while simultaneously it falls in a vertical direction the distance $x = g t^2/2$. Different motion-determinative circumstances exercise no mutual effect on one another, and the motions determined by them take place *independently of each other.* Galileo was led to this assumption by the attentive observation of the phenomena; and the assumption proved itself true.

For the curve which a body describes when the two motions in question are compounded, we find, by employing the two equations above given, the expression $y = \sqrt{(2 c^2/g) x}$. It is the parabola of Appolonius having its parameter equal to c^2/g and its axis vertical, as Galileo knew (Fig. 100).

We readily perceive with Galileo, that *oblique* projection involves nothing new. The velocity c imparted to a body at the angle a with the horizon is resolvable into the horizontal component $c \cdot \cos a$ and the vertical

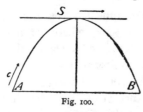

Fig. 100.

component $c \cdot \sin a$. With the later velocity the body ascends during the same interval of time t which it would take to acquire this velocity in falling vertically downwards. Therefore, $c \cdot \sin a = g t$. When it has reached its greatest height the vertical component of its initial velocity has vanished, and from the point S onward it continues its motion as a horizontal projection. If we examine any two epochs equally distant in time, before and after the transit through S, we shall see that the body at these two epochs is equally distant from the perpendicular through S and situated the same distance below the horizontal line through S. The curve is therefore symmetrical with respect to the vertical line through S. It is a parabola with vertical axis and the parameter $(c \cos \alpha)^2/g$.

To find the so-called range of projection, we have simply to consider the horizontal motion during the time of the rising and falling of the body. For the ascent this time is, according to the equations above given, $t = c \sin \alpha/g$, and the same for the descent. With the horizontal velocity $c \cdot \cos \alpha$, therefore, the distance is traversed

$$w = c \cos \alpha \cdot 2 \, \frac{c \sin \alpha}{g} = \frac{c^2}{g} \, 2 \sin \alpha \cos \alpha = \frac{c^2}{g} \sin 2 \, \alpha.$$

The range of projection is greatest accordingly when $\alpha = 45°$, and equally great for any two angles $\alpha = 45° \pm \beta°$.

19. We cannot adequately appreciate the extent of Galileo's achievement in the analysis of the motion of projectiles until we examine his predecessors' endeavors in this field. Santbach (1561) is of the opinion that a cannon-ball speeds onward in a straight line until its velocity is exhausted and then drops to the ground in a vertical direction. Tartaglia (1537) compounds the path of a projectile out of a straight line, the arc of a circle, and lastly the vertical tangent to the arc. He is perfectly aware, as Rivius later (1582) more distinctly states, that accurately viewed the path is curved at all points, since the deflective action of gravity never ceases; but he is yet unable to arrive at a complete analysis. The initial portion of the path is well calculated to arouse the illusive impression that the action

of gravity has been annulled by the velocity of the projection—an illusion to which even Benedetti fell a victim. We fail to observe any *descent* in the initial part of the curve, and forget to take into account the shortness of the corresponding *time* of the descent. By a similar oversight a jet of water may assume the appearance of a solid body suspended in the air, if one is unmindful of the fact that it is made up of a mass of rapidly alternating minute particles. The same illusion is met with in the centrifugal pendulum, in the top, in Aitken's flexible chain rendered rigid by rapid rotation (*Philosophical Magazine,* 1878) in the locomotive which rushes safely across a defective bridge, through which it would have crashed if at rest, but which, owing to the insufficient time of descent and of the period in which it can do work, leaves the bridge intact. On thorough analysis none of these phenomena are more surprising than the most ordinary events. As Vailati remarks, the rapid spread of firearms in the fourteenth century gave a distinct impulse to the study of the motion of projectiles, and indirectly to that of mechanics generally. Essentially the same conditions occur in the case of the ancient catapults and in the hurling of missiles by hand, but the new and imposing

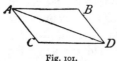
Fig. 101.

form of the phenomenon doubtless exercised a great fascination on the curiosity of the people.

20. The recognition of the mutual *independence* of the forces, or motion-determinative circumstances occurring in nature, which was reached and found expression in the investigations relating to projection, is important. A body may move (Fig. 101) in the direction *A B,* while the space in

which this motion occurs is displaced in the direction
$A\,C$. The body then goes from A to D. Now, this also
happens if the two circumstances that simultaneously
determine the motions $A\,B$ and $A\,C$, have no influence
on one another. It is easy to see that we may compound by the parallelogram not only displacements that
have taken place but also velocities and accelerations
that simultaneously take place.

Galileo's conception of the motion of a projectile
as a process compounded of two distinct and independent motions, is suggestive of an entire group of similar important epistemological processes. We may say
that it is as important to perceive the *non dependence*
of two circumstances A and B on each other, as it is
to perceive the *dependence* of two circumstances A
and C on each other. For the first perception alone
enables us to pursue the second relation with compo-
sure. Think only of how serious an obstacle the
assumption of non-existing causal relations constituted
to the research of the Middle Ages. Similar to Galileo's discovery is that of the parallelogram of forces
by Newton, the composition of the vibrations of
strings by Sauveur, the composition of thermal disturbances by Fourier. Through this latter inquirer
the method of compounding a phenomenon out of mutually independent partial phenomena by means of
representing a general integral as the sum of particular
integrals has penetrated into every nook and corner
of physics. The decomposition of phenomena into
mutually independent parts has been aptly characterized by P. Volkmann as *isolation,* and the composition
of a phenomenon out of such parts, *superposition.* The
two processes combined enable us to comprehend, or
reconstruct in thought, piecemeal, what, *as a whole,* it
would be impossible for us to grasp.

"Nature with its myriad phenomena assumes a unified aspect only in the rarest cases; in the majority of instances it exhibits a thoroughly composite character . . . ; it is accordingly one of the duties of science to conceive phenomena as made up of sets of partial phenomena, and at first to study these partial phenomena in their purity. Not until we know to what extent each circumstance shares in the phenomenon as an entirety do we acquire a command over the whole. . . ." (*Cf.* P. Volkmann, *Erkenntnisstheoretische Grundzüge der Naturwissenschaft,* 1896, p. 70. *Cf.* also my *Principles of Heat,* German edition, pp. 123, 151, 452).

21. If, now, we ask what views into the nature of things Galileo has bequeathed to us, or at least facilitated in a lasting manner by classically simple examples, we find:

(1) The emphasis upon the conception of work in a statical connection. There is no saving work with machines;

(2) The advancement of the conception of work in a dynamical connection. The velocity attained by falling, when resistance is neglected, only depends on the distance fallen through;

(3) The law of inertia;

(4) The principle of the superposition of motions.

22. Galileo's creative activity extends far beyond the limits of mechanics; we will only call to mind his founding of thermometry, his sketch of a method for the determination of the velocity of light,[1] his direct proof of the numerical ratio of the vibrations of the musical interval and his explanation of synchronous vibrations. He heard of the telescope, and that was

---

[1] [See Mach's *Popular Scientific Lectures,* 3rd ed., Chicago and London, 1898, pp. 50-54.]

enough for him to rediscover and to improvise one with two lenses and an organ-pipe. In quick succession he discovered, by the help of his instrument, the mountains of the moon, whose height he measured, Jupiter with his satellites, like a small model of the solar system, the peculiar form of Saturn, the phases of Venus, and the spots and rotation of the sun. These were new and very strong arguments for Copernicus. Also his thoughts on geometrically similar animals and machines and on the form and firmness of bones must be considered to be stimuli to the development of new mathematical methods. Besides Wohlwill, E. Goldbeck ("Galilei's Atomistik," *Biblioth. Math.,* 3rd series, III, 1902, part 1) has recently shown that this revolutionizing thinker was not wholly independent of ancient and mediæval influences. In particular, the first day of the *Discorsi* contains a lengthy exposition of Galileo's atomistic reflections which clearly stand in opposition to Aristotle, and as clearly approximate to Hero's position. These reflections led him to extraordinary discussions on the continuum and to speculations, in which mysticism and mathematics were combined, on the finite and the infinite, which remind us, on the one hand, of Nicolas of Cusa, and, on the other hand, of many modern mathematical researches which are hardly free from mysticism.[1] That Galileo could not attain complete clearness in all his thoughts need surprise us no more than his occupation with paradoxes, whose disturbing and clarifying force every thinker must have experienced.

---

[1] [See the German translation of the first two days of the *Discorsi* in *Ostwald's Klassiker,* No. 11 (the other days are translated in Nos. 24 and 25), especially pp. 30-32. Besides the articles of Goldbeck mentioned in the text above, there is an article by E. Kasner on "Galileo and the Modern Concept of Infinity," which is noticed in the *Jahrbuch über die Fortschritte der Mathematik,* vol. xxxvi, 1905, p. 49. See also Crew and de Salvio's translation of the *Discorsi,* pp. 26-40.]

23. With respect to the knowledge of accelerated motion Galileo has done the greatest service. For the sake of completeness we will refer to P. Duhem's researches ("De l'accélération produite par une force constante; notes pour servir à l'histoire de la dynamique," *Congrès international de philosophie*, Geneva, 1905, p. 859). Without entering into the many historically interesting details communicated by Duhem, we will here only add the following. According to the literal Aristotelian doctrine, a constant force conditions a constant velocity. But since the increasing velocity of falling can hardly escape even rough observations, the difficulty arises of bringing this acceleration into harmony with the doctrine that held the field. On approaching the ground, the body, in the opinion of Aristotle, becomes heavier. The traveller hastens when approaching his destination, as Tartaglia expresses it. The air which at one time was viewed as a hindrance and at another time as a motive power must, in order to make the contradictions more supportable, play at one time the one part and at another time the other. The hindering space of air between the body and the ground is, according to the commentator Simplicius, greater at the beginning of the motion of falling than at the end of this motion. The "forerunner" of Leonardo found that air which has once been set in motion is less of a hindrance for the body moved. The naïf observer of a stone projected obliquely or horizontally and describing an initial line which is almost straight must receive the natural impression that gravity is removed by the impulse to motion (*Mech.*, p. 151-3). Hence the distinction between natural and forced motion. The considerations of Leonardo, Tartaglia, Cardano, Galileo, and Torricelli on projectiles showed

how the idea of an alteration of two motions which were considered to be fundamentally different gradually yields to that of a mixture and simultaneity of them. Leonardo was acquainted with the accelerated motion of falling, and conjectured the increase of velocity proportionally to the time, which he ascribed to the successively diminished resistance of the air, but did not know how to determine the correct dependence of the space fallen through on the time. It was first at about the middle of the sixteenth century that the thought appeared that gravity continually communicates impulses to the falling body, and these impulses are added to the impressed force which is already present and which gradually decreases. This view was embraced by A. Piccolomini, J. C. Scaliger, and J. B. Benedetti. Already Leonardo remarked, quite by the way, that the arrow is not projected only at the greatest tension of the bow, but also in the other positions by the touching string (Duhem, *loc. cit.*, p. 882). But it was only when Galileo gave up this supposition of a gradual and spontaneous decrease of the impressed force and reduced this decrease to resisting forces, and investigated the motion of falling experimentally and without taking its causes into consideration, could the laws of the uniformly accelerated motion of falling appear in a purely quantitative form.

Further, from Duhem's historical exposition the fact results that Descartes rendered, independently of Galileo, more important services in the development of modern dynamics than is usually supposed, and than I too have supposed (*Mech.*, Chap. III). I am very grateful for this instruction. Descartes busied himself during his residence in Holland (1617-19), in cooperation with Beeckmann and in connection with the researches of Cardano and probably also of Scaliger and Benedetti, with the acceleration of falling bodies. He thoroughly recognized the law of inertia, as results from letters written to Mersenne in 1629, before Galileo's publication (E. Wohlwill, in *Die Entdeckung des Beharrungsgesetzes,* pp. 142, 143, considered it possible that Galileo indirectly stimulated him). Descartes also recognized the law of uniformly accelerated motion under the influence of a constant force, and was only mistaken with respect to the law of dependence of the path described on the time. The thoughts of Galileo and Descartes mutually complete each other. Galileo investigated the motion of descent phenomenologically, and without inquiring into its causes, while Descartes derived this motion from the constant force. Naturally in both investigations, a constructive and speculative element was active, but this element with Galileo kept close to the concrete case, while with Descartes it came in earlier with more general experiences. Certainly Descartes, in his *Principles of Philosophy,* observed the transference of motion and the loss of motion of the impinging body and the general philosophical consequences that (1) without the giving of motion to other bodies there can be no loss of motion (inertia); (2) every motion is either original or transferred from somewhere; (3) the original quantity of motion is indestructible. From this standpoint he could imagine that every apparently spontaneous motion whose origin was not perceptible was introduced by invisible impacts.

The great advantage which I—perhaps in opposition to Duhem—ascribe to the method of Galileo consists in the careful and complete exposition of the mere facts. In this exposition nothing remains concealed

CHRISTIANUS HUGENIUS
natus 14 Aprilis 1629.
denatus 8 Junii 1695.

behind the expression "force" which could be conjectured or disentangled by speculation. On this point opinions are divided even at the present time.

II.   THE ACHIEVEMENTS OF HUYGENS.

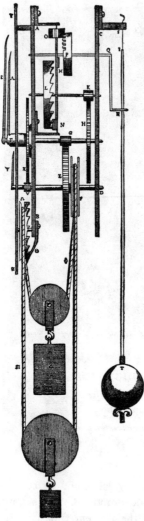

Huygens's Pendulum Clock.

1. HUYGENS in every respect must be ranked as Galileo's peer. If, perhaps, his philosophical endowments were less splendid than those of Galileo, this deficiency was compensated by the superiority of his geometrical powers. Huygens not only continued the researches which Galileo had begun, but he also solved the first problems in the *dynamics of several masses,* whereas Galileo had throughout restricted himself to the dynamics of a *single* body.

The plenitude of Huygens's achievements is best seen in his *Horologium Oscillatorium,* which appeared in 1673. The most important subjects there treated for the first time, are: the theory of the center of oscillation, the invention and construction of the pendulum-clock, the invention of the escapement, the determination of the acceleration of gravity, g, by pendulum-observations, a proposition regarding the employment of the length of the seconds pendulum as the unit of length, the theorems respecting centrifugal force, the mechanical and geometrical properties of cycloids, the doctrine of evolutes, and the theory of the circle of curvature.

2. With respect to the form of presentation of his work, it is to be remarked that Huygens shares with Galileo, in all its perfection, the latter's exalted and inimitable candor. He is frank without reserve in the presentment of the methods that led him to his discoveries, and thus always brings his reader to a full comprehension of his performances. Nor had he cause to conceal these methods. If, a thousand years from now, it shall be found that he was a man, it will likewise be seen what manner of man he was. In our discussion of the achievements of Huygens, however, we shall have to proceed in a somewhat different manner from that which we pursued in the case of Galileo. Galileo's views, in their classical simplicity, could be given in an almost unmodified form. With Huygens this is not possible. The latter deals with more complicated problems; his mathematical methods and notations become inadequate and cumbrous. For reasons of brevity, therefore, we shall reproduce all the conceptions of which we treat, in modern form, retaining, however, Huygens's essential and characteristic ideas.

3. We begin with the investigations concerning centrifugal force. When once we have recognized with Galileo that force determines acceleration, we are impelled, unavoidably, to ascribe every *change* of velocity and consequently also every change in the *direction* of a motion (since the direction is determined by three velocity-components perpendicular to one another). If, therefore, any body attached to a string, say a stone, is

swung uniformly round in a circle, the curvilinear motion which it performs is intelligible only on the supposition of a constant force that deflects the body from the rectilinear path. The tension of the string is in this force; by it the body is constantly deflected from the rectilinear path and made to move towards the center of the circle. This tension, accordingly, represents a centripetal force. On the other hand, the axis also, or the fixed center, is acted on by the tension of the string, and in this aspect the tension of the string appears as a centrifugal force.

Let us suppose that we have a body to which a velocity has been imparted and which is maintained in uniform motion in a circle by an acceleration constantly directed towards the center. It is our purpose to investigate the conditions on which this acceleration depends. We imagine (Fig. 102) two equal circles uniformly travelled round by two bodies; the velocities in the circles I and II bear to each other the proportion $1 : 2$. If in the two circles we consider any same arc-element corresponding to some very small angle a,

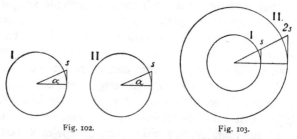

Fig. 102.   Fig. 103.

then the corresponding element s of the distance that the bodies in consequence of the centripetal acceleration have departed from the rectilinear path (the tangent), will also be the same. If we call φ_1 and φ_2 the respective accelerations, and τ and $\tau/2$ the time-elements for the angle α, we find by Galileo's law

$$\varphi_1 = \frac{2s}{\tau^2}, \; \varphi_2 = 4\frac{2s}{\tau^2}, \text{ that is to say } \varphi_2 = 4\varphi_1.$$

Therefore, by generalization, in equal circles the centripetal acceleration is proportional to the square of the velocity of the motion.

Let us now consider the motion in the circles I and II (Fig. 103), the radii of which are to each other as 1 : 2, and let us take for the ratio of the velocities of the motions also 1 : 2, so that like arc-elements are travelled through in equal times. φ_1, φ_2, s, $2s$ denote the accelerations and the elements of the distance traversed; τ is the element of the time, equal for both cases. Then

$$\varphi_1 = \frac{2s}{\tau^2}, \; \varphi_2 = \frac{4s}{\tau^2}, \text{ that is to say } \varphi_2 = 2\varphi_1.$$

If now we reduce the velocity of the motion in II one-half, so that the velocities in I and II become equal, φ_2 will thereby be reduced one-fourth, that is to say to $\varphi_1/2$. Generalizing, we get this rule: when the velocity of the circular motion is the *same*, the centripetal acceleration is inversely proportional to the radius of the circle described.

4. The early investigators, owing to their following the conceptions of the ancients, generally obtained their propositions in the cumbersome form of proportions. We shall pursue a different method. On a movable object having the velocity v let a force act during the element of time τ which imparts to the object perpendicularly to the direction of its motion the acceleration φ (Fig. 104). The new velocity-component thus becomes $\varphi\tau$, and its composition with the first velocity produces a new direction of the motion, making the

angle α with the original direction. From this results, by conceiving the motion to take place in a circle of radius r, and on account of the *smallness of the angu-*

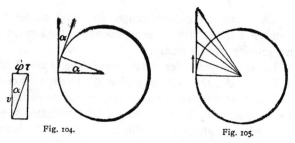

Fig. 104.   Fig. 105.

lar element putting tan $\alpha = \alpha$, the following, as the complete expression for the centripetal acceleration of a uniform motion in a circle,

$$\frac{\varphi\tau}{v} = \tan \alpha = \alpha = \frac{v\tau}{r} \text{ or } \varphi = \frac{v^2}{r}.$$

The idea of uniform motion in a circle conditioned by a constant centripetal acceleration is a little paradoxical. The paradox lies in the assumption of a constant acceleration towards the center without actual approach thereto and without increase of velocity. This is lessened when we reflect that without this centripetal acceleration the body would be continually moving away from the center (Fig. 105); that the direction of the acceleration is constantly changing; and that a change of velocity (as will appear in the discussion of the principle of *vis viva*) is connected with an approach of the bodies that accelerate each other, which does not take place here. The more complex case of elliptical central motion is elucidative in this direction.

The perspicuous deduction of the expression for centrifugal force based on the principle of Hamilton's hodograph may also be mentioned. If a body move uniformly in a circle of radius r (Fig. 105a), the velocity v at the point A of the path is transformed by the traction of the string into the velocity v of like

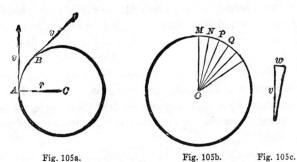

Fig. 105a. Fig. 105b. Fig. 105c.

magnitude but different direction at the point B. If from O as center (Fig. 105b) we lay off as to magnitude and direction all the velocities the body successively acquires, these lines will represent the sum of the radii v of the circle. For OM to be transformed into ON, the perpendicular component to it, MN, must be added. During the period of revolution T the velocity is uniformly increased in the directions of the radii r by an amount $2\pi v$. The numerical measure of the radial acceleration is therefore

$$\varphi = \frac{2\pi v}{T}, \text{ and since } vT = 2\pi r, \text{ therefore also } \varphi = \frac{v^2}{r}.$$

If to $OM = v$ the very small component w is added (Fig. 105c), the resultant will strictly be a greater velocity $\sqrt{v^2 + w^2} = v + \dfrac{w^2}{2v}$, as the approximate extraction of the square root will show. But on *continuous* deflection $\dfrac{w^2}{2v}$ vanishes with respect to v; hence, only the direction, but not the magnitude, of the velocity changes.

5. The expression for the centripetal or centrifugal acceleration, $\varphi = v^2/r$, can easily be put in a somewhat different form. If T denote the periodic time of the circular motion, the time occupied in describing the circumference, then $vT = 2\,r\,\pi$, and consequently $\varphi = 4\,r\,\pi^2/T^2$, in which form we shall employ the expression later on. If several bodies moving in circles have the same periodic times, the respective centripetal accelerations by which they are held in their paths, as is apparent from the last expression, are proportional to the radii.

6. We shall take it for granted that the reader is familiar with the phenomena that illustrate the considerations here presented: as the rupture of strings of insufficient strength on which bodies are whirled about, the flattening of soft rotating spheres, and so on. Huygens was able, by the aid of his conception, to explain at once whole series of phenomena. When a pendulum-clock, for example, which had been taken from Paris to Cayenne by Richer (1671-1673), showed a retardation of its motion, Huygens deduced the apparent diminution of the acceleration of gravity g thus established, from the greater centrifugal acceleration of the rotating earth at the equator; an explanation that at once rendered the observation intelligible.

An experiment instituted by Huygens may here be noticed, on account of its historical interest. When Newton brought out his theory of universal gravitation, Huygens belonged to the great number of those who were unable to reconcile themselves to the idea of action at a distance. He was of the opinion that gravitation could be explained by a vortical medium. If we enclose

in a vessel filled with a liquid a number of lighter bodies, say wooden balls in water, and set the vessel rotating about its axis, the balls will at once rapidly move
towards the axis. If for instance (Fig. 106), we place
the glass cylinders *RR* containing the wooden balls *KK*
by means of a pivot *Z* on a rotatory apparatus, and rotate the latter about its vertical axis, the balls will immediately run up the cyl-
inders in the direction away
from the axis. But if the
tubes be filled with water,
each rotation will force the
balls floating at the extremities *EE* towards the axis.

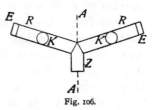

Fig. 106.

The phenomenon is easily explicable by analogy with
the principle of Archimedes. The wooden balls receive
a centripetal impulsion, comparable to buoyancy, which
is equal and opposite to the centrifugal force acting on
the displaced liquid. Even Descartes thought of explaining the centripetal impulsion of floating bodies in
a vortical medium, in this manner. But Huygens correctly remarked that on this hypothesis we should have
to assume that the *lightest* bodies received the *greatest*
centripetal impulsion, and that all heavy bodies would
without exception have to be lighter than the vortical
medium. Huygens observes further that like phenomena are also necessarily presented in the case of
bodies, be they what they may, that do *not* participate
in the whirling movement, that is to say, such as might
exist without centrifugal force in a vortical medium
affected with centrifugal force. For example, a sphere
composed of any material whatsoever but movable only
along a *stationary* axis, say a wire, is impelled toward
the axis of rotation in a whirling medium.

In a closed vessel containing water Huygens placed small particles of sealing wax which are slightly *heavier* than water and hence touch the bottom of the vessel. If the vessel be rotated, the particles of sealing wax will flock toward the outer rim of the vessel. If the vessel be then suddenly brought to rest, the water will continue to rotate while the particles of sealing wax which touch the bottom and are therefore more rapidly arrested in their movement, will now be impelled toward the axis of the vessel. In this process Huygens saw an exact replica of gravity. An ether whirling in one direction only, did not appear to fulfil his requirements. Ultimately, he thought, it would sweep everything with it. He accordingly assumed ether-particles that sped rapidly about in all directions, it being his theory that in a closed space, circular, as contrasted with radial, motions would of themselves preponderate. This ether appeared to him adequate to explain gravity. The detailed exposition of this kinetic theory of gravity is found in Huygens's tract *On the Cause of Gravitation* (German trans. by Mewes, Berlin, 1893). See also Lasswitz, *Geschichte der Atomistik,* 1890, II., p. 344.

7. Before we proceed to Huygens's investigations on the center of oscillation, we shall present to the reader a few considerations concerning pendulous and oscillatory motion generally, which will make up in obviousness for what they lack in rigor.

Many of the properties of pendulum motion were known to GALILEO. That he had formed the conception which we shall now give, or that at least he was on the verge of so doing, may be inferred from many scattered allusions to the subject in his *Dialogues.* The bob of a simple pendulum of length *l* moves in a circle

(Fig. 107) of raduis l. If we give the pendulum a very small excursion, it will travel in its oscillations over a very small arc which coincides approximately with the chord belonging to it. But this chord is described by a falling particle, moving on it as on an inclined plane (see SECT. I of this Chapter, § 7), in the same time as the vertical diameter $BD = 2l$. If the time of descent be called t, we shall have $2l = \frac{1}{2}gt^2$, that is $t = 2\sqrt{l/g}$. But since the continued movement from B up the line BC' occupies an equal inter-

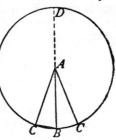

Fig. 107.

val of time, we have to put for the time T of an oscillation from C to C', $T = 4\sqrt{l/g}$. It will be seen that even from so crude a conception as this the correct *form* of the pendulum-laws is obtainable. The exact expression for the time of very small oscillations is, as we know, $T = \pi\sqrt{l/g}$.

Again, the motion of a pendulum bob may be viewed as a motion of descent on a succession of inclined planes. If the string of the pendulum makes the angle α with the perpendicular, the pendulum bob receives in the direction of the position of equilibrium the acceleration $g \cdot \sin \alpha$. When α is small, $g \cdot \alpha$ is the expression of this acceleration; in other words, the acceleration is always proportional and oppositely directed to the excursion. When the excursions are small the curvature of the path may be neglected.

8. From these preliminaries, we may proceed to the study of oscillatory motion in a simpler manner. A body is free to move on a straight line OA (Fig. 108), and constantly receives in the direction towards the

202   *THE SCIENCE OF MECHANICS*

point *O* an acceleration proportional to its distance from
O. We will represent these accelerations by ordinates
erected at the positions considered. Ordinates upwards
denote accelerations towards the left; ordinates downwards represent accelerations towards the right. The

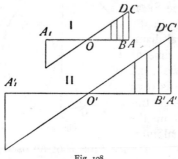

Fig. 108.

body, left to itself at
A, will move towards
O with varied accel-
eration, pass through
O to A_1, where OA_1
$= OA$, come back to
O, and so again con-
tinue its motion. It is
in the first place easily
demonstrable that the
period of oscillation
(the time of the motion through AOA_1) is independent
of the amplitude of the oscillation (the distance *OA*).
To show this, let us imagine in I and II the same oscillation performed, with single and double amplitudes
of oscillation. As the acceleration varies from point
to point, we must divide *OA* and $O'A' = 2OA$ into a
very large equal number of elements. Each element
A'B' of *O'A'* is then twice as large as the corresponding element *AB* of *OA*.

The initial accelerations φ and φ' stand in the relation $\varphi' = 2\varphi$. Accordingly, the elements *AB* and *A'B'*
$= 2AB$ are described with their respective accelerations
φ and 2φ in the same time ι. The final velocities *v*
and *v'* in I and II, for the first element, will be $v = \varphi\iota$
and $v' = 2\varphi\iota$, that is $v' = 2v$. The accelerations and
the initial velocities at *B* and *B'* are therefore again as
1:2. Accordingly, the corresponding elements that
next succeed will be described in the same time. And
of every succeeding pair of elements the same asser-

tion holds true. Therefore, generalizing, it will be readily perceived that the period of oscillation is independent of its amplitude or breadth.

Next, let us conceive two oscillatory motions, I and II, that have equal excursions (Fig. 109); but in II let a fourfold acceleration correspond to the same distance from O. We divide the amplitudes of the oscillations AO and $O'A' = OA$ into a very large equal number of parts. These parts are then equal in I and II. The initial accelerations at A and A' are φ and 4φ; the elements of the distance described are $AB = A'B' = s$; and the times are respectively ι and ι'. We obtain, then, $\iota = \sqrt{2s/\varphi}$, $\iota' = \sqrt{2s/4\varphi} = \iota/2$. The element $A'B'$ is accordingly travelled through in one-half the time as the element AB. The final velocities v and v' at B and

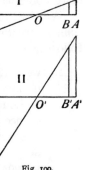

Fig. 109.

B' are found by the equations $v = \varphi\iota$ and $v' = 4\varphi(\iota/2)$ $= 2v$. Since, therefore, the initial velocities at B and B' are to one another as $1:2$, and the accelerations are again as $1:4$, the element of II succeeding the first will again be traversed in half the time of the corresponding one in I. Generalizing, we get: For equal excursions the time of oscillation is inversely proportional to the square root of the accelerations.

9. The considerations last presented may be put in a very much abbreviated and very obvious form by a method of conception first employed by Newton. Newton calls those material systems *similar* that have geometrically similar configurations and whose homologous masses bear to one another the same ratio. He says further that systems of this kind execute similar movements when the homologous points describe simi-

lar paths in proportional times. Conformably to the geometrical terminology of the present day we should not be permitted to call mechanical structures of this kind (of five dimensions) *similar* unless their homologous linear dimensions as well as the times and the masses bore to one another the *same* ratio. The structures might more appropriately be termed *affined* to one another.

We shall retain, however, the name phoronomically *similar* structures, and in the consideration that is to follow leave the masses entirely out of account.

In two such similar motions, then, let
the homologous paths be s and αs,
the homologous times be t and βt;
whence the homologous
velocities are............... $v = \dfrac{s}{t}$ and $\gamma v = \dfrac{\alpha}{\beta} \dfrac{s}{t}$,
the homologous accel-
erations .... $\varphi = \dfrac{2s}{t^2}$ and $\varepsilon\varphi = \dfrac{\alpha}{\beta^2} \dfrac{2s}{t^2}$.

Now all oscillations which a body performs under the conditions above set forth with any two different amplitudes 1 and α, will be readily recognized as *similar* motions. Noting that the ratio of the homologous accelerations in this case is $\varepsilon = \alpha$, we have $\alpha = \alpha/\beta^2$. Wherefore the ratio of the homologous times, that is to say of the times of oscillation, is $\beta = \pm 1$. We obtain thus the law, that the period of oscillation is independent of the amplitude.

If in two oscillatory motions we put for the ratio between the amplitudes $1 : \alpha$, and for the ratio between the accelerations $1 : \alpha\mu$, we shall obtain for this case $\varepsilon = \alpha\mu = \alpha/\beta^2$, and therefore $\beta = \dfrac{\pm 1}{\sqrt{\mu}}$; wherewith the second law of oscillating motion is obtained.

Two uniform circular motions are always phoronomically similar. Let the ratio of their radii be 1 : a and the ratio of their velocities 1 : γ. The ratio of their accelerations is then $\varepsilon = a/\beta^2$, and since $\gamma = a/\beta$, also $\varepsilon = \gamma^2/a$; whence the theorems relative to centripetal acceleration are obtained.

It is a pity that investigations of this kind respecting mechanical and phoronomical *affinity* are not more extensively cultivated, since they promise the most beautiful and most elucidative extensions of insight imaginable.

10. Between uniform motion in a circle and oscillatory motion of the kind just discussed an important relation exists which we shall now consider. We assume a system of rectangular coordinates, having its origin at the center, O, of the circle (Fig. 110), about the circumference of which we conceive a body to move uniformly. The centripetal acceleration φ which conditions this motion, we resolve in the directions

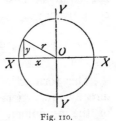

Fig. 110.

of X and Y; and observe that the X-components of the motion are affected only by the X-components of the acceleration. We may regard both the motions and both the accelerations as independent of each other.

Now, the two components of the motion are oscillatory motions to and fro about O. The acceleration-component $\varphi(x/r)$ or $(\varphi/\mathrm{r})\,x$ in the direction O corresponds to the excursion X. The acceleration is *proportional*, therefore, to the excursion. And accordingly the motion is of the kind just investigated. The time T of a complete to and fro movement is also the periodic time of the circular motion. With respect to the latter, however, we know that $\varphi = 4\,r\,\pi^2/T^2$, or, what is the same, that $T = 2\,\pi\sqrt{r/\varphi}$. Now φ/r is the acceleration for $x = 1$, the acceleration that corresponds to unit of excursion, which we shall briefly designate by f. For the oscillatory motion we may put, therefore, $T = 2\pi\sqrt{1/f}$. or a single movement to, or a single movement fro—the common method of reckoning the time of oscillation—we get, then, $T = \pi\sqrt{1/f}$.

11. Now this result is directly applicable to pendulum vibrations of *very small* excursions, where neglecting the curvature of the path, it is possible to adhere to the conception developed. For the angle of elongation a we obtain as the distance of the pendulum bob from the position of equilibrium, $l\,a$; and as the corresponding acceleration, $g\,a$; whence

$$f = \frac{g\,a}{l\,a} = \frac{g}{l} \text{ and } T = \pi\sqrt{\frac{l}{g}}.$$

This formula tells us, that the time of vibration is directly proportional to the square root of the length of the pendulum, and inversely proportional to the square root of the acceleration of gravity. A pendulum that is four times as long as the seconds pendulum, therefore, will perform its oscillation in two seconds. A seconds pendulum removed a distance equal to the earth's radius from the surface of the earth, and subjected therefore to the acceleration $g/4$, will likewise perform its oscillation in two seconds.

12. The dependence of the time of oscillation on the length of the pendulum is very easily verifiable by experiment. If (Fig. 111) the pendulums a, b, c, which to maintain the plane of oscillation invariable are suspended by double threads, have the lengths 1, 4, 9, then a will execute two oscillations to one oscillation of b, and three to one of c.

The verification of the dependence of the time of

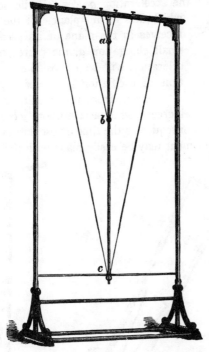

Fig. 111.

oscillation on the acceleration of gravity g is somewhat more difficult; since the latter cannot be arbitrarily altered. But the demonstration can be effected
by allowing one component only of g to act on the
pendulum. If we imagine the axis of oscillation of
the pendulum AA fixed in the vertically placed plane
of the paper, EE will be the intersection of the plane
of oscillation with the plane of the paper and likewise
the position of equilibrium of the pendulum. The axis
makes with the horizontal plane, and the plane of os-

cillation makes with the vertical plane, the angle β; wherefore the acceleration $g \cdot \cos \beta$ is the acceleration which acts in this plane. If the pendulum receives in the plane of its oscillation the small elongation α, the corresponding acceleration will be $(g \cos \beta) \alpha$; whence the time of oscillation is $T = \pi \sqrt{l/g \cos \beta}$.

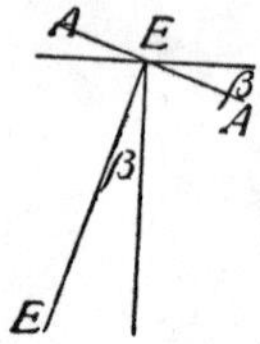

Fig. 112.

We see from this result, that as β is increased the acceleration $g \cos \beta$ diminishes, and consequently the time of oscillation increases. The experiment may be easily made with the apparatus

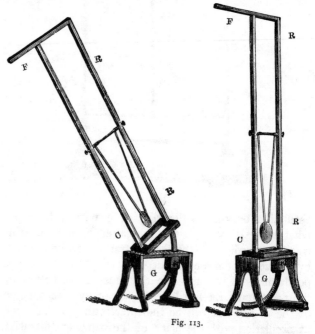

Fig. 113.

represented in Fig. 113. The frame RR is free to turn about a hinge at C; it can be inclined and placed on its side. The angle of inclination is fixed by a graduated arc *G* held by a set-screw. Every increase of β increases the time of oscillation. If the plane of oscillation is made horizontal, in which position *R* rests on the foot *F*, the time of oscillation becomes infinitely great. The pendulum in this case no longer returns to any definite position but describes several complete revolutions in the same direction until its entire velocity has been destroyed by friction.

13. If the movement of the pendulum does not take place in a plane, but is performed in space, the thread of the pendulum will describe the surface of a cone. The motion of the conical pendulum was also investigated by Huygens. We shall examine a simple case of this motion. We imagine (Fig. 114) a pendulum of length *l* removed from the vertical by the angle α, a velocity *v* imparted to the bob of the pendulum at right angles to the plane of elongation, and the pendulum released. The bob of the pendulum will move in a horizontal circle if the centrifugal acceleration φ developed exactly equilibrates the acceleration of gravity *g*; that is, if the resultant acceleration falls in the direction of the pendulum thread. But in that case $\varphi/g = \tan \alpha$. If T stands for the time taken to describe one revolution, the periodic time, then $\varphi = 4 r \pi^2/T^2$ or $T = 2 \pi \sqrt{r/\varphi}$. Introducing, now, in the place of r/φ the value $l \sin \alpha/g \tan \alpha = l \cos \alpha/g$, we get for the periodic time of the pendulum, $T = 2 \pi \sqrt{l \cos \alpha/g}$. For the velocity *v* of the revolution we find $v = \sqrt{r \varphi}$, and since $\varphi = g \tan \alpha$ it follows that $v = \sqrt{g l \sin \alpha \tan \alpha}$. For very small elongations of the conical pendulum we may put $T = 2 \pi \sqrt{l/g}$, which coincides with the regular formula for

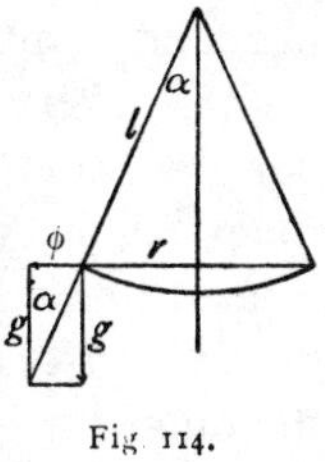

Fig. 114.

the pendulum, when we reflect that a single revolution of the conical pendulum corresponds to *two* vibrations of the common pendulum.

14. Huygens was the first to undertake the exact determination of the acceleration of gravity g by means of pendulum observations. From the formula $T = \pi\sqrt{l/g}$ for a simple pendulum with small bob we obtain directly $g = \pi^2 l/T^2$. For latitude 45° we obtain as the value of g, in meters and seconds, 9.806.

For provisional mental calculations it is sufficient to remember that the acceleration of gravity amounts in round numbers to 10 meters a second.

15. Every thinking beginner puts to himself the question how the duration of an oscillation, that is a *time*, can be found by dividing a number that is the measure of a *length* by a number that is the measure of an *acceleration* and extracting the square root of the quotient. But the fact here to be borne in mind is that $g = 2s/t^2$, that is a length divided by the square of a time. In reality therefore the formula we have is $T = \pi\sqrt{(l/2s)t^2}$. And since $l/2s$ is the ratio of two lengths, and therefore a number, what we have under the radical sign is consequently the square of a time. It stands to reason that we shall find T in seconds only when, in determining g, we also take the second as unit of time.

In the formula $g = \pi^2 l/T^2$ we see directly that g is a length divided by the square of a time, according to the nature of an acceleration.

16. The most important achievement of Huygens is his solution of the problem to determine the center of oscillation. So long as we have to deal with the dynamics of a *single* body, the Galilean principles amply suffice. But in the problem just mentioned we have to

determine the motion of *several* bodies that mutually
influence each other. This cannot be done without
resorting to a *new* principle. Such a one Huygens
actually discovered.

We know that long pendulums perform their oscillations more slowly than short ones. Let us imagine a
heavy body, free to rotate about an axis, the center of
gravity of which lies outside of the axis; such
a body will represent a compound pendulum.
Every material particle of a pendulum of this
kind would, if it were situated alone at the
same distance from the axis, have its own period of oscillation. But owing to the connec-   Fig. 115.
tions of the parts the whole body can vibrate with only
a single, determinate period of oscillation. If we picture to ourselves several pendulums of unequal lengths
(Fig. 115) the shorter ones will swing more quickly,
the longer ones more slowly. If all be joined together
so as to form a single pendulum, it is to be presumed
that the longer ones will be accelerated, the shorter
ones retarded, and that a sort of mean time of oscillation will result. There must exist therefore a simple
pendulum, intermediate in length between the shortest
and the longest, that has the same time of oscillation
as the compound pendulum. If we lay off the length
of this pendulum on the compound pendulum, we shall
find a point that preserves the same period of oscillation in its connection with the other points as it would
have if detached and left to itself. This point is the
center of oscillation. MERSENNE was the first to pro-
pound the problem of determining the center of oscillation. The solution of DESCARTES, who attempted it,
was, however, precipitate and insufficient.

17. Huygens was the first who gave a general solution. Besides Huygens nearly all the great inquirers

of that time employed themselves on the problem, and we may say that the most important principles of modern mechanics were developed in connection with it.

The *new* idea from which Huygens set out, and which is more important by far than the whole problem, is this. In whatsoever manner the material particles of a pendulum may by mutual interaction modify each other's motions, in every case the velocities acquired in the descent of the pendulum can be such only that by virtue of them, the center of gravity of the particles, whether still in connection or with their connections dissolved, is able to rise just as *high* as the point from which it *fell*. Huygens found himself compelled, by the doubts of his contemporaries as to the correctness of this principle, to remark, that the only assumption implied in the principle is, that heavy bodies of themselves do not move upwards. If it were possible for the center of gravity of a connected system of falling material particles to rise higher after the dissolution of its connections than the point from which it had fallen, then by repeating the process heavy bodies could, by virtue of their own weights, be made to rise to any height we wished. If after the dissolution of the connections the center of gravity should rise to a height less than that from which it had fallen, we should only have to reverse the motion to produce the same result. What Huygens asserted, therefore, no one had ever really doubted; on the contrary, every one had *instinctively* perceived it. Huygens, however, gave this instinctive perception an *abstract, conceptual* form. He does not omit, moreover, to point out, on the ground of this view, the fruitlessness of endeavors to establish a perpetual motion. The principle just developed will be recognized as a *generalization of one of Galileo's ideas*.

18. Let us now see what the principle accomplishes in the determination of the center of oscillation. Let

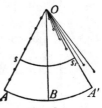

Fig. 116.

OA (Fig. 116), for simplicity's sake, be a linear pendulum, made up of a large number of masses indicated in the diagram by points. Set free at *OA*, it will swing through *B* to *OA'*, where $AB = BA'$. Its center of gravity *S* will ascend just as high on the second side as it fell on the first. From this, so far, nothing would follow. But also, if we should suddenly, at the position *OB*, release the individual masses from their connections, the masses could, by virtue of the velocities impressed on them by their connections, only attain the same height with respect to center of gravity. If we arrest the free outward-swinging masses at the *greatest heights* they severally attain, the shorter pendulums will be found below the line *OA'*, the longer ones will have passed beyond it, but the center of gravity of the system will be found on *OA'* in its former position.

Now let us note that the enforced velocities are proportional to the distances from the axis; therefore, *one* being given, all are determined, and the height of ascent of the center of gravity given. Conversely, therefore, the velocity of any material particle also is determined by the known height of the center of gravity. But if we know in a pendulum the velocity corresponding to a given distance of descent, we know its whole motion.

19. Premising these remarks, we proceed to the problem itself. On a compound linear pendulum (Fig. 117) we cut off, measuring from the axis, the portion $= 1$. If the pendulum move from its position of

greatest excursion to the position of equilibrium, the point at the distance $= 1$ from the axis will fall through the height k. The masses m, m', m'' . . . at the distances r, r', r'' . . . will fall in this case the distances, rk, $r'k$, $r''k$ . . ., and the distance of the descent of the center of gravity will be:

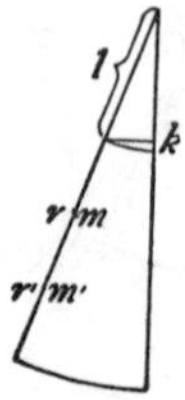

Fig. 117.

$$\frac{m r k + m' r' k + m'' r'' k + \ldots}{m + m' + m'' + \ldots} = k \frac{\Sigma m r}{\Sigma m}.$$

Let the point at the distance 1 from the axis acquire, on passing through the position of equilibrium, the velocity, as yet unascertained, v. The height of its ascent, after the dissolution of its connections, will be $v^2/2g$. The corresponding heights of ascent of the other material particles will then be $(rv)^2/2g$, $(r'v)^2/2g$, $(r''v)^2/2g$ . . . . The height of ascent of the center of gravity of the liberated masses will be

$$\frac{m \dfrac{(rv)^2}{2g} + m' \dfrac{(r'v)^2}{2g} + m'' \dfrac{(r''v)^2}{2g} + \ldots}{m + m' + m'' + \ldots} = \frac{v^2 \Sigma m r^2}{2g \Sigma m}.$$

By Huygens's fundamental principle, then,

$$k \frac{\Sigma m r}{\Sigma m} = \frac{v^2}{2g} \frac{\Sigma m r^2}{\Sigma m} \ldots \ldots (a).$$

From this a relation is deducible between the distance of descent k and the velocity v. Since, however, all pendulum motions of the same excursion are phoronomically similar, the motion here under consideration is, in this result, completely determined.

To find the length of the simple pendulum that has the same period of oscillation as the compound pendulum considered, be it noted that the same relation must obtain between the distance of its descent and its velocity, as in the case of its unimpeded fall. If y is the length of this pendulum, $k y$ is the distance of its descent, and $v y$ its velocity; wherefore

$$\frac{(vy)^2}{2g} = ky, \text{ or}$$

$$y \cdot \frac{v^2}{2g} = k \ldots \ldots \ldots (b).$$

Multiplying equation (a) by equation (b) we obtain

$$y = \frac{\Sigma m r^2}{\Sigma m r}.$$

Employing the principle of phoronomic similitude, we may also proceed in this way. From (a) we get

$$v = \sqrt{2gk} \sqrt{\frac{\Sigma m r}{\Sigma m r^2}}.$$

A simple pendulum of length 1, under corresponding circumstances, has the velocity

$$'_1 = \sqrt{2gk}.$$

Calling the time of oscillation of the compound pendulum T, that of the simple pendulum of length 1 $T_1 = \pi \sqrt{1/g}$, we obtain, adhering to the supposition of equal excursions,

$$\frac{T}{T_1} = \frac{'_1}{v'}; \text{ wherefore } T = \pi \sqrt{\frac{\Sigma m r^2}{g \Sigma m r}}.$$

20. We see without difficulty in the Huygenian principle the recognition of *work* as the condition *determinative of velocity*, or, more exactly, the condition determinative of the so-called *vis viva*. By the *vis viva* or living force of a system of masses m, $m_{,,}$

$m_{,,}$ . . . ., affected with the velocities v, $v_{,}$ $v_{,,}$ . . . ., we understand the sum*

$$\frac{m\,v^2}{2} + \frac{m_{,}\,v_{,}{}^2}{2} + \frac{m_{,,}\,v_{,,}{}^2}{2} + \ldots$$

The fundamental principle of Huygens is identical with the principle of *vis viva*. The additions of later inquirers were made not so much to the idea as to the form of its expression.

If we picture to ourselves generally any system of weights p, $p_{,}$ $p_{,,}$ . . . ., which fall connected or unconnected through the heights h, $h_{,}$ $h_{,,}$ . . . ., and attain thereby the velocities v, $v_{,}$ $v_{,,}$ . . . ., then, by the Huygenian conception, a relation of equality exists between the distance of *descent* and the distance of *ascent* of the center of gravity of the system, and, consequently, the equation holds

$$\frac{p\,h + p'h' + p''h'' + \ldots}{p + p' + p'' + \ldots} = \frac{p\dfrac{v^2}{2g} + p'\dfrac{v'^2}{2g} + p''\dfrac{v''^2}{2g} + \ldots}{p + p' + p'' + \ldots}$$

$$\text{or } \Sigma p\,h = \frac{1}{g}\,\Sigma\,\frac{p\,v^2}{2}.$$

If we have reached the concept of "mass," which Huygens did not yet possess in his investigations, we may substitute for p/g the mass m and thus obtain the form $\Sigma p\,h = \frac{1}{2}\Sigma\,m\,v^2$, which is very easily generalized for non-constant forces.

21. With the aid of the principle of living forces we can determine the duration of the infinitely small oscillations of any pendulum whatsoever. We let fall from the center of gravity s (Fig. 118) a perpendicular on the axis; the length of the perpendicular is, say, a.

---

* This is not the usual definition of English writers, who follow the older authorities in making the *vis viva* twice this quantity.—*Trans.*

We lay off on this, measuring from the axis, the length
= 1. Let the distance of descent of the point in question to the position of equilibrium be k, and v the
velocity acquired. Since the work
done in the descent is determined by
the motion of the center of gravity, we
have

work done in descent = *vis viva*:

$$a k g M = \frac{v^2}{2} \Sigma m r^2.$$

Fig. 118.

Here we call M the total mass of the pendulum and
anticipate the expression *vis viva*. By an inference
similar to that in the preceding case, we obtain $T = \pi \sqrt{\Sigma m r^2 / a g M}$.

22. We see that the duration of infinitely small
oscillations of any pendulum is determined by two factors—by the value of the expression $\Sigma m r^2$, which
Euler called the *moment of inertia* and which Huygens
had employed without any particular designation, and
by the value of $a g M$. The latter expression, which we
shall briefly term the *statical moment,* is the product
$a P$ of the weight of the pendulum into the distance of
its center of gravity from the axis. If these two values
be given, the length of the simple pendulum of the
same period of oscillation (the isochronous pendulum)
and the position of the center of oscillation are determined.

For the determination of the lengths of the pendulums referred to, Huygens, in the lack of the analytical
methods later discovered, employed a very ingenious
geometrical procedure, which we shall illustrate by one
or two examples. Let the problem be to determine the
time of oscillation of a homogeneous, material, and

heavy rectangle *ABCD*, which swings on the axis *AB*

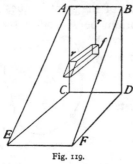

Fig. 119.

(Fig. 119). Dividing the rectangle into minute elements of area $f, f_{,}, f_{,,}, \ldots$ having the distances $r, r_{,}, r_{,,}, \ldots$ from the axis, the expression for the length of the isochronous simple pendulum, or the distance of the center of oscillation from the axis, is given by the equation

$$\frac{fr^2 + f_{,}r_{,}^2 + f_{,,}r_{,,}^2 + \ldots}{fr + f_{,}r_{,} + f_{,,}r_{,,} + \ldots}.$$

Let us erect on *ABCD* at *C* and *D* the perpendiculars $CE = DF = AC = BD$ and picture to ourselves a homogeneous wedge *ABCDEF*. Now find the distance of the center of gravity of this wedge from the plane through *AB* parallel to *CDEF*. We have to consider, in so doing, the tiny columns $fr, f_{,}r_{,}, f_{,,}r_{,,}, \ldots$ and their distances $r, r_{,}, r_{,,} \ldots$ from the plane referred to. Proceeding thus, we obtain for the required distance of the center of gravity the expression

$$\frac{fr \cdot r + f_{,}r_{,} \cdot r_{,} + f_{,,}r_{,,} \cdot r_{,,} + \ldots}{fr + f_{,}r_{,} + f_{,,}r_{,,} + \ldots},$$

that is, the same expression as before. The center of oscillation of the rectangle and the center of gravity of the wedge are consequently at the same distance from the axis, $\frac{2}{3} AC$.

Following out this idea, we readily perceive the correctness of the following assertions. For a homogeneous rectangle of height *h* swinging about one of its sides, the distance of the center of gravity from the axis is $h/2$, the distance of the center of oscillation $\frac{2}{3} h$. For a homogeneous triangle of height h, the axis of which passes through the vertex parallel to the base, the distance of the center of gravity from the axis is $\frac{2}{3} h$, the distance of the center of oscillation $\frac{3}{4} h$. Calling the moments of inertia of the rectangle and of the triangle $\varDelta_1$, $\varDelta_2$, and their respective masses M_1, M_2, we get

$$\tfrac{2}{3} h = \frac{\varDelta_1}{\frac{h}{2} M_1}, \quad \tfrac{3}{4} h = \frac{\varDelta_2}{\frac{2h}{3} M_2}.$$

Consequently $\varDelta_1 = \dfrac{h^2 M_1}{3}, \quad \varDelta_2 = \dfrac{h^2 M_2}{2}.$

By this pretty geometrical conception many problems can be solved that are today treated—more conveniently it is true—by routine forms.

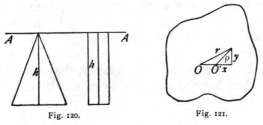

Fig. 120. Fig. 121.

23. We shall now discuss a proposition relating to moments of inertia, that Huygens made use of in a somewhat different form. Let O (Fig. 121) be the center of gravity of any given body. Make this the origin of a system of rectangular coördinates, and suppose the moment of inertia with reference to the Z-axis determined. If m is the element of mass and r its distance from the Z-axis, then this moment of inertia is $\varDelta = \varSigma m r^2$. We now displace the axis of rotation

parallel to itself to O', the distance a in the X-direction. The distance r is transformed, by this displacement, into the new distance ρ, and the new moment of inertia is

$$\Theta = \Sigma m \rho^2 = \Sigma m \left[(x-a)^2 + y^2 \right] = \Sigma m (x^2 + y^2) - 2a \Sigma m x + a^2 \Sigma m, \text{ or, since } \Sigma m (x^2 + y^2) = \Sigma m r^2 = \Delta,$$

calling the total mass $M = \Sigma m$, and remembering the property of the center of gravity $\Sigma m x = 0$,

$$\Theta = \Delta + a^2 M.$$

From the moment of inertia for one axis through the center of gravity, therefore, that for any other axis *parallel* to the first is easily derivable.

24. An additional observation presents itself here. The distance of the center of oscillation is given by the equation $l = \overline{\Delta + a^2 M}/a M$, where Δ, M, and a have their previous significance. The quantities Δ and M are invariable for any one given body. So long therefore as a retains the same value, l will also remain invariable. For all *parallel* axes situated at the *same* distance from the center of gravity, the same body as pendulum has the same period of oscillation. If we put $\Delta/M = \varkappa$, then

$$l = \frac{\varkappa}{a} + a.$$

Now since l denotes the distance of the center of oscillation, and a the distance of the center of gravity from the axis, therefore the center of oscillation is always farther away from the axis than the center of gravity by the distance $\varkappa/a$. Therefore $\varkappa/a$ is the distance of the center of oscillation from the center of gravity. If through the center of oscillation we place

a second axis parallel to the original axis, a passes
thereby into $\varkappa/a$, and we obtain the new pendulum
length

$$l' = \frac{\varkappa}{\dfrac{\varkappa}{a}} + \frac{\varkappa}{a} = a + \frac{\varkappa}{a} = l.$$

The time of oscillation remains the same therefore
for the second parallel axis through the center of oscillation, and consequently the same also for every parallel axis that is at the same distance $\varkappa/a$ from the
center of gravity as the center of oscillation.

The totality of all parallel axes corresponding to
the same period of oscillation and having the distances a
and $\varkappa/a$ from the center of gravity, is consequently realized in two coaxial cylinders. Each generating line
is interchangeable as axis with every other generating
line without affecting the period of oscillation.

25. To obtain a clear view of the relations subsisting between the two axial cylinders, as we shall briefly
call them, let us institute the following considerations.
We put $\varDelta = k^2 M$, and then

$$l = \frac{k^2}{a} + a.$$

If we seek the a that corresponds to a given l, and
therefore to a given time of oscillation, we obtain

$$a = \frac{l}{2} \pm \sqrt{\frac{l^2}{4} - k^2}.$$

Generally therefore to one value of l there correspond
two values of a. Only where $\sqrt{l^2/4 - k^2} = 0$, that is
in cases in which $l = 2\,k$, do both values coincide in
$a = k$.

If we designate the two values of a that correspond to every l, by α and β, then

$$l = \frac{k^2 + \alpha^2}{\alpha} = \frac{k^2 + \beta^2}{\beta}, \text{ or}$$
$$\beta(k^2 + \alpha^2) = \alpha(k^2 + \beta^2),$$
$$k^2(\beta - \alpha) = \alpha\beta(\beta - \alpha),$$
$$k^2 = \alpha \cdot \beta.$$

If, therefore, in any pendulous body we know two parallel axes that have the same time of oscillation and different distances α and β from the center of gravity, as is the case for instance where we are able to give the center of oscillation for any point of suspension, we can construct k. We lay off (Fig. 122) α and β consecutively on a straight line, describe a semicircle on $\alpha + \beta$ as diameter, and erect a perpendicular at the point of junction of the two divisions α and β. On this perpendicular the semicircle cuts off k. If on the other hand we know k, then for every value of α, say λ, a value μ is obtainable that will give the same period

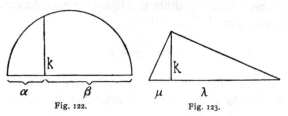

Fig. 122.   Fig. 123.

of oscillation as λ. We construct (Fig. 123) with λ and k as sides a right angle, join their extremities by a straight line on which we erect at the extremity of k a perpendicular which cuts off on λ produced the portion μ.

Now let us imagine any body whatsoever (Fig. 124) with the center of gravity O. We place it in the plane of the drawing, and make it swing about all possible parallel axes at right angles to the plane of the paper. All the axes that pass through the circle α are, we find, with respect to period of oscillation, interchangeable with each other and also with those that pass

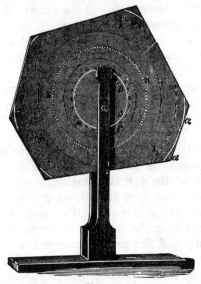

Fig. 124.

through the circle β. If instead of α we take a smaller circle λ, then in the place of β we shall get a larger circle μ. Continuing in this manner, both circles ultimately meet in one with the radius k.

26. We have dwelt at such length on the foregoing matters for good reasons. In the first place, they have served our purpose of displaying in a clear light the splendid results of the investigations of Huygens. For all that we have given is virtually contained, though in somewhat different form, in the writings of Huygens, or is at least so approximately presented in them that

it can be supplied without the slightest difficulty. Only a very small portion of it has found its way into our modern elementary text-books. One of the propositions that has thus been incorporated in our elementary treatises is that referring to the convertibility of the point of suspension and the center of oscillation. The usual presentation, however, is not exhaustive. Captain KATER, as we know, employed this principle for determining the exact length of the seconds pendulum.

The points raised in the preceding paragraphs have also rendered us the service of supplying enlightenment as to the nature of the conception "moment of inertia." This notion affords us no insight, in point of principle, that we could not have obtained without it. But since we *save* by its aid the individual consideration of the particles that make up a system, or dispose of them once for all, we arrive by a shorter and easier way at our goal. This idea, therefore, has a high import in the *economy* of mechanics. Poinsot, after Euler and Segner had attempted a similar object with less success, developed the ideas that belong to this subject further, and by his ellipsoid of inertia and central ellipsoid introduced further simplifications.

27. The investigations of Huygens concerning the geometrical and mechanical properties of cycloids are of less importance. The cycloidal pendulum, a contrivance in which Huygens realized, not an approximate, but an exact independence of the time and amplitude of oscillation, has been dropped as unnecessary from the practice of modern horology. We shall not, therefore, enter into these investigations here, however much of the geometrically beautiful they may present.

Great as the merits of Huygens are with respect to the most different physical theories, the art of horology, practical dioptrics, and mechanics in particular, his chief performance, the one that demanded the greatest intellectual courage and that was also accompanied with the greatest results, remains his enunciation of the principle by which he solved the problem of the center of oscillation. This very principle, however, was the only one he enunciated that was not adequately appreciated by his less far seeing contemporaries; nor was it appreciated for a long period thereafter. We hope we have placed this principle here in its right light as identical with the principle of *vis viva*.

28. It has been impossible for us to enter upon the signal achievements of Huygens in physics proper. But a few points may be briefly indicated. He is the creator of the wave-theory of light, which ultimately overthrew the emission theory of Newton. His attention was drawn, in fact, to precisely those features of luminous phenomena that had escaped Newton. With respect to physics he took up with great enthusiasm the idea of Descartes that all things were to be explained mechanically, though without being blind to its errors, which he acutely and correctly criticized. His predilection for mechanical explanations rendered him also an opponent of Newton's action at a distance, which he wished to replace by pressures and impacts, that is, by action due to contact. In his endeavor to do so he lighted upon some peculiar conceptions, like that of magnetic currents, which at first could not compete with the influential theory of Newton, but which has recently been reinstated in its full rights in the unbiassed efforts of Faraday and Maxwell. As a geometer and mathematician also Huygens is to be ranked high, and in this connection reference need be made only to his theory of games of chance. His astronomical observations, his achievements in theo-

retical and practical dioptrics advanced these departments very considerably. As a technicist he is the inventor of the powder-machine, the idea of which has found actualization in the modern gas-machine. As a physiologist he surmised the accommodation of the eye by deformation of the lens. All these things can scarcely be mentioned here. Our opinion of Huygens grows as his labors are made better known by the complete edition of his works. A brief and reverential sketch of his scientific career in all its phases is given by J. Bosscha in a pamphlet entitled *Christian Huyghens, Rede am 200. Gedächtnisstage seines Lebensendes,* German trans. by Engelmann, Leipzig, 1895.

III.

THE ACHIEVEMENTS OF NEWTON.

1. The merits of NEWTON with respect to our subject are twofold. First, he greatly extended the range of mechanical physics by his discovery of *universal gravitation.* Second, he *completed the formal enunciation of the mechanical principles now generally accepted.* Since his time no essentially new principle has been stated. All that has been accomplished in mechanics since his day, has been a deductive, formal, and mathematical development of mechanics on the basis of Newton's laws.

2. Let us first cast a glance at Newton's achievement in the domain of *physics.* Kepler had deduced from the observations of Tycho Brahe and his own, three empirical laws for the motion of the planets about the sun, which Newton by his new view rendered intelligible. The laws of KEPLER are as follows:

1) The planets move about the sun in ellipses, in one focus of which the sun is situated.

2) The radius vector joining each planet with the sun describes equal areas in equal times.

3) The cubes of the mean distances of the planets from the sun are proportional to the squares of their times of revolution.

He who clearly understands the doctrine of Galileo and Huygens, must see that a *curvilinear* motion implies deflective *acceleration*. Hence, to explain the phenomena of planetary motion, an acceleration must be supposed constantly directed towards the concave side of the planetary orbits.

Now Kepler's second law, the law of areas, is explained at once by the assumption of a constant planetary acceleration towards the sun; or rather, this acceleration is another form of expression for the same fact. If a radius vector describes in an element of time the area *ABS* (Fig. 125), then in the next equal element of time, assuming no acceleration, the area *BCS* will be described, where *BC* = *AB* and lies in the prolongation of *AB*. But if the central

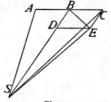

Fig. 125.

acceleration during the first element of time produces a velocity by virtue of which the distance *BD* will be traversed in the same interval, the next-succeeding area swept out is not *BCS,* but *BES,* where *CE* is parallel and equal to *BD*. But it is evident that *BES* = *BCS* = *ABS*. Consequently, the law of the areas constitutes, in another aspect, a central acceleration.

Having thus ascertained the fact of a central acceleration, the *third* law leads us to the discovery of its character. Since the planets move in ellipses slightly different from circles, we may assume, for the sake of simplicity, that their orbits actually are circles. If R_1,

R_2, R_3 are the radii and T_1, T_2, T_3 the respective times of revolution of the planets, Kepler's third law may be written as follows:

$$\frac{R_1{}^3}{T_1{}^2} = \frac{R_2{}^3}{T_2{}^2} = \frac{R_3{}^3}{T_3{}^2} = \ldots = \text{a constant.}$$

But we know that the expression for the central acceleration of motion in a circle is $\varphi = 4 R \pi^2 / T^2$, or $T^2 = 4 \pi^2 R / \varphi$. Substituting this value we get

$$\varphi_1 R_1{}^2 = \varphi_2 R_2{}^2 = \varphi_3 R_3{}^2 = \text{constant; or}$$

$$\varphi = \text{constant} / R^2 ;$$

that is to say, on the assumption of a central acceleration inversely proportional to the square of the distance, we get, from the known laws of central motion, Kepler's third law; and *vice versa*.

Moreover, though the demonstration is not easily put in an elementary form, when the idea of a central acceleration inversely proportional to the square of the distance has been reached, the demonstration that this acceleration is another expression for the motion in conic sections, of which the planetary motion in ellipses is a particular case, is a mere affair of mathematical analysis.

3. But in addition to the *intellectual* performance just discussed, the way to which was fully prepared by Kepler, Galileo, and Huygens, still another achievement of Newton remains to be estimated which in no respect should be underrated. This is an achievement of the *imagination*. We have, indeed, no hesitation in saying that this last is the most important of all. Of what nature is the acceleration that conditions the curvilinear motion of the planets about the sun, and of the satellites about the planets?

Newton perceived, with great audacity of thought, and first in the instance of the moon, that this acceleration differed in no substantial respect from the acceleration of gravity so familiar to us. It was probably the principle of continuity, which accomplished so much in Galileo's case, that led him to his discovery. He was wont, and this habit appears to be common to all truly great investigators, to adhere as closely as possible, even in cases presenting altered conditions, to a conception once formed, to preserve the same uniformity in his conceptions that nature teaches us to see in her processes. That which is a property of nature at any one time and in any one place, constantly and everywhere recurs, though it may not be with the same prominence. If the attraction of gravity is observed to prevail, not only on the surface of the earth, but also on high mountains and in deep mines, the physical inquirer, accustomed to continuity in his beliefs, conceives this attraction as also operative at greater heights and depths than those accessible to us. He asks himself: Where lies the limit of this action of terrestrial gravity? Should its action not extend to the moon? With this question the great flight of fancy was taken, of which the great scientific achievement, of Newton's intellectual genius, was but a necessary consequence.

Rosenberger is correct in his statement (*Newton und seine physikalischen Principien,* 1895) that the idea of universal gravitation did not originate with Newton, but that Newton had many highly deserving predecessors. But it may be safely asserted that it was, with all of them, a question of conjecture, of a groping and imperfect grasp of the problem, and that no one before Newton grappled with the notion so comprehensively and energetically; so that above and beyond the great mathematical problem, which Rosenberger concedes, there still remains to Newton the credit of a colossal feat of the imagination.

Among Newton's forerunners may first be mentioned Copernicus, who (in 1543) says: "I am at least of opinion that gravity is nothing more than a natural tendency implanted in particles by the divine providence of the Master of the Universe, by virtue of which, they, collecting together in the shape of a sphere, do form their own proper unity and integrity. And it is to be assumed that this propensity is inherent also in the sun, the moon, and the other planets." Similarly, Kepler (1609), like Gilbert before him (1600), conceives of gravity as the analogue of magnetic attraction. By this analogy, Hooke, it seems, is led to the notion of a *diminution* of gravity with the distance; and in picturing its action as due to a kind of radiation, he even hits upon the idea of its acting inversely as the square of the distance. He even sought to determine the diminution of its effect (1686) by weighing bodies hung at different heights from the top of Westminster Abbey (precisely after the more modern method of Jolly), by means of spring-balances and pendulum clocks, but of course without results. The conical pendulum appeared to him admirably adapted for illustrating the motion of the planets. Thus Hooke really approached nearest to Newton's conception, though he never completely reached the latter's altitude of view.

In two instructive writings (*Kepler's Lehre von der Gravitation, Halle,* 1896: *Die Gravitation bei Galileo u. Borelli,* Berlin, 1897) E. Goldbeck investigates the early history of the doctrine of gravitation with Kepler on the one hand and Galileo and Borelli on the other. Despite his adherence to scholastic, Aristotelian notions, Kepler has sufficient insight to see that there is a real physical problem presented by the phenomena of the planetary system; the moon, in his view, is swept along with the earth in its motion round the sun, and in its turn drags the tidal wave along with it, just as the earth attracts heavy bodies. Also, for the planets the source of motion is sought in the sun, from which immaterial levers extend that rotate with the sun and carry the distant planets around more slowly than the near ones. By this view, Kepler was enabled to guess that the period of rotation of the sun was less than eighty-eight days, the period of revolution of Mercury. At times, the sun is also conceived as a revolving magnet, over against which are placed the magnetic planets. In Galileo's conception of the universe, the formal, mathematical, and esthetical point of view predominates. He rejects each and every assumption of attraction, and even scouted the idea as childish in Kepler. The planetary system had not yet taken the shape of a genuine physical problem for him. Yet he assumed with Gilbert that an immaterial geometric point can exercise no physical action, and he did very much toward demonstrating the terrestrial nature of the heavenly bodies. Borelli (in his work on the satellites of the Jupiter) conceives the planets as floating between layers of ether of differing densities. They have a *natural* tendency to approach their central body, (the term attraction is avoided,) which is offset by the centrifugal force set up by the revolution. Borelli illustrates his theory by an experiment very similar to that described by us in Fig. 106, p. 199. As will be seen, he approaches very closely to Newton. His theory is, though, a combination of Descartes's and Newton's.

Newton discovered first in the case of the moon that the same acceleration that controls the descent of a stone also prevented this heavenly body from moving away in a rectilinear path from the earth, and that, on the other hand, its tangential velocity prevented it from falling towards the earth. The motion of the moon thus suddenly appeared to him in an entirely new light, but withal under quite familiar points of view. The new conception was attractive in that it embraced objects that previously were very remote, and it was convincing in that it involved the most familiar elements. This explains its prompt application in other fields and the sweeping character of its results.

Newton not only solved by his new conception the thousand years' puzzle of the planetary system, but also furnished by it the key to the explanation of a number of other important phenomena. In the same way that the acceleration due to terrestrial gravity extends to the moon and to all other parts of space, so do the accelerations that are due to the other heavenly bodies, to which we must, by the principle of continuity, ascribe the same properties, extend to all parts of space, including also the earth. But if gravitation is not peculiar to the earth, its seat is not exclusively in the *center* of the earth. Every portion of the earth, however small, shares it. Every part of the earth attracts, or determines an acceleration of, every other part. Thus an amplitude and freedom of physical view were reached of which men had no conception previously to Newton's time.

A long series of propositions respecting the action of spheres on other bodies situated beyond, upon, or within the spheres; inquiries as to the shape of the earth, especially concerning its flattening by rotation, sprang, as it were, spontaneously from this view. The riddle of the tides, the connection of which with the moon had long before been guessed, was suddenly explained as due to the acceleration of the mobile masses of terrestrial water by the moon.

Newton illustrated the identity of terrestrial gravity with the universal gravitation that determined the motions of the celestial bodies, as follows: He conceived a stone to be hurled with successive increases of horizontal velocity from the top of a high mountain. Neglecting the resistance of the air, the parabolas successively described by the stone will increase in length until finally they will fall clear of the earth altogether, and the stone will be converted into a satellite circling round the earth. Newton begins with the *fact* of universal gravity. An explanation of the phenomenon was not forthcoming, and it was not his wont, he says, to frame hypotheses. Nevertheless he could not set his thoughts at rest so easily, as is apparent from his well-known letter to Bentley. That gravity was immanent and innate in matter, so that one body could act on another directly through empty space, appeared to him absurd. But he is unable to decide whether the intermediary agency is material or immaterial (spiritual?). Like all his predecessors and successors, Newton felt the need of explaining gravitation, by some such means as actions of contact. Yet the great success which Newton achieved in astronomy with forces acting at a distance as the basis of deduction, soon changed the situation very considerably. Inquirers accustomed themselves to these forces as points of departure for their explanations and the impulse to inquire after their origin soon disappeared almost completely. The attempt was now made to introduce these forces into all the departments of physics, by conceiving bodies to be composed of particles separated by vacuous interstices and thus acting on one another at a distance. Finally even, the resistance of bodies to pressure and impact, this is to say, even forces of contact, were explained by forces acting at a distance between particles. As a fact, the functions representing the former are more complicated than those representing the latter.

The doctrine of forces acting at a distance doubtless stood in highest esteem with Laplace and his contemporaries. Faraday's unbiassed and ingenuous conceptions and Maxwell's mathematical formulation of them again turned the tide in favor of the forces of contact. Divers difficulties had raised doubts in the minds of astronomers as to the exactitude of Newton's law, and slight quantitative variations of it were looked for. After it had been demonstrated, however, that electricity travelled with finite velocity, the question of a like state of affairs in connection with the analogous action of gravitation again naturally arose. As a fact, gravitation bears a close resemblance to electrical forces acting at a distance, save in the single respect that so far as we know, attraction only, and not repulsion, takes place in the case of gravitation. Föppl ("Ueber eine Erweiterung des Gravitationsgesetzes," *Sitzungsber. d. Münch. Akad.*, 1897, p. 6 et seq.) believes that we may, without becoming involved in contradictions, assume also with respect to gravitation negative masses, which attract one another but repel positive masses, and assume therefore also *finite* fields of gravitation, similar to the electric fields. Drude (in his report on actions at a distance made for the German Naturforscherversammlung of 1897) enumerates many experiments for establishing a velocity of propagation for gravitation, which go back as far as Laplace. The result is to be regarded as a negative one, for the velocities which are possible to be considered as such, do not accord with one another, though they are all very large multiples of the velocity of light. Paul Gerber alone ("Ueber die räumliche u. zeitliche Ausbreitung der Gravitation," *Zeitschrift f. Math. u. Phys.*, 1898, II), from the perihelial motion of Mercury, forty-one seconds in a century, finds the velocity of propagation of gravitation to be the same as that of light. This would speak in favor of the ether as the medium of gravitation. (Compare W. Wien, "Ueber die Möglichkeit einer elektromagnetischen Begründung der Mechanik," *Archives Néerlandaises*, The Hague, 1900, V, p. 96.)

4. The reaction of the new ideas on mechanics was a result which speedily followed. The greatly varying accelerations which by the new view the *same* body became affected with according to its position in space, suggested at once the idea of *variable* weight, yet also pointed to *one* characteristic property of bodies which was constant. The notions of *mass* and *weight* were thus first clearly distinguished. The recognized variability of acceleration led Newton to determine by special experiments the fact that the acceleration of gravity is independent of the chemical constitution of bodies; whereby new positions of vantage were gained for the elucidation of the relation of mass and weight, as will presently be shown more in detail. Finally, the *universal applicability* of Galileo's *idea of force* was more palpably impressed on the mind by Newton's performances than it ever had been before. People could no longer believe that this idea was alone applicable to the phenomenon of falling bodies and the processes most immediately connected therewith. The generalization was effected as of itself, and without attracting particular attention.

5. Let us now discuss, more in detail, the achievements of Newton as they bear upon the *principles of mechanics*. In so doing, we shall first devote ourselves exclusively to Newton's ideas, seek to bring them forcibly home to the reader's mind, and restrict our criticisms wholly to preparatory remarks, reserving the criticism of details for a subsequent section. On perusing Newton's work (*Philosophiae Naturalis Principia Mathematica*. London, 1687), the following things strike us at once as the chief advances beyond Galileo and Huygens:

1) The generalization of the idea of force.
2) The introduction of the concept of mass.
3) The distinct and general formulation of the principle of the parallelogram of forces.
4) The statement of the law of action and reaction.

6. With respect to the first point little is to be added to what has already been said. Newton conceives all circumstances determinative of motion, whether terrestrial gravity or attractions of planets, or the action of magnets, and so forth, as circumstances determinative of *acceleration*. What considerations led to this great and immediate generalization is hard to prove. Special experiments on every kind of force cannot well be made. On the other hand the thought suggests that all forces that manifest themselves by pressure or pull, will also be equal in regard to acceleration. (*Cf. Erkenntnis und Irrtum,* 2nd ed., 1906, pp. 140, 315).

Newton's reiterated and emphatic protestations that he is not concerned with hypotheses as to the causes of phenomena, but has simply to do with the investigation and transformed statement of *actual facts,*—a direction of thought that is distinctly and tersely uttered in his words "hypotheses non fingo," (I do not frame hypotheses)—stamps him as a philosopher of the *highest* rank. He is not desirous to astound and startle, or to impress the imagination by the originality of his ideas: his aim is to know *Nature*.*

7. With regard to the concept of "mass," it is to be observed that the formulation of Newton, which defines mass to be the quantity of matter of a body as measured by the product of its volume and density, is unfortunate. As we can only define density as the mass of unit of volume, the circle is manifest. Newton felt distinctly that in every body there was inherent a property whereby the amount of its motion was determined and perceived that this must be different from weight. He called it, as we still do, mass; but he did not succeed in correctly stating this perception. We shall revert later on to this point, and shall stop here only to make the following preliminary remarks.

---

* This is conspicuously shown in the rules that Newton formed for the conduct of natural inquiry (the *Regulæ Philosophandi*):

"Rule I.   No more causes of natural things are to be admitted than such as truly exist and are sufficient to explain the phenomena of these things.

"Rule II.   Therefore, to natural effects of the same kind we must, as far as possible, assign the same causes; e. g., to respiration in man and animals; to the descent of stones in Europe and in America; to the light of our kitchen fire and of the sun; to the reflection of light on the earth and on the planets.

"Rule III.   Those qualities of bodies that can be neither increased nor diminished, and which are found to belong to all bodies within the reach of our experiments, are to be regarded as the universal qualities of all bodies. [Here follows the enumeration of the properties of bodies which has been incorporated in all text-books.]

"If it universally appear, by experiments and astronomical observations, that all bodies in the vicinity of the earth are heavy with respect to the earth, and this in proportion to the quantity of matter which they severally contain; that the moon is heavy with respect to the earth in the proportion of its mass, and our seas with respect to the moon; and all the planets with respect to one another, and the comets also with respect to the sun; we must, in conformity with this rule, declare, that *all* bodies are heavy with respect to one another.

"Rule IV.   In experimental physics propositions collected by induction from phenomena are to be regarded either as accurately true or very nearly true, notwithstanding any contrary hypotheses, till other phenomena occur, by which they are made more accurate, or are rendered subject to exceptions.

"This rule must be adhered to, that the results of induction may not be annulled by hypotheses."

8. Numerous experiences, of which a sufficient number stood at Newton's disposal, point clearly to the existence of a property distinct from weight, whereby the quantity of motion of the body to which it belongs is determined. Baliani, in his preface to *De motu gravium* (1638), distinguished, according to G. Vailati, between the weight as *agens* and the weight as *patiens*,

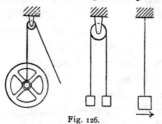

Fig. 126.

and is therefore a fore-runner of Newton. If (Fig. 126) we tie a fly-wheel to a rope and attempt to lift it by means of a pulley, we feel the *weight* of the fly-wheel. If, however, the wheel be placed on a perfectly cylindrical axle and well balanced, it will no longer assume by virtue of its weight any determinate position. Nevertheless, we are sensible of a powerful resistance the moment we endeavor to set the wheel in motion or attempt to stop it when in motion. This is the phenomenon that led to the enunciation of a distinct property of matter termed inertia, or "force" of inertia—a step which, as we have already seen and shall further explain below, is unnecessary. Two equal loads simultaneously raised, offer resistance by their weight. Tied to the extremities of a cord that passes over a pulley, they offer resistance to any motion, or rather to any change of velocity of the pulley, by their mass. A large weight hung as a pendulum on a very long string can be held at an angle of slight deviation from the line of equilibrium with very little effort. The weight-component that forces the pendulum into the position of equilibrium, is very small. Yet notwithstanding this we shall experience a considerable resistance if we suddenly attempt to move or stop the weight. A weight that is just supported by a balloon, although we have no longer to overcome its gravity, opposes a perceptible resistance to motion. Add to this the fact that the same body experiences in different geographical latitudes and in different parts of space very unequal gravitational accelerations and we shall clearly recognize that mass exists as a property wholly distinct from weight determining the amount of acceleration which a given force communicates to the body to which it belongs.

It should be observed that the notion of mass as quantity of matter was *psychologically* a very natural conception for Newton, with his peculiar development. Critical inquiries as to the origin of the concept of matter could not possibly be expected of a scientist in Newton's day. The concept developed quite instinctively; it is discovered as a datum perfectly complete, and is adopted with absolute ingenuousness. The same is the case with the concept of force. But force appears conjoined with matter. And, inasmuch as Newton invested all material particles with precisely identical gravitational forces, inasmuch as he regarded the forces exerted by the heavenly bodies on one another as the sum of the forces of the individual particles composing them, naturally these forces appear to be inseparably conjoined with the quantity of matter. Rosenberger has called attention to this fact in his book, *Newton und seine physikalischen Principien* (Leipzig, 1895, especially page 192).

I have endeavored to show elsewhere (*Analysis of the Sensations,* Chicago, 1897) how starting from the constancy of the *connection* between different sensations we have been led to the assumption of an *absolute constancy,* which we call *substance,* the most obvious and prominent example being that of a movable

body distinguishable from its environment. And seeing that such bodies are divisible into homogeneous
parts, of which each presents a constant complexus
of properties, we are induced to form the notion of a
substantial something that is quantitatively variable,
which we call *matter*. But that which we take away
from one body, makes its appearance again at some
other place. The quantity of matter in its entirety,
thus, proves to be *constant*. Strictly viewed, however, we are concerned with precisely as many substantial quantities as bodies have properties, and there
is no other function left for *matter* save that of representing the constancy of connection of the several
properties of bodies, of which *mass* is *one* only. (Compare my *Principles of Heat*, German edition, 1896,
page 425.)

9. Important is Newton's demonstration that the
mass of a body may, nevertheless, under certain conditions, be measured by its weight. Let us suppose a
body to rest on a support, on
which it exerts by its weight a
pressure. The obvious inference
is that 2 or 3 such bodies, or one-
half or one-third of such a body,

Fig. 127.

will produce a corresponding pressure 2, 3, $\frac{1}{2}$, or $\frac{1}{3}$ times
as great. If we imagine the acceleration of descent
increased, diminished, or wholly removed, we shall expect that the pressure also will be increased, diminished, or wholly removed. We thus *see,* that the pressure attributable to weight increases, decreases, and
vanishes along with the "quantity of matter" and the
magnitude of the acceleration of descent. In the simplest manner imaginable we conceive the pressure p as
quantitatively representable by the product of the
quantity of matter m into the acceleration of descent

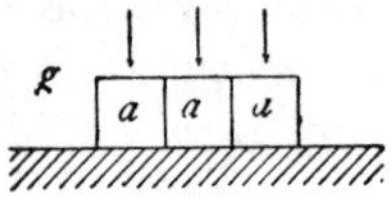

g—by $p = m\,g$. Suppose now we have two bodies that exert respectively the weight-pressures p, p', to which we ascribe the "quantities of matter" m, m', and which are subjected to the accelerations of descent g, g'; then $p = m\,g$ and $p' = m'\,g'$. If, now, we were able to prove, that, independently of the material (chemical) composition of bodies, $g = g'$ at every same point on the earth's surface, we should obtain $m/m' = p/p'$; that is to say, on the same spot of the earth's surface, it would be possible to *measure mass* by *weight*.

Now Newton established this fact, that g is independent of the chemical composition of bodies, by experiments with pendulums of equal lengths but different material, which exhibited equal times of oscillation. He carefully allowed, in these experiments, for the disturbances due to the resistance of the air; this last factor being eliminated by constructing from different materials spherical pendulum-bobs of exactly the same size, the weights of which were equalized by appropriately hollowing the spheres. Accordingly, all bodies may be regarded as affected with the same g, and their quantity of matter or mass can, as Newton pointed out, be measured by their weight.

If we imagine a rigid partition placed between an assemblage of bodies and a magnet, the bodies, if the magnet be powerful enough, or at least the majority of the bodies, will exert a pressure on the partition. But it would occur to no one to employ this magnetic pressure, in the manner we employed pressure due to weight, as a measure of mass. The strikingly noticeable inequality of the accelerations produced in the different bodies by the magnet excludes any such idea. The reader will furthermore remark that this whole argument possesses an additional dubious feature, in that the concept of mass which up to this point has simply been *named* and *felt as a necessity,* but not *defined,* is assumed by it.

10. To Newton we owe the distinct formulation of the principle of the composition of forces.* If a body is simultaneously acted on by two forces (Fig. 128), of which one would produce the motion *AB* and the other the motion *AC* in the same interval of time, the body, since the two forces and the motions produced

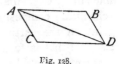

Fig. 128.

by them are *independent of each other,* will move in that interval of time to *AD.* This conception is in every respect natural, and distinctly characterizes the essential point involved. It contains none of the artificial and forced characters that were afterwards imported into the doctrine of the composition of forces.

We may express the proposition in a somewhat different manner, and thus bring it nearer its modern form. The accelerations that different forces impart to the same body are at the same time the measure of these forces. But the paths described in equal times are proportional to the accelerations. Therefore the latter also may serve as the measure of the forces. We may say accordingly: If two forces, which are proportional to the lines *AB* and *AC,* act on a body *A* in the directions *AB* and *AC,* a motion will result that could also be produced by a third force acting alone in the direction of the diagonal of the parallelogram constructed on *AB* and *AC* and proportional to that diagonal. The latter force, therefore, may be substituted for the other two. Thus, if φ and ψ are the two accelerations set up in the directions *AB* and *AC,* then for any definite interval of time t, $AB = \varphi t^2/2$, $AC =$

* Roberval's (1668) and Lami's (1687) achievements with respect to the doctrine of the composition of forces are also to be mentioned here. Varignon has already been referred to. (See the text, page 47.)

$\psi\,t^2/2$. If, now, we imagine AD produced in the same interval of time by a single force determining the acceleration χ, we get

$$AD = \chi t^2/2, \text{ and } AB:AC:AD = \varphi:\psi:\chi.$$

As soon as we have perceived the fact that the forces are independent of each other, the principle of the parallelogram of forces is easily reached from Galileo's notion of force. Without the assumption of this independence any effort to arrive abstractly and philosophically at the principle, is in vain.

11. Perhaps the most important achievement of Newton with respect to the principles is the distinct and general formulation of the law of the *equality of action and reaction,* of pressure and counter-pressure. Questions respecting the motions of bodies that exert a reciprocal influence on each other, cannot be solved by Galileo's principles alone. A new principle is necessary that will define this mutual action. Such a principle was that resorted to by Huygens in his investigation of the center of oscillation. Such a principle also is Newton's law of action and reaction.

A body that presses or pulls another body is, according to Newton, pressed or pulled in exactly the same degree by that other body. Pressure and counter-pressure, force and counter-force, are always equal to each other. As the measure of force is defined by Newton to be the quantity of motion or momentum (mass $\times$ velocity) generated in a unit of time, it consequently follows that bodies that act on each other communicate to each other in equal intervals of time equal and opposite quantities of motion (momenta), or receive contrary velocities reciprocally proportional to their masses.

Now, although Newton's law, in the form here expressed, appears much more simple, more immediate, and at first glance more admissible than that of Huygens, it will be found that it by no means contains less unanalyzed experience or fewer instinctive elements. Unquestionably the original incitation that prompted the enunciation of the principle was of a purely instinctive nature. We know that we do not experience any resistance from a body until we seek to set it in motion. The more swiftly we endeavor to hurl a heavy stone from us, the more our body is forced back by it. Pressure and counter-pressure go hand in hand. The assumption of the equality of pressure and counter-pressure is quite immediate if, using Newton's own illustration, we imagine a rope stretched between two bodies, or a distended or compressed spiral spring between them.

There exist in the domain of statics very many instinctive perceptions that involve the equality of pressure and counter-pressure. The trivial experience that one cannot lift one's self by pulling on one's chair is of this character. In a scholium in which he cites the physicists Wren, Huygens, and Wallis as his predecessors in the employment of the principle, Newton puts forward similar reflections. He imagines the earth, the single parts of which gravitate towards one another, divided by a plane. If the pressure of the one portion on the other were not equal to the counter-pressure, the earth would be compelled to move in the direction of the greater pressure. But the motion of a body can, so far as our experience goes, only be determined by other bodies external to it. Moreover, we might place the plane of division referred to at any point we choose, and the direction of the resulting motion, therefore, could not be exactly determined.

12. The indistinctness of the concept of mass takes a very palpable form when we attempt to employ the principle of the equality of action and reaction dynamically. Pressure and counter-pressure may be equal. But whence do we know that equal pressures generate velocities in the inverse ratio of the masses? Newton, indeed, actually felt the necessity of an experimental corroboration of this principle. He cites in a scholium, in support of his proposition, Wren's experiments on impact, and made independent experiments himself. He enclosed in one sealed vessel a magnet and in another a piece of iron, placed both in a tub of water, and left them to their mutual action. The vessels approached each other, collided, clung together, and afterwards remained at rest. This result is proof of the equality of pressure and counter-pressure and of equal and opposite momenta (as we shall learn later on, when we come to discuss the laws of impact).

13. The reader has already felt that the various enunciations of Newton with respect to mass and the principle of reaction, hang consistently together, and that they support one another. The experiences that lie at their foundation are: the instinctive perception of the connection of pressure and counter-pressure; the discernment that bodies offer resistance to change of velocity independently of their weight, but proportionately thereto; and the observation that bodies of greater weight receive under equal pressure smaller velocities. Newton's sense of *what* fundamental concepts and principles were required in mechanics was admirable. The *form* of his enunciations, however, as we shall later indicate in detail, leaves much to be desired. But we have no right to underrate on this account the magnitude of his achievements; for the dif-

ficulties he had to conquer were of a formidable kind, and he shunned them less than any other investigator.

14. Newton's achievements are not limited to the domain which is the subject of this book. Even his *Principia* treats questions which do not belong to mechanics proper. Motion in resisting media and the motion of fluids—even under the influence of friction —are treated there, and the velocity of the propagation of sound is theoretically deduced for the first time. The optical works of Newton contain a series of the most important discoveries. He demonstrated the prismatic decomposition of light and the compounding of white light from rays of light of different colors and unequal refrangibilities, and, in this connection, gave a proof of the periodicity of light and determined the length of period as a function of the color and refrangibility. Also it was Newton who first grasped the essential point in the polarization of light. Other studies led him to establish his law of cooling and the thermometric or pyrometric principle founded on this law.[1] In his papers and book on optics[2] Newton showed the paths which led to his discoveries quite frankly and without any restraint. Apparently the unpleasant controversies in which these first publications of his involved him had an influence on his exposition

[1] [Cf. Mach, *Die Principien der Wärmelehre*, 2nd ed., Leipzig, 1900, pp. 58-61.]

[2] [Newton's *Opticks: or a Treatise of the Reflexions, Refractions, Inflexions, and Colours of Light; also Treatises of the Species and Magnitude of Curvilinear Figures* was published at London in 1704, and again, with additions but without the mathematical appendices, in 1717, 1718, 1721, and 1730. A Latin translation, by Samuel Clarke, was first published at London in 1706; and a useful annotated German translation by W. Abendroth was published as Nos. 96 and 97 of *Ostwald's Klassiker der exakten Wissenschaften* in 1898. Newton's *Optical Lectures read in the Publick Schools of the University of Cambridge Anno Domini, 1669*, was translated into English from the original Latin and published at London in 1728, after Newton's death. The Latin was published at London in 1729. Newton's papers on optics are printed in vols. vi-xi of the *Philosophical Transactions*, and begin in the year 1672.]

in the *Principia*. In the *Principia* he gave the proofs of the theorems that he had discovered in a synthetic form, and did not disclose the methods which had led him to these theorems. The acrimonious controversy between Newton and Leibniz, and between their respective followers, on the priority of the discovery of the infinitesimal calculus, was chiefly caused by the late publication of Newton's method of fluxions. To-day it is quite clear that both Newton and Leibniz were stimulated by their predecessors and had no need to borrow from one another, and also that the discoveries were sufficiently prepared for to enable them to appear in different forms. The preparatory works of Kepler, Galileo, Descartes, Fermat, Roberval, Cavalieri, Guldin, Wallis, and Barrow were accessible to both Newton and Leibniz.[1]

IV.

DISCUSSION AND ILLUSTRATION OF THE PRINCIPLE OF
REACTION.

1. We shall now devote ourselves a moment exclusively to the Newtonian ideas, and seek to bring the principle of reaction more clearly home to our mind and feeling. If two masses M and m (Fig. 129) act on

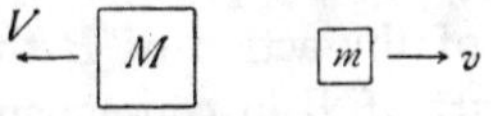

Fig. 129. 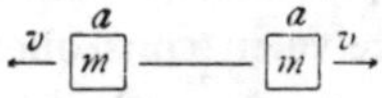   Fig. 130.

one another, they impart to each other, according to Newton, *contrary* velocities V and v, which are in-

---

[1] [On Newton's mathematical and physical achievements, we may refer to M. Cantor's *Vorlesungen über Geschichte der Mathematik*, III, 2nd ed., Leipzig, 1901, pp. 156-328, and F. Rosenberger's excellent compilation, *Isaac Newton und seine physikalischen Principien*, Leipzig, 1895.]

versely proportional to their masses, so that

$$MV + mv = 0.$$

The appearance of greater evidence may be imparted to this principle by the following consideration. We imagine first two absolutely *equal* bodies *a* (Fig. 130), also absolutely alike in chemical constitution. We set these bodies opposite each other and put them in mutual action; then, on the supposition that the influences of any third body and of the spectator are excluded, the communication of *equal* and contrary velocities in the direction of the line joining the bodies is the sole *uniquely* determined interaction.

Now let us group together in *A* (Fig. 131) *m* such bodies *a,* and put at *B* over against them *m'* such bodies *a.* We have then before us bodies whose quantities of matter or masses bear to each other the pro-

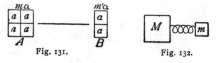

Fig. 131.   Fig. 132.

portion *m : m'.* The distance between the groups we assume to be so great that we may neglect the extension of the bodies. Let us regard now the accelerations *a,* that every two bodies *a* impart to each other, as independent of each other. Every part of *A,* then, will receive in consequence of the action of *B* the acceleration *m'a,* and every part of *B* in consequence of the action of *A* the acceleration *m a*—accelerations which will therefore be inversely proportional to the masses.

2. Let us picture to ourselves now a mass *M* (Fig. 132) joined by some elastic connection with a mass *m,*

both masses made up of bodies *a* equal in all respects. Let the mass *m* receive from some *external* source an acceleration φ. At once a distortion of the connection is produced, by which on the one hand *m* is retarded and on the other *M* accelerated. When both masses have begun to move with the same acceleration, all *further* distortion of the connection ceases. If we call α the acceleration of *M* and β the diminution of the acceleration of *m*, then $\alpha = \varphi - \beta$, where agreeably to what precedes $\alpha M = \beta m$. From this follows

$$\alpha + \beta = \alpha + \frac{\alpha M}{m} = \varphi, \text{ or } \alpha = \frac{m \varphi}{M + m}.$$

If we were to enter more exhaustively into the details of this last occurrence, we should discover that the two masses, in addition to their motion of progression, also generally perform with respect to each other motions of oscillation. If the connection on slight distortion develops a powerful tension, it will be impossible for any great amplitude of vibration to be reached, and we may entirely neglect the oscillatory motions, as we actually have done.

If the expression $\alpha = m\varphi/\overline{M + m}$, which determines the acceleration of the entire system, be examined, it will be seen that the product $m \varphi$ plays a decisive part in its determination. Newton therefore invested this product of the mass into the acceleration imparted to it, with the name of "moving force." On the other hand, $M + m$, represents the entire mass of the rigid system. We obtain, accordingly, the acceleration of any mass m' on which the moving force p acts, from the expression p/m'.

3. To reach this result, it is not at all necessary that the two connected masses should act directly on each other in all their parts. We have, connected together, let us say, the three masses m_1, m_2, m_3, where m_1 is supposed to act only on m_2, and m_3 only on m_2. Let the mass m_1 receive from some external source the acceleration φ. In the distortion that follows, the

Fig. 133.

$$
\begin{array}{ccc}
m_3 & m_2 & m_1 \\
+\,\delta & +\,\beta & +\,\varphi \\
& -\,\gamma & -\,\alpha.
\end{array}
$$

masses receive the accelerations

Here all accelerations to the right are reckoned as positive, those to the left as negative, and it is obvious that the distortion ceases to increase

when $\delta = \beta - \gamma$, $\delta = \varphi - \alpha$,
where $\delta\, m_3 = \gamma\, m_2$, $\alpha\, m_1 = \beta\, m_2$.

The resolution of these equations yields the common acceleration that all the masses receive; namely,

$$ \delta = \frac{m_1\,\varphi}{m_1 + m_2 + m_3}, $$

—a result of exactly the same form as before. When therefore a magnet acts on a piece of iron which is joined to a piece of wood, we need not trouble ourselves about ascertaining what particles of the wood are distorted directly or indirectly (through other particles of the wood) by the motion of the piece of iron.

The considerations advanced will, in some measure, perhaps, have contributed towards clearly impressing on us the great importance for mechanics of the Newtonian enunciations. They will also serve, in a subsequent place, to render more readily obvious the defects of these enunciations.

4. Let us now turn to a few illustrative physical examples of the principle of reaction. We consider, say, a load L on a table T (Fig. 134). The table is pressed by the load *just so much,* and so much only, as it in return presses the load,

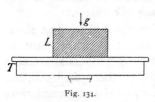

Fig. 134.

that is *prevents* the same from falling. If p is the weight, m the mass, and g the acceleration of gravity, then by Newton's conception $p = mg$. If the table be let fall vertically downwards with the acceleration of free descent g, all pressure on it ceases. We discover thus, that the pressure on the table is determined by the relative acceleration of the load with respect to the table. If the table fall or rise with the acceleration γ, the pressure on it is respectively $m(g - \gamma)$ and $m(g + \gamma)$. Be it noted, however, that no change of the relation is produced by a *constant velocity* of ascent or descent. The relative *acceleration* is determinative.

Galileo knew this relation of things very well. He not only refuted the doctrine of the Aristotelians, that bodies of greater weight fall faster than bodies of less weight, by experiments, but cornered his adversaries by logical arguments. Heavy bodies fall faster than light bodies, the Aristotelians said, because the upper parts weigh down on the under parts and accelerate their descent. In that case, returned Galileo, a small body tied to a larger body must, if it possesses *in se* the property of less rapid descent, retard the larger. Therefore, a larger body falls more slowly than a smaller body. The entire fundamental assumption iş wrong, Galileo says, because *one* portion of a *falling* body cannot by its weight under any circumstances press *another* portion.

A pendulum with the time of oscillation $T = \pi \sqrt{l/g}$, would aquire, if its axis received the downward acceleration γ, the time of oscillation $T = \pi \sqrt{l/g - \gamma}$, and if let fall freely would acquire an infinite time of oscillation, that is, would cease to oscillate.

We ourselves, when we jump or fall from an elevation, experience a peculiar sensation, which must be due to the discontinuance of the gravitational pressure of the parts of our body on one another—the blood, and so forth. A similar sensation, as if the ground were sinking beneath us, we should have on a smaller planet, to which we were suddenly transported. The sensation of constant ascent, like that felt in an earthquake, would be produced on a larger planet.

5. The conditions referred to are very beautifully illustrated by an apparatus (Fig. 135c) constructed by Poggendorff. A string loaded at both extremities

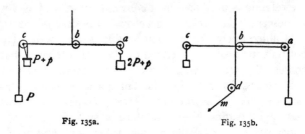

Fig. 135a.   Fig. 135b.

by a weight P (Fig. 135a) is passed over a pulley c, attached to the end of a scale-beam. A weight p is laid on one of the weights first mentioned and tied by a fine thread to the axis of the pulley. The pulley now supports the weight $2P + p$. Burning away the thread that holds the over-weight, a uniformly accelerated motion begins with the acceleration γ, with which $P + p$ descends and P rises. The load on the

pulley is thus lessened, as the turning of the scales indicates. The descending weight P is counterbalanced
by the rising weight P, while the added over-weight,
instead of weighing p, now weighs $(p/g)(g-\gamma)$. And
since $\gamma=(p/\overline{2P+p})g$, we have now to regard the
load on the pulley, not as p, but as $p(2P/\overline{2P+p})$. The

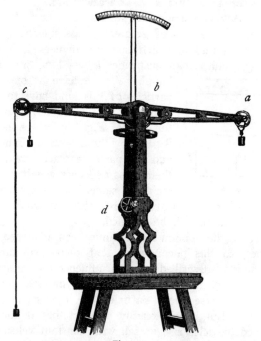

Fig. 135c.

descending weight, only partially impeded in its motion
of descent, exerts only a partial pressure on the pulley.

We may vary the experiment. We pass a thread
loaded at one extremity with the weight P over the
pulleys a, b, d, of the apparatus as indicated in Fig.
135b., tie the unloaded extremity at m, and equilibrate

the balance. If we pull on the string at *m,* this cannot *directly* affect the balance since the direction of the string passes exactly through its axis. But the side *a* immediately falls. The slackening of the string causes *a* to rise. An *unaccelerated* motion of the weights would not disturb the equilibrium. But we cannot pass from rest to motion *without* acceleration.

6. A phenomenon that strikes us at first glance is, that minute bodies of greater or less specific gravity than the liquid in which they are immersed, if sufficiently small, remain suspended a very long time in the

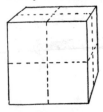

liquid. We perceive at once that particles of this kind have to overcome the friction of the liquid. If the cube of Fig. 136 be divided into 8 parts by the 3 sections indicated, and the parts be placed in a row, their mass and over-weight will remain the same, but their cross-section and superficial area, with which the friction goes hand in hand, will be doubled.

Fig. 136.

Now, the opinion has at times been advanced with respect to this phenomenon that suspended particles of the kind described have no influence on the specific gravity indicated by an areometer immersed in the liquid, because these particles are themselves areometers. But it will readily be seen that if the suspended particles rise or fall with constant velocity, as in the case of very small particles immediately occurs, the effect on the balance and the areometer must be the same. If we imagine the areometer to oscillate about its position of equilibrium, it will be evident that the liquid with all its contents will be moved with it. Applying the principle of virtual displacements, therefore, we can be no longer in doubt that the areometer must indicate the mean specific gravity. We may convince ourselves of the untenability of the rule by which the areometer is supposed to indicate only the specific gravity of the liquid and not that of the suspended particles, by the following consideration. In a liquid A a smaller quantity of a heavier liquid B is introduced and distributed in fine drops. The areometer, let us assume, indicates only the specific gravity of A. Now, take more and more of the liquid B, finally just as much of it as we have of A: we can, then, no longer say which liquid is suspended in the other, and which specific gravity, therefore, the areometer must indicate.

7. A phenomenon of an imposing kind, in which the relative acceleration of the bodies concerned is seen to be determinative of their mutual pressure, is that of the tides. We will enter into this subject here only in so far as it may serve to illustrate the point we are considering. The connection of the phenomenon of the tides with the motion of the moon asserts itself in the coincidence of the tidal and lunar periods, in the augmentation of the tides at the full and new moons, in the daily retardation of the tides (by about 50 minutes), corresponding to the retardation of the culmination of the moon, and so forth. As a matter of fact, the connection of the two occurrences was thought of very early. In Newton's time people imagined to themselves a kind of wave of atmospheric pressure, by means of which the moon in its motion was supposed to create the tidal wave.

The phenomenon of the tides makes, on every one that sees it for the first time in its full proportions, an overpowering impression. We must not be surprised, therefore, that it is a subject that has actively engaged the investigators of all times. The warriors of Alexander the Great had, from their Mediterranean homes, scarcely the faintest idea of the phenomenon of the tides, and they were, therefore, not a little taken aback by the sight of the powerful ebb and flow at the mouth of the Indus; as we learn from the account of Curtius Rufus (*De Rebus Gestis Alexandri Magni*), whose words we here quote literally:

"34. Proceeding, now, somewhat more slowly in their course, owing to the current of the river being slackened by its meeting the waters of the sea, they at last reached a second island in the middle of the river. Here they brought the vessels to the shore, and, landing, dispersed to seek provisions, wholly unconscious of the great misfortune that awaited them.

"35. It was about the third hour, when the ocean, in its constant tidal flux and reflux, began to turn and press back upon the river. The latter, at first merely checked, but then more vehemently repelled, at last set back in the opposite direction with a force greater than that of a rushing mountain torrent. The nature of the ocean was unknown to the multitude, and grave portents and evidences of the wrath of the Gods were seen in what happened. With ever-increasing vehemence the sea poured in, completely covering the fields which shortly before were dry. The vessels were lifted and the entire fleet dispersed before those who had been set on shore, terrified and dismayed at this unexpected calamity, could return. But the more haste, in times of great disturbance, the less speed. Some pushed the ships to the shore with poles; others, not waiting to adjust their oars, ran aground. Many, in their great haste to get away, had not waited for their companions, and were barely able to set in motion the huge, unmanageable barks; while some of the ships were too crowded to receive the multitudes that struggled to get aboard. The unequal division impeded all. The cries of some clamoring to be taken aboard, of others crying to put off, and the conflicting commands of men, all desirous of different ends, deprived every one of the possibility of seeing or hearing. Even the steersmen were powerless; for neither could their cries be heard by the struggling masses nor were their orders noticed by the terrified and distracted crews. The vessels collided, they broke off each other's oars, they plunged against one another. One would think it was not the fleet of one and the same army that was here in motion, but two hostile fleets in combat. Prow struck stern; those that had thrown the foremost in confusion were themselves thrown into confusion by those that followed; and the desperation of the struggling mass sometimes culminated in hand-to-hand combats.

"36. Already the tide had overflown the fields surrounding the banks of the river, till only the hillocks jutted forth from above the water, like islands. These were the point towards which all that had given up hope of being taken on the ships, swam. The scattered vessels rested in part in deep water, where there were depressions in the land, and in part lay aground in shallows, according as the waves had covered the unequal surface of the country. Then, suddenly, a new and greater terror took possession of them. The sea began to retreat, and its waters flowed back in great long swells, leaving the land which shortly before had been immersed by the salt waves, uncovered and clear. The ships, thus forsaken by the water, fell, some on their prows, some on their sides. The fields were strewn with luggage, arms, and pieces of broken planks and oars. The soldiers dared neither to venture on the land nor to remain in the ships, for every moment they expected something new and worse than had yet befallen them. They could scarcely believe that that which they saw had really happened—a shipwreck on dry land, an ocean in a river. And of their misfortune there seemed no end. For wholly ignorant that the tide would shortly bring back the sea and again set their vessels afloat, they prophesied hunger and direst distress. On the fields horrible animals crept about, which the subsiding floods had left behind.

"37. The night fell, and even the king was sore distressed at the slight hope of rescue. But his solicitude could not move his unconquerable spirit. He remained during the whole night on the watch, and despatched horsemen to the mouth of the river, that, as soon as they saw the sea turn and flow back, they might return and announce its coming. He also commanded that the damaged vessels should be repaired and that those that had been overturned by the tide should be set upright, and ordered all to be near at hand when the sea should again inundate the land. After he had thus passed the entire night in watching and in exhortation, the horsemen came back at full speed and the tide as quickly followed. At first, the approaching waters, creeping in light swells beneath the ships, gently raised them, and, inundating the fields, soon set the entire fleet in motion. The shores resounded with the cheers and clappings of the soldiers and sailors, who celebrated with immoderate joy their unexpected rescue. 'But whence,' they asked, in wonderment, 'had the sea so suddenly given back these great masses of water? Whither had they, on the day previous, retreated? And what was the nature of this element, which now opposed and now obeyed the dominion of the hours?' As the king concluded from what had happened that the fixed time for the return of the tide was after sunrise, he set out, in order to anticipate it, at midnight, and proceeding down the river with a few ships he passed the mouth and, finding himself at last at the goal of his wishes, sailed out 400 stadia into the ocean. He then offered a sacrifice to the divinities of the sea, and returned to his fleet."

8. The essential point to be noted in the explanation of the tides is, that the earth as a rigid body can receive but *one* determinate acceleration towards the moon, while the mobile particles of water on the sides nearest to and remotest from the moon can acquire various accelerations.

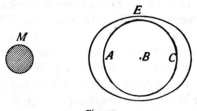

Fig. 137.

Let us consider on the earth E, opposite which stands the moon M, three points A, B, C (Fig. 137). The accelerations of the three points in the direction of the moon, if we regard them as free points, are respectively $\varphi + \Delta\varphi, \varphi, \varphi - \Delta\varphi$. The earth as a whole, however, has, as a rigid body, the acceleration φ. The acceleration towards the center of the earth we will call g. Designating now all accelerations to the left as negative, and all to the right as positive, we get the following table:

A	B	C
$-(\varphi + \varDelta\varphi),$	$-\varphi,$	$-(\varphi - \varDelta\varphi)$
$+g$		$-g.$
$-\varphi,$	$-\varphi,$	$-\varphi$
$g - \varDelta\varphi,$	$0,$	$-(g - \varDelta\varphi),$

where the symbols of the first and second lines represent the accelerations which the *free* points that head the columns receive, those of the third line the acceleration of corresponding rigid points of the earth, and those of the fourth line, the difference, or the resultant accelerations of the free points towards the earth. It will be seen from this result that the weight of the water at A and C is diminished by exactly the same amount. The water will rise at A and C (Fig. 137). A tidal wave will be produced at these points twice every day.

It is a fact not always sufficiently emphasized, that the phenomenon would be an essentially different one if the moon and the earth were not affected with accelerated motion towards each other but were relatively fixed and at rest. If we modify the considerations presented to comprehend this case, we must put for the rigid earth in the foregoing computation, $\varphi = 0$ simply. We then obtain for

the free points .... A C

the accelerations ..$- (\varphi + \varDelta\varphi),$ $-(\varphi - \varDelta\varphi),$
$+g$ $-g$

or ....$(g - \varDelta\varphi) - \varphi,$ $-(g - \varDelta\varphi) - \varphi$

or ....$g' - \varphi,$ $-(g' + \varphi),$

where $g' = g - \varDelta\varphi$. In such case, therefore, the weight of the water at A would be diminished, and the weight at C increased; the height of the water at A would be increased, and the height at C diminished.

The water would be elevated only on the side facing the moon. (Fig. 138.)

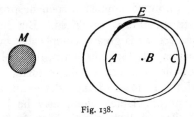

Fig. 138.

9. It would hardly be worth while to illustrate propositions best reached deductively, by experiments that can only be performed with difficulty. But such experiments are not beyond the limits of possibility. If we imagine a small iron sphere K to swing as a conical pendulum about the pole of a magnet N (Fig. 139), and cover the sphere with a solution of magnetic sulphate of iron, the fluid drop should, if the magnet is sufficiently powerful, represent the phenomenon of the tides. But if we imagine the sphere to be fixed and at rest with respect to the pole of the magnet, the fluid drop will certainly not be found tapering to a point *both* on the side facing and the side opposite to the pole of the magnet, but will remain suspended only on the side of the sphere towards the pole of the magnet.

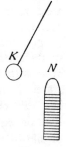

Fig. 139.

10. We must not, of course, imagine, that the entire tidal wave is produced at once by the action of the moon. We have rather to conceive the tide as an oscillatory movement *maintained* by the moon. If, for example, we should sweep a fan uniformly and continuously along over the surface of the water of a

circular canal, a wave of considerable magnitude following in the wake of the fan would by this gentle and constantly continued impulsion soon be produced. In like manner the tide is produced. But in the latter case the occurrence is greatly complicated by the irregular formation of the continents, by the periodical variation of the disturbance, and so forth.

11. Of the theories of the tides enunciated before Newton, that of Galileo alone may be briefly mentioned. Galileo explains the tides as due to the relative motion of the solid and liquid parts of the earth, and regards this fact as direct evidence of the motion of the earth and as a cardinal argument in favor of the Copernican system. If the earth (Fig. 140) rotates from the west to the east, and is affected at the same time with a progressional motion, the parts

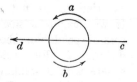

Fig. 140.

of the earth at *a* will move with the sum, and the parts at *b* with the difference, of the two velocities. The water in the bed of the ocean, which is unable to follow this change in velocity quickly enough, behaves like the water in a plate swung rapidly back and forth, or like that in the bottom of a skiff which is rowed with rapid alterations of speed: it piles up now in the front and now at the back. This is substantially the view that Galileo set forth in the *Dialogue on the Two World Systems*. Kepler's view, which supposes attraction by the moon, appears to him mystical and childish. He is of the opinion that it should be relegated to the category of explanations by "sympathy" and "antipathy," and that it admits as easily of refutation as the doctrine according to which the tides are created by radiation and the consequent expansion of the water. That on his theory the tides rise only once a day, did not, of course, escape Galileo's attention. But he deceived himself with regard to the difficulties involved, believing himself able to explain the daily, monthly, and yearly periods by considering the natural oscillations of the water and the alterations to which its motions are subject. The principle of relative motion is a correct feature of this theory, but it is so infelicitously applied that only an extremely illusive theory could result. We will first convince ourselves that the conditions supposed to be involved would not have the effect ascribed to them. Conceive a homogeneous sphere of water; any other effect due to rotation than that of a corresponding oblateness we should not expect. Now, suppose the ball to acquire in addition a uniform motion of progression. Its various parts will now as before remain at relative rest with respect to one another. For the case in question does not differ, according to our view, in any essential respect from the preceding, inasmuch as the progressive motion of the sphere may be conceived to be replaced by a motion in the opposite direction of all surrounding bodies. Even for the person who is inclined to regard the motion as an "absolute" motion, no change is produced in the relation of the parts to one another by uniform motion of progression. Now, let us cause the sphere, the parts of which have no tendency to move with respect to one another, to congeal at certain points, so that sea-beds with liquid water in them are produced. The undisturbed uniform rotation will continue, and consequently Galileo's theory is erroneous.

But Galileo's idea appears at first blush to be extremely plausible; how is the paradox explained? It is due entirely to a *negative* conception of the law of inertia. If we ask what acceleration the water experiences, everything is clear. Water having no weight would be hurled off at the beginning of rotation; water having weight, on the other hand, would describe a central motion around the center of the earth. With its slight velocity of rotation it would be forced more and more toward the center of the earth, with just enough of its centripetal acceleration counteracted by the resistance of the mass lying beneath, as to make the remainder, conjointly with the given tangential velocity, sufficient for motion in a circle. Looking at it from this point of view, all doubt and obscurity vanishes. But it must in justice be added that it was almost impossible for Galileo, unless his genius were supernatural, to have gone to the bottom of the matter. He would have been obliged to anticipate the great intellectual achievements of Huygens and Newton.

It is noteworthy that Galileo in his theory of the tides treats the first dynamic problem of space without troubling himself about the new system of coördinates. In the most naïve manner he considers the fixed stars as the new system of reference.

V.

CRITICISM OF THE PRINCIPLE OF REACTION AND OF THE

CONCEPT OF MASS.

1. Now that the preceding discussions have made us familiar with Newton's ideas, we are sufficiently prepared to enter on a critical examination of them. We shall restrict ourselves primarily in this, to the

consideration of the concept of mass and the principle of reaction. The two cannot, in such an examination, be separated; in them is contained the gist of Newton's achievement.

2. In the first place we do not find the expression "quantity of matter" adapted to explain and elucidate the concept of mass, since that expression itself is not possessed of the requisite clearness. And this is so, though we go back, as many authors have done, to an enumeration of the hypothetical atoms. We only complicate, in so doing, indefensible conceptions. If we place together a number of equal, chemically homogeneous bodies, we can, it may be granted, connect some clear idea with "quantity of matter," and we perceive, also, that the resistance the bodies offer to motion increases with this quantity. But the moment we suppose chemical heterogeneity, the assumption that there is still something that is measurable by the same standard, which something we call quantity of matter, may be suggested by mechanical experiences, but is an assumption nevertheless that needs to be justified. When therefore, with Newton, we make the assumptions, respecting pressure due to weight, that $p = m\,g$, $p' = m'g$, and put in conformity with such assumptions $p/p' = m/m'$, we have made actual use in the operation thus performed of the *supposition,* yet to be justified, that different bodies are measurable by the *same* standard.

We might, indeed, *arbitrarily posit,* that $m/m' = p/p'$; that is, might define the ratio of mass to be the ratio of pressure due to weight when g was the same. But we should then have to *substantiate* the use that is made of this notion of mass in the principle of reaction and in other relations.

Fig. 140 a.   Fig. 140 b.

3. When two bodies (Fig. 140 a), perfectly equal in all respects, are placed opposite each other, we expect, agreeably to the principle of symmetry, that they will produce in each other in the direction of their line of junction equal and opposite accelerations. But if these bodies exhibit any difference, however slight, of form, of chemical constitution, or are in any other respects different, the principle of symmetry forsakes us, *unless we assume or know beforehand* that sameness of form or sameness of chemical constitution, or whatever else the thing in question may be, is not determinative. If, however, mechanical experiences clearly and indubitably point to the existence in bodies of a special and distinct property determinative of *accelerations*, nothing stands in the way of our arbitrarily establishing the following definition:

All those bodies are bodies of equal mass, which, mutually acting on each other, produce in each other equal and opposite accelerations.

We have, in this, simply designated, or *named*, an actual relation of things. In the general case we proceed similarly. The bodies A and B receive respectively as the result of their mutual action (Fig. 140 b) the accelerations $-\varphi$ and $+\varphi'$, where the senses of the accelerations are indicated by the signs. We say then, B has φ/φ' times the mass of A. *If we take A as our unit, we assign to that body the mass m which imparts to A m times the acceleration that A in the reaction imparts to it.* The ratio of the masses is the negative inverse ratio of the counter-accelerations. That these accelerations always have opposite signs, that there are therefore, by our definition, only positive masses, is a point that experience teaches, and experience alone can teach. In our concept of mass no theory is involved; "quantity of matter" is wholly unnecessary in it; all it contains is the exact establishment, designation, and denomination of a fact.

H. Streintz's objection (*Die physikalischen Grundlagen der Mechanik*, Leipzig, 1883, p. 117), that a comparison of masses satisfying my definition can be effected only by astronomical means, I am unable to admit. The expositions on these pages and on page 248 amply refute this. Masses produce in each other accelerations in impact, as well as when subject to electric and magnetic forces, and when connected by a string on Atwood's machine. In my *Elements of Physics* (second German edition, 1891, page 27) I have shown how mass-ratios can be experimentally determined on a centrifugal machine, in a very elementary and popular manner. The criticism in question, therefore, may be regarded as refuted.

My definition is the outcome of an endeavor to establish *the interdependence of phenomena* and to remove all metaphysical obscurity, without accomplishing on this account less than other definitions have done. I have pursued exactly the same course with respect to the ideas, "quantity of electricity" ("On the Fundamental Concepts of Electrostatics," *Popular Scientific Lectures*, "temperature," "quantity of heat" (*Zeitschrift für den physikalischen und chemischen Unterricht*, Berlin, 1888, No. 1), and so forth. With the view here taken of the concept of mass is associated, however, another difficulty, which must also be carefully noted, if we would be rigorously critical in our analysis of other concepts of physics, as for example the concepts of the theory of heat. Maxwell made reference to this point in his investigations of the concept of temperature, about the same time as I did with respect to the concept of heat. I would refer here to the discussions on this subject in my *Principles of Heat* (German edition, Leipzig, 1896), particularly page 41 and page 190.

4. We now wish to consider this difficulty, since its removal is absolutely necessary to the formation of a perfectly clear concept of mass. We consider a set of bodies, A, B, C, D . . ., and compare them all with A as unit.

$$A, \quad B, \quad C, \quad D, \quad E, \quad F.$$
$$1, \quad m, \quad m', \quad m'', \quad m''', \quad m'''',$$

We find thus the respective mass-values, 1, m, m', m''. . . ., and so forth. The question now arises: If we select B as our standard of comparison (as our unit), shall we obtain for C the mass-value m'/m, and for D the value m''/m, or will perhaps wholly different values result? More simply, the question may be put thus: Will two bodies B, C, which in mutual action with A have acted as equal masses, also act as equal masses in mutual action with each other? No *logical* necessity exists whatsoever, that two masses that are equal to a third mass should also be equal to each other. For we are concerned here, not with a mathematical, but with a *physical* question. This will be rendered quite clear by recourse to an analogous relation. We place by the side of each other the bodies A, B, C in the proportions of weight a, b, c in which they enter into the chemical combinations AB and AC. There exists, now, no *logical* necessity at all for assuming that the same proportions of weight b, c of the bodies B, C will also enter into the chemical combination BC. Experience, however, informs us that they do. If we place by the side of each other any set of bodies in the proportions of weight in which they combine with the body *A*, they will also unite with each other in the same proportions of weight. But no one can know this who has not tried it. And this is precisely the case with the mass-values of bodies.

If we were to assume that the order of combination of the bodies, by which their mass-values are determined, exerted any influence on the mass-values, the consequences of such an assumption would, we should find, lead to conflict with experience. Let us suppose, for instance, that we have three elastic bodies, *A, B, C,* movable on an absolutely smooth and rigid ring (Fig. 141). We presuppose that *A* and *B* in their mutual relations comport themselves like equal masses and that *B* and *C* do the same. We are then also obliged to assume, if we wish to avoid conflicts with experience, that *C* and *A* in their mutual relations act like equal

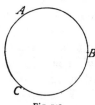

Fig. 141.

masses. If we impart to *A* a velocity, *A* will transmit this velocity by impact to *B*, and *B* to *C*. But if *C* were to act towards *A*, say, as a greater mass, *A* on impact would acquire a greater velocity than it originally had while *C* would still retain a residue of what it had. With every revolution in the direction of the hands of a watch the *vis viva* of the system would be increased. If *C* were the smaller mass as compared with *A*, reversing the motion would produce the same result. But a constant increase of *vis viva* of this kind is at decided variance with our *experience*.

5. The concept of mass when reached in the manner just developed renders unnecessary the special enunciation of the principle of reaction. In the con-

cept of mass and the principle of reaction, as we have stated in a preceding page, the same fact is *twice* formulated; which is redundant. If two masses 1 and 2 act on each other, our very definition of mass asserts that they impart to each other contrary accelerations which are to each other respectively as $2:1$.

6. The fact that *mass* can be *measured* by *weight*, where the acceleration of gravity is invariable, can also be deduced from our definition of mass. We are sensible at once of any increase or diminution of a pressure, but this feeling affords us only a very inexact and indefinite measure of magnitudes of pressure. An exact, serviceable measure of pressure springs from the observation that every pressure is replaceable by the pressure of a number of like and commensurable weights. Every pressure can be counterbalanced by the pressure of

Fig. 142.

weights of this kind. Let two bodies m and m' (Fig. 142) be respectively affected in opposite directions with the acceleration φ and φ', determined by external circumstances. And let the bodies be joined by a string. If equilibrium prevails, the acceleration φ in m and the acceleration φ' in m' are exactly balanced by *interaction*. For this case, accordingly, $m\varphi = m'\varphi'$. When, therefore, $\varphi = \varphi'$, as is the case when the bodies are abandoned to the acceleration of gravity, we have, in the case of equilibrium, also $m = m'$. It is obviously immaterial whether we make the bodies act on each other directly by means of a string, or by means of a string passed over a pulley, or by placing them on the two pans of a balance. The fact that mass can be measured by weight is evident from our definition without recourse or reference to "quantity of matter."

7. As soon therefore as we, our attention being

drawn to the fact by experience, have *perceived* in bodies the existence of a special property determinative of accelerations, our task with regard to it ends with the recognition and unequivocal designation of this *fact*. Beyond the recognition of this fact we shall not get, and every venture beyond it will only be productive of obscurity. All uneasiness will vanish when once we have made clear to ourselves that in the concept of mass no theory of any kind whatever is contained, but simply a fact of experience. The concept has hitherto held good. It is very improbable, but not impossible, that it will be shaken in the future, just as the conception of a constant quantity of heat, which also rested on experience, was modified by new experiences.

This passage appeared already in the first edition of 1883, long before the discussion concerning electromagnetic mass had started.

8. I would like to refer, here, to A. Lampa, *"Eine Ableitung des Massenbegriffs,"* in the periodical *Lotos,* (Prague), 1911, p. 303, especially to the excellent explanations of the universal method of the treatment of such questions on p. 306 seq.

VI.

NEWTON'S VIEWS OF TIME, SPACE AND MOTION.

1. In a scholium which he appends immediately to his definitions, Newton presents his views regarding time and space which we must examine more in detail. We shall literally cite, to this end, only the passages that are absolutely necessary to the characterization of Newton's views.

"So far, my object has been to explain the senses in which certain words little known are to be used in the sequel. Time, space, place, and motion, being words well known to everybody, I do not define. Yet it is to be remarked, that the vulgar conceive these quantities only in their relation to sensible objects. And hence certain prejudices with respect to them have arisen, to remove which it will be convenient to distinguish them into absolute and relative, true and apparent, mathematical and common, respectively.

"I. Absolute, true and mathematical time, of itself, and by its own nature, flows uniformly on, without regard to anything external. It is also called *duration*.

"Relative, apparent, and common time, is some sensible and external measure of absolute time (duration), estimated by the motions of bodies, whether accurate or inequable, and is commonly employed in place of true time; as an hour, a day, a month, a year. . .

"The natural days, which, commonly, for the purpose of the measurement of time, are held as equal, are in reality unequal. Astronomers correct this inequality, in order that they may measure by a truer time the celestial motions. It may be that there is no equable motion, by which time can accurately be measured. All motions can be accelerated and retarded. But the flow of *absolute* time cannot be changed. Duration, or the persistent existence of things, is always the same, whether motions be swift or slow or null."

2. It would appear as though Newton in the remarks here cited still stood under the influence of the mediæval philosophy, as though he had grown unfaithful to his resolves to investigate only actual facts. When we say a thing A changes with the time, we mean simply that the conditions that determine a thing A depend on the conditions that determine another thing B. The vibrations of a pendulum take place *in time* when its excursion *depends* on the position of the earth. Since, however, in the observation of the pendulum, we are

not under the necessity of taking into account its dependence on the position of the earth, but may compare it with any other thing (the conditions of which of course also depend on the position of the earth), the illusory notion easily arises that *all* the things with which we compare it are unessential. Nay, we may, in attending to the motion of a pendulum, neglect entirely other external things, and find that for every position of it our thoughts and sensations are different. Time, accordingly, appears to be some particular and independent thing, on the progress of which the position of the pendulum depends, while the things that we resort to for comparison and choose at random appear to play a wholly collateral part. But we must not forget that all things in the world are connected with one another and depend on one another, and that we ourselves and all our thoughts are also a part of nature. It is utterly beyond our power to *measure* the changes of things by *time*. Quite the contrary, time is an abstraction, at which we arrive by means of the changes of things; made because we are not restricted to any one *definite* measure, all being interconnected. A motion is termed uniform in which equal increments of space described correspond to equal increments of space described by some motion with which we form a comparison, as the rotation of the earth. A motion may, with respect to another motion, be uniform. But the question whether a motion is *in itself* uniform, is senseless. With just as little justice, also, may we speak of an "absolute time"—*of a time independent of* change. This absolute time can be measured by comparison with no motion; it has therefore neither a practical nor a scientific value; and no one is justified in saying that he knows aught about it. It is an idle metaphysical conception.

It would not be difficult to show from the points of view of psychology, history, and the science of language (by the names of the chronological divisions), that we reach our ideas of time in and through the interdependence of things on one another. In these ideas the profoundest and most universal connection of things is expressed. When a motion takes place in time, it depends on the motion of the earth. This is not refuted by the fact that mechanical motions can be reversed. A number of variable quantities may be so related that one set can suffer a change without the others being affected by it. Nature behaves like a machine. The individual parts reciprocally determine one another. But while in a machine the position of one part determines the position of *all* the other parts, in nature more complicated relations obtain. These relations are best represented under the conception of a number, n, of quantities that satisfy a lesser number, n', of equation. Were $n = n'$, nature would be invariable. Were $n' = n - 1$, then with one quantity all the rest would be controlled. If this latter relation obtained in nature, time could be reversed the moment this had been accomplished with any one single motion. But the true state of things is represented by a different relation between n and n'. The quantities in question are partially determined by one another; but they retain a greater indeterminateness, or freedom, than in the case last cited. We ourselves feel that we are such a partially determined, partially undetermined element of nature. In so far as a portion only of the changes of nature depends on us and can be reversed by us, does time appear to us irreversible, and the time that is past as irrevocably gone.

We arrive at the idea of time—to express it briefly and popularly—by the connection of that which is contained in the province of our memory with that which is contained in the province of our sense-perception. When we say that time flows on in a definite direction or sense, we mean that physical events generally (and therefore also physiological events) take place only in a definite sense.* Differences of temperature, electrical differences, differences of level generally, if left to themselves, all grow less and not greater. If we contemplate two bodies of different temperatures, put in contact and left wholly to themselves, we shall find that it is possible only for greater differences of temperature in the field of memory to exist with lesser ones in the field of sense-perception, and not the reverse. In all this there is simply expressed a peculiar and profound connection of things. To demand at the present time a full elucidation of this matter, is to anticipate, in the manner of speculative philosophy, the results of all future special investigation, that is, a perfected physical science.

As in the study of thermal phenomena we take as our measure of temperature an *arbitrarily chosen indicator of volume,* which varies in almost parallel correspondence with our sensation of heat, and which is not liable to the uncontrollable disturbances of our organs of sensation, so, for similar reasons, we select as our measure of time an *arbitrarily chosen motion,* (the angle of the earth's rotation, or path of a free body,) which proceeds in almost parallel correspondence with our sensation of time. If we have once made clear to ourselves that we are concerned only with the ascertainment of the *interdependence* of phenomena, as I pointed out as early as 1865 (*Ueber den Zeitsinn des*

---

* On the physiological nature of the sensations of time and space. Cf. *Analyse der Empfindungen*, 6th ed. [English ed. The Analysis of Sensations (1914)]; *Erkenntnis und Irrtum*, 2nd ed.

Ohres, Sitzungsberichte der Wiener Akademie) and 1866 (Fichte's *Zeitschrift für Philosophie*), all metaphysical obscurities disappear. (Compare J. Epstein, *Die logischen Principien der Zeitmessung*, Berlin, 1887.)

I have endeavored also (*Principles of Heat*, German edition, page 51) to point out the reason for the natural tendency of man to hypostatize the concepts which have great value for him, particularly those at which he arrives instinctively, without a knowledge of their development. The considerations which I there adduced for the concept of temperature may be easily applied to the concept of time, and render the origin of Newton's concept of "absolute" time intelligible. Mention is also made there (page 338) of the connection obtaining between the concept of energy and the irreversibility of time, and the view is advanced that the entropy of the universe, if it could ever possibly be determined, would actually represent a species of absolute measure of time. I have finally to refer here to the discussions of Petzoldt ("Das Gesetz der Eindeutigkeit," *Vierteljahrsschrift für wissenschaftliche Philosophie*, 1894, p. 146) and to my own volume: *Erkenntnis und Irrtum*, 2nd ed., Leipzig, 1906, pp. 434-448.

3. Views similar to those concerning time, are developed by Newton with respect to space and motion. We extract here a few passages which characterize his position.

"II. Absolute space, in its own nature and without regard to anything external, always remains similar and immovable.

"Relative space is some movable dimension or measure of absolute space, which our senses determine by its position with respect to other bodies, and which is commonly taken for immovable [absolute] space. . . .

"IV. Absolute motion is the translation of a body from one absolute place* to another absolute place; and relative motion, the translation from one relative place to another relative place. . . .

". . . . And thus we use, in common affairs, instead of *absolute* places and motions, *relative* ones; and that without any inconvenience. But in physical disquisitions, we should abstract from the senses. For it may be that there is no body really at rest, to which the places and motions of others can be referred . . . .

"The effects by which absolute and relative motions are distinguished from one another, are centrifugal forces, or those forces in circular motion which produces a tendency of recession from the axis. For in a circular motion which is purely relative no such forces exist; but in a true and absolute circular motion they do exist, and are greater or less according to the quantity of the [absolute] motion.

"For instance. If a bucket, suspended by a long cord, is so often turned about that finally the cord is strongly twisted, then is filled with water, and held at rest together with the water; and afterwards by the action of a second force, it is suddenly set whirling about the contrary way, and continues, while the cord is untwisting itself, for some time in this motion; the surface of the water will at first be level, just as it was before the vessel began to move; but, subsequently, the vessel, by gradually communicating its motion to the water, will make it begin sensibly to rotate, and the water will recede little by little from the middle and rise up at the sides of the vessel, its surface assuming a concave form. (This experiment I have made myself.)

* The place, or *locus* of a body, according to Newton, is not its position, but the *part of space* which it occupies. It is either absolute or relative.— *Trans.*

" . . . . At first, when the *relative* motion of the water in the vessel was *greatest,* that motion produced no
tendency whatever of recession from the axis; the water made no endeavor to move towards the circumference, by rising at the sides of the vessel, but remained
level, and for that reason its *true* circular motion had
not yet begun.   But afterwards, when the relative motion of the water had decreased, the rising of the water
at the sides of the vessel indicated an endeavor to recede from the axis; and this endeavor revealed the
real motion of the water, continually increasing, till it
had reached its greatest point, when *relatively* the water
was at rest in the vessel . . . .

"It is indeed a matter of great difficulty to discover,
and effectually to distinguish, the *true* from the apparent motions of particular bodies; for the parts of
that immovable space in which bodies actually move,
do not come under the observation of our senses.

"Yet the case is not altogether desperate; for there
exist to guide us certain marks, abstracted partly from
the apparent motions, which are the differences of the
true motions, and partly from the forces that are the
causes and effects of the true motions.   If, for instance,
two globes, kept at a fixed distance from one another
by means of a cord that connects them, be revolved
about their common center of gravity, one might, from
the simple tension of the cord, discover the tendency
of the globes to recede from the axis of their motion,
and on this basis the quantity of their circular motion
might be computed.   And if any equal forces should be
simultaneously impressed on alternate faces of the
globes to augment or diminish their circular motion, we
might, from the increase or decrease of the tension of
the cord, deduce the increment or decrement of their
motion; and it might also be found thence on what faces

forces would have to be impressed, in order that the motion of the globes should be most augmented; that is, their real faces, or those which, in the circular motion, follow. But as soon as we knew which faces followed, and consequently which preceded, we should likewise know the direction of the motion. In this way we might find both the quantity and the direction of the circular motion, considered even in an immense vacuum, where there was nothing external or sensible with which the globes could be compared . . . ."

If, in a material spatial system, there are masses with different velocities, which can enter into mutual relations with one another, these masses present to us forces. We can only decide how great these forces are when we know the velocities to which those masses are to be brought. *Resting* masses too are forces if *all* the masses do not rest. Think, for example, of Newton's rotating bucket in which the water is not yet rotating. If the mass m has the velocity v_1 and it is to be brought to the velocity v_2, the force which is to be spent on it is $p = m(v_1 - v_2)/t$, or the work which is to be expended is $ps = m(v_1^2 - v_2^2)$. *All* masses and *all* velocities, and consequently *all* forces, are relative. There is no decision about relative and absolute which we can possibly meet, to which we are forced, or from which we can obtain any intellectual or other advantage. When quite modern authors let themselves be led astray by the Newtonian arguments which are derived from the bucket of water, to distinguish between relative and absolute motion, they do not reflect that the system of the world is only given *once* to us, and the Ptolemaic or Copernican view is *our* interpretation, but both are equally actual. Try to fix Newton's bucket and rotate the heaven of fixed stars and then prove the absence of centrifugal forces.

4. It is scarcely necessary to remark that in the reflections here presented Newton has again acted contrary to his expressed intention only to investigate *actual facts*. No one is competent to predicate things about absolute space and absolute motion; they are pure things of thought, pure mental constructs, that cannot be produced in experience. All our principles of mechanics are, as we have shown in detail, experimental knowledge concerning the relative positions and motions of bodies. Even in the provinces in which they are now recognized as valid, they could not, and were not, admitted without previously being subjected to experimental tests. No one is warranted in extending these principles beyond the boundaries of experience. In fact, such an extension is meaningless, as no one possesses the requisite knowledge to make use of it.

We must suppose that the change in the point of view from which the system of the world is regarded which was initiated by Copernicus, left deep traces in the thought of Galileo and Newton. But while Galileo, in his theory of the tides, quite naïvely chose the sphere of the fixed stars as the basis of a new system of coördinates, we see doubts expressed by Newton as to whether a given fixed star is at rest only apparently or really (*Principia*, 1687, p. 11). This appeared to him to cause the difficulty of distinguishing between true (absolute) and apparent (relative) motion. By this he was also impelled to set up the conception of *absolute space*. By further investigations in this direction —the discussion of the experiment of the rotating spheres which are connected together by a cord and that of the rotating water-bucket (pp. 9, 11)—he believed that he could prove an absolute rotation, though he could not prove any absolute translation. By absolute rotation he understood a rotation relative to the fixed stars, and here centrifugal forces can always be found. "But how we are to collect," says Newton in the Scholium at the end of the Definitions, "the true motions from their causes, effects, and apparent differences, and *vice versa*; how from the motions, either true or apparent, we may come to the knowledge of their causes and effects, shall be explained more at large in the following Tract." The resting sphere of fixed stars seems to have made a certain impression on Newton as well. The natural system of reference is for him that which has any uniform motion or translation without rotation (relatively to the sphere of fixed stars).[1] But do not the words quoted in inverted commas give the impression that Newton was glad to be able now to pass over to less precarious questions that could be tested by experience?

Let us look at the matter in detail. When we say that a body K alters its direction and velocity solely through the influence of another body K', we have asserted a conception that it is impossible to come at unless other bodies $A, B, C \ldots$ are present with reference to which the motion of the body K has been estimated. In reality, therefore, we are simply cognizant of a relation of the body K to $A, B, C \ldots$ If now we suddenly neglect $A, B, C \ldots$ and attempt to speak of the deportment of the body K in absolute space, we implicate ourselves in a twofold error. In the first place, we cannot know how K would act in the absence of $A, B, C \ldots$; and in the second place, every means would be wanting of forming a judgment of the behavior of K and of putting to the test what we had predicated—which latter therefore would be bereft of all scientific significance.

[1] *Principia*, p. 19, Coroll. V: "The motions of bodies included in a given space are the same among themselves, whether that space is at rest or moves uniformly forwards in a right line without any circular motion."

Two bodies K and K', which gravitate toward each other, impart to each other in the direction of their line of junction accelerations inversely proportional to their masses m, m'. In this proposition is contained, not only a relation of the bodies K and K' to one another, but also a relation of them to other bodies. For the proposition asserts, not only that K and K' suffer with respect to one another the acceleration designated by $\varkappa(\overline{m + m'}/r^2)$, but also that K experiences the acceleration $- \varkappa \, m'/r^2$ and K' the acceleration $+ \varkappa \, m/r^2$ in the direction of the line of junction; facts which can be ascertained only by the presence of other bodies.

The motion of a body K can only be estimated by reference to other bodies $A, B, C \ldots$. But since we always have at our disposal a sufficient number of bodies, that are as respects each other relatively fixed, or only slowly change their positions, we are, in such reference, restricted to no one *definite* body and can alternately leave out of account now this one and now that one. In this way the conviction arose that these bodies are indifferent generally.

It might be, indeed, that the isolated bodies $A, B, C \ldots$. play merely a collateral rôle in the determination of the motion of the body K, and that this motion is determined by a *medium* in which K exists. In such a case we should have to substitute this medium for Newton's absolute space. Newton certainly did not entertain this idea. Moreover, it is easily demonstrable that the atmosphere is not this motion-determinative medium. We should, therefore, have to picture to ourselves some other medium, filling, say, all space, with respect to the constitution of which and its kinetic relations to the bodies placed in it we have at present no adequate knowledge. In itself such a state of things would not belong to the impossibilities. It is known, from recent hydrodynamical investigations, that a rigid body experiences resistance in a frictionless fluid only when its velocity *changes*. True, this result is derived theoretically from the notion of inertia; but it might, conversely, also be regarded as the primitive fact from which we have to start. Although, practically, and at present, nothing is to be accomplished with this conception, we might still hope to learn more in the future concerning this hypothetical medium; and from the point of view of science it would be in every respect a more valuable acquisition than the forlorn idea of absolute space. When we reflect that we cannot abolish the isolated bodies $A, B, C \ldots$, that is, cannot determine by experiment whether the part they play is fundamental or collateral, that hitherto they have been the sole and only competent means of the orientation of motions and of the description of mechanical facts, it will be found expedient provisionally to regard all motions as determined by these bodies.

5. Let us now examine the point on which Newton, apparently with sound reasons, rests his distinction of absolute and relative motion. If the earth is affected with an *absolute* rotation about its axis, centrifugal forces are set up in the earth: it assumes an oblate form, the acceleration of gravity is diminished at the equator, the plane of Foucault's pendulum rotates, and so on. All these phenomena disappear if the earth is at rest and the other heavenly bodies are affected with absolute motion round it, such that the same *relative* rotation is produced. This is, indeed, the case, if we start *ab initio* from the idea of absolute space. But if we take our stand on the basis of facts, we shall find we have knowledge only of *relative* spaces and motions. *Relatively*, not considering the unknown and neglected medium of space, the motions of the uni-

verse are the same whether we adopt the Ptolemaic or the Copernican mode of view. Both views are, indeed, equally *correct*; only the latter is more simple and more *practical*. The universe is not *twice* given, with an earth at rest and an earth in motion; but only *once*, with its *relative* motions, alone determinable. It is, accordingly, not permitted us to say how things would be if the earth did not rotate. We may interpret the one case that is given us, in different ways. If, however, we so interpret it that we come into conflict with experience, our interpretation is simply wrong. The principles of mechanics can, indeed, be so conceived, that even for relative rotations centrifugal forces arise.

Newton's experiment with the rotating vessel of water simply informs us, that the relative rotation of the water with respect to the sides of the vessel produces *no* noticeable centrifugal forces, but that such forces *are* produced by its relative rotation with respect to the mass of the earth and the other celestial bodies. No one is competent to say how the experiment would turn out if the sides of the vessel increased in thickness and mass till they were ultimately several leagues thick. The one experiment only lies before us, and our business is, to bring it into accord with the other facts known to us, and not with the arbitrary fictions of our imagination.

6. When Newton examined the principles of mechanics discovered by Galileo, the great value of the simple and precise law of inertia for deductive derivations could not possibly escape him. He could not think of renouncing its help. But the law of inertia, referred in such a naïve way to the earth supposed to be at rest, could not be accepted by him. For, in Newton's case, the rotation of the earth was not a debatable point; it rotated without the least doubt. Galileo's

happy discovery could only hold approximately for small times and spaces, during which the rotation did not come into question. Instead of that, Newton's conclusions about planetary motion, referred as they were to the fixed stars, appeared to conform to the law of inertia. Now, in order to have a generally valid system of reference, Newton ventured the fifth corollary of the *Principia* (p. 19 of the first edition). He imagined a momentary terrestrial system of coördinates, for which the law of inertia is valid, held fast in space without any rotation relatively to the fixed stars. Indeed he could, without interfering with its usability, impart to this system any initial position and any uniform translation relatively to the above momentary terrestrial system. The Newtonian laws of force are not altered thereby; only the initial positions and initial velocities—the constants of integration—may alter. By this view Newton gave the *exact* meaning of his hypothetical extension of Galileo's law of inertia. We see that the reduction to absolute space was by no means necessary, for the system of reference is just as relatively determined as in every other case. In spite of his metaphysical liking for the absolute, Newton was correctly led by the *tact of the natural investigator*. This is particularly to be noticed, since, in former editions of this book, it was not sufficiently emphasized. How far and how accurately the conjecture will hold good in future is of course undecided.

The comportment of terrestrial bodies with respect to the earth is reducible to the comportment of the earth with respect to the remote heavenly bodies. If we were to assert that we knew more of moving objects than this their last-mentioned, experimentally-given comportment with respect to the celestial bodies, we should render ourselves culpable of a falsity. When, accordingly, we say, that a body preserves unchanged its direction and velocity *in space*, our assertion is nothing more or less than an abbreviated reference to *the entire universe*. The use of such an abbreviated expression is permitted the original author of the principle, because he knows, that as things are no difficulties stand in the way of carrying out its implied directions. But no remedy lies in his power, if difficulties of the kind mentioned present themselves; if, for example, the requisite, relatively fixed bodies are wanting.

7. Instead, now, of referring a moving body K to space, that is to say to a system of coördinates, let us view directly its relation to the bodies of the universe, by which alone such a system of coördinates can be determined. Bodies very remote from each other, moving with constant direction and velocity with respect to other distant fixed bodies, change their mutual distances proportionately to the time. We may also say, all very remote bodies—all mutual or other forces neglected—alter their mutual distances proportionately to those distances. Two bodies, which, situated at a short distance from one another, move with constant direction and velocity with respect to other fixed bodies, exhibit more complicated relations. If we should regard the two bodies as dependent on one another, and call r the distance, t the time, and a a constant dependent on the directions and velocities, the formula would be obtained: $d^2r/dt^2 = (1/r)[a^2 - (dr/dt)^2]$. It is manifestly much *simpler* and *clearer* to regard the two bodies as independent of each other and to consider the constancy of their direction and velocity with respect to other bodies.

Instead of saying, the direction and velocity of a mass μ in space remain constant, we may also employ the expression, the mean acceleration of the mass μ

with respect to the masses m, m', m''. . . . at the distances r, r', r''. . . . is $= 0$, or $d^2(\Sigma\, m\, r/\Sigma\, m)/dt^2 = 0$. The latter expression is equivalent to the former, as soon as we take into consideration a sufficient number of sufficiently distant and sufficiently large masses. The mutual influence of more proximate small masses, which are apparently not concerned about each other, is eliminated of itself. That the constancy of direction and velocity is given by the condition adduced, will be seen at once if we construct through μ as vertex cones that cut out different portions of space, and set up the condition with respect to the masses of these separate portions. We may put, indeed, for the *entire* space encompassing μ, $d^2(\Sigma\, m\, r/\Sigma\, m)/dt^2 = 0$. But the equation in this case asserts nothing with respect to the motion of μ, since it holds good for all species of motion where μ is uniformly surrounded by an infinite number of masses. If two masses μ_1, μ_2 exert on each other a force which is dependent on their distance r, then $d^2r/dt^2 = (\mu_1 + \mu_2)f(r)$. But, at the same time, the acceleration of the center of gravity of the two masses or the mean acceleration of the mass-system with respect to the masses of the universe (by the principle of reaction) remains $= 0$; that is to say,

$$\frac{d^2}{dt^2}\left[\mu_1\,\frac{\Sigma\, m\, r_1}{\Sigma\, m} + \mu_2\,\frac{\Sigma\, m\, r_2}{\Sigma\, m}\right] = 0.$$

When we reflect that the time-factor that enters into the acceleration is nothing more than a quantity that is the measure of the distances (or angles of rotation) of the bodies of the universe, we see that even in the simplest case, in which apparently we deal with the mutual action of only *two* masses, the neglecting of the rest of the world is *impossible*. Nature does not begin with elements, as we are obliged to begin with them. It is certainly fortunate for us, that we can, from time to time, turn aside our eyes from the overpowering unity of the All, and allow them to rest on individual details. But we should not omit, ultimately to complete and correct our views by a thorough consideration of the things which for the time being we left out of account.

8. The considerations just presented show, that it is not necessary to refer the law of inertia to a special absolute space. On the contrary, it is perceived that the masses that in the common phraseology exert forces on each other as well as those that exert none, stand with respect to acceleration in quite similar relations. We may, indeed, regard *all* masses as related to each other. That *accelerations* play a prominent part in the relations of the masses, must be accepted as a fact of experience; which does not, however, exclude attempts to *elucidate* this fact by a comparison of it with other facts, involving the discovery of new points of view. In all the processes of nature the *differences* of certain

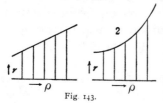

Fig. 143.

quantities *u* play a determinative rôle. Differences of temperature, of potential function, and so forth, induce the natural processes, which consist in the equalization of these differences. The familiar expressions d^2u/dx^2, d^2u/dy^2, d^2u/dz^2, which are determinative of the character of the equalization, may be regarded as the measure of the departure of the condition of any point from the mean of the conditions of its environment—to which mean the point tends. The accelerations of masses may be analogously conceived. The great distances between masses that stand in no especial force-relation to one another, change *proportionately to each other*. If we lay off, therefore, a certain distance ϱ as abscissa, and another r as ordinate, we obtain a straight line (Fig. 143). Every r-ordinate corresponding to a definite ϱ-value represents, accordingly, the mean of the adjacent ordinates. If a force-relation exists between the bodies, some value d^2r/dt^2 is determined by it which conformably to the remarks above we may replace by an expression of the form $d^2r/d\varrho^2$. By the force-relation, therefore, a *departure* of the r-ordinate from the *mean of the adjacent ordinates* is produced, which would not exist if the supposed force-relation did not obtain. This intimation will suffice here.

9. We have attempted in the foregoing to give the law of inertia a different expression from that in ordinary use. This expression will, so long as a sufficient number of bodies are apparently fixed in space, accomplish the same as the ordinary one. It is as easily applied, and it encounters the same difficulties. In the one case we are unable to come at an absolute space, in the other a limited number of masses only is within the reach of our knowledge, and the summation indicated can consequently not be fully carried out. It is impossible to say whether the new expression would still represent the true condition of things if the stars were to perform rapid movements among one another. The general experience cannot be constructed from the particular case given us. We must, on the contrary, *wait* until such an experience presents itself. Perhaps when our physico-astronomical knowledge has been extended, it will be offered somewhere in celestial space, where more violent and complicated motions take place than in our environment. The most important result of our reflections is, however, *that precisely*

the apparently simplest mechanical principles are of a very complicated character, that these principles are founded on uncompleted experiences, nay on experiences that never can be fully completed, that practically, indeed, they are sufficiently secured, in view of the tolerable stability of our environment, to serve as the foundation of mathematical deduction, but that they can by no means themselves be regarded as mathematically established truths but only as principles that not only admit of constant control by experience but actually require it.

I do not believe that the writings of the advocates of absolute space which have appeared during the last ten years can assert anything else than the italicized passage, which appeared in the first German edition of 1883 (pp. 221, 222). This perception is valuable in that it is propitious to the advancement of science.

10. The law of inertia has often been discussed in ancient and modern times, and almost always the empty conception of absolute space, which is open to such grave objections in point of principle, has been mixed up with it in a disturbing manner. Here we will limit ourselves to the mention of the more modern discussions of this subject.

In the first place we must mention the writings of C. Neumann: *Ueber die Principien der Galilei-Newton'schen Theorie*, of 1870, and "Über den Körper Alpha" (*Ber. der königl. sächs. Ges. der Wiss.*, 1910, III). The author denotes, on p. 22 of the former treatise, the relation to the body Alpha as a relation to a system of axes which proceeds uniformly in a straight line without rotation, and thus his statement coincides with the fifth corollary of Newton which we have already mentioned. However, I do not believe that the fiction of the body Alpha and the preservation of the distinction between absolute and relative motion and the paradoxes (pp. 27, 28) connected with this distinction have particularly contributed to the clarification of the matter. In the publication of 1910 (p. 70, note 1) Neumann calls what he has brought forward purely hypothetical, and in this lies an essential progress in the knowledge of Newton's fifth corollary. In the same publication, Lange's standpoint is exposed as in essentials coinciding with his own.

H. Streintz (*Die physikalischen Grundlagen der Mechanik*, 1883) accepts the Newtonian distinction between absolute and relative motion, but also comes to the view expressed in Newton's fifth corollary. What I had to say against Streintz's criticism of my views was contained in the former editions of this work and shall not be repeated here.

We will now consider L. Lange: "Über die wissenschaftliche Fassung der Galilei'schen Beharrungsgesetzes," Wundt's *Philos. Studien*, II, 1885, pp. 266-297, 539-545; *Ber. d. königl. sächs. Ges. der Wiss., math.-physik. Klasse*, 1885, pp. 333-351; *Die geschichtliche Entwicklung des Bewegungsbegriffs*, Leipzig, 1886; *Das Inertialsystem vor dem Forum der Naturforschung*, Leipzig, 1902.

L. Lange sets out from the supposition that the general Newtonian law of inertia subsists and seeks the system of coördinates to which it is to be referred (1885). With respect to any moving point P_1 — which can even move in a curve — we can so move a system of coördinates that the point P_1 describes a straight line G_1 in this system. If we have also a second moving point P_2, the system can still be moved so that a second straight line G_2, in general warped with respect to G_1, is described by P_2, if only the shortest distance $G_1 G_2$ does not surpass the shortest distance which

P_1 P_2 can ever have. Still the system can rotate about $P_1 P_2$. If we choose a third straight line G_3, such that all the triangles P_1, P_2, P_3 which can arise by means of any third moving point P_3 are representable by points on G_1, G_2, G_3, then P_3 can also advance on G_3. Thus, for at most three points, a system of coördinates in which these points proceed in a straight line is a mere convention. Now, Lange sees the essential contents of the law of inertia in that, by the help of three material points which are left to themselves, a system of coördinates can be found with respect to which four or arbitrarily many material points which are left to themselves move in a straight line and describe paths which are proportional to one another. The process in nature is thus a simplification and limitation of the kinematically possible variety of cases.

This promising thought and its consequences found much recognition with mathematicians, physicists, and astronomers. (*Cf.* H. Seeliger's account of Lange's works in the *Vierteljahrsschrift der astronom. Ges.*, XXII, p. 252; H. Seeliger, "Über die sogenannte absolute Bewegung," *Sitzungsber. der Münchener Akad. der Wiss.*, 1906, p. 85.) Now, J. Petzoldt ("Die Gebiete der absoluten und der relativen Bewegung," Ostwald's *Annalen der Naturphilosophie*, VII, 1908, pp. 29-62) has found certain difficulties in Lange's thoughts, and these difficulties have also disturbed others and are not quickly to be put on one side. On this account we will here break off our remarks on Lange's system of coördinates or inertial systems till the clouds pass away. Seeliger has attempted to determine the relation of the inertial system to the empirical astronomical system of coördinates which is in use, and believes that he can say that the empirical system cannot rotate about the inertial system by more than some seconds of arc in a century. *Cf.* also A. Anding, "Über Koordinaten und Zeit," in the *Encyklopädie* der *mathematischen Wissenschaften,* VI, 2, 1.

11. The view that "absolute motion" is a conception which is devoid of content and cannot be used in science struck almost everybody as strange thirty years ago, but at the present time it is supported by many and worthy investigators. Some "relativists" are: Stallo, J. Thomson, Ludwig Lange, Love, Kleinpeter, J. G. MacGregor, Mansion, Petzoldt, Pearson. The number of relativists has grown very quickly, and the above list is certainly already incomplete. Probably there will soon be no important supporter of the opposite view. But, if the inconceivable hypotheses of absolute space and absolute time cannot be accepted, the question arises: In what way can we give a comprehensible meaning to the law of inertia? MacGregor shows in an excellent paper (*Phil. Mag.,* XXXVI, 1893, p. 223),[1] which is very clearly written and shows great recognition of Lange's work, that there are two ways that we can take: (1) the historical and critical way, which considers anew the facts on which the law of inertia rests and which draws its limits of validity and finally considers a new formulation; (2) the supposition that the law of inertia in its old form teaches us the motions sufficiently, and the derivation of the correct system of coördinates *from* these motions.

For the first method it seems to me that Newton himself gave the first example with his system of reference indicated in the fifth corollary, which has been often mentioned above. It ·is obvious that we must take account of modifications of expression which have become necessary by extension of our experience.

[1] [This paper, "On the Hypotheses of Dynamics," was occasioned by some remarks of O. Lodge on a former paper of MacGregor's.]

The second way is very closely connected *psychologically* with the great trust which mechanics, as the most exact natural science, enjoys.   Indeed, this way has often been followed with more or less success.   W. Thomson and P. G. Tait (*Treatise on Natural Philosophy*, I, part 1, 1879, § 249)[2] remark that two material points which are simultaneously projected from the same place and then left to themselves move in such a way that the line joining them remains parallel to itself.   Thus, if four points O, P, Q, and R are projected simultaneously from the same place and then subject to no further force, the lines OP, OQ, and OR always give fixed directions.   J. Thomson attempts, in two articles (*Proc. Roy. Soc. Edinb.*, 1884, pp. 568, 730), to construct the system of reference corresponding to the law of inertia, and in this recognizes that the suppositions about uniformity and rectilinearity are *partly conventional.*   Tait (loc. cit., p. 743), stimulated by J. Thomson, takes part in the solution of the same problem by quaternions.   We find also MacGregor in the same path ("The Fundamental Hypotheses of Abstract Dynamics," *Trans. Roy. Soc. of Canada*, X, 1892, § iii, especially pp. 5 and 6).

The same psychological motives were certainly active in the case of Ludwig Lange, who has been most fortunate in his efforts to interpret correctly the Newtonian law of inertia.   This he did in two articles in Wundt's *Philos. Studien* of 1885.

More recently Lange (*Philos. Studien*, XX, 1902) published a critical paper in which he also worked out the method of obtaining a *new* system of coördinates according to his principles, when the usual rough reference to the fixed stars shall be, in consequence of more accurate astronomical observations, no longer suf-

[2] [*Cf.* §§ 267, 245.]

ficient. There is, I think, no difference of meaning between Lange and myself about the *theoretical* and formal value of Lange's expressions, and about the fact that, at the present time, the heaven of fixed stars is the only practically usable system of reference, and about the method of obtaining a new system of reference by gradual corrections. The difference which still subsists, and perhaps will always do so, lies in the fact that Lange approaches the question as a *mathematician,* while I was concerned with the *physical* side of the subject.

Lange supposes with some confidence that *his* expression would remain valid for celestial motions on a large scale. I cannot share this confidence. The surroundings in which we live, with their almost constant angles of direction to the fixed stars, appear to me to be an extremely special case, and I would not dare to conclude from this case to a very different one. Although I expect that astronomical observation will only as yet necessitate very small corrections, I consider it possible that the law of inertia in its simple Newtonian form has only, for us human beings, a meaning which depends on space and time. Allow me to make a more general remark. We measure time by the angle of rotation of the earth, but could measure it just as well by the angle of rotation of any other planet. But, on that account, we would not believe that the *temporal* course of all physical phenomena would have to be disturbed if the earth or the distant planet referred to should suddenly experience an abrupt variation of angular velocity. We consider the dependence as not immediate, and consequently the temporal orientation as *external.* Nobody would believe that the chance disturbance—say by an impact—of one body in a system of uninfluenced bodies which are left to themselves and move uniformly in a straight line, where all the bodies combine to fix the system of coördinates, will immediately cause a disturbance of the others as a consequence. The orientation is external here also. Although we must be very thankful for this, especially when it is purified from meaninglessness, still the natural investigator must feel the need of further insight —of knowledge of the *immediate* connections, say, of the masses of the universe. There will hover before him as an ideal an insight into the principles of the whole matter, from which accelerated and inertial motions result in the *same* way. The progress from Kepler's discovery to Newton's law of gravitation, and the impetus given by this to the finding of a physical understanding of the attraction in the manner in which electrical actions at a distance have been treated, may here serve as a model. We must even give rein to the thought that the masses which we see, and by which we by chance orientate ourselves, are perhaps not those which are really decisive. On this account we must not underestimate even experimental ideas like those of Friedländer[1] and Föppl,[2] even if we do not yet see any immediate result from them. Although the investigator gropes with joy after what he can immediately reach, a glance from time to time into the depths of what is uninvestigated cannot hurt him.

12. A small elementary paper of J. R. Schütz ("Prinzip der absoluten Erhaltung der Energie," *Göttinger Nachrichten, math.-physik. Klasse,* 1897) shows, by simple examples, that Newton's laws can be obtained from the principle spoken of. The term "absolute" is only meant to express that the principle is to

---

[1] B. and J. Friedländer, *Absolute und relative Bewegung,* Berlin, 1896.

[2] "Über einen Kreiselversuch zur Messung der Umdrehungsgeschwindigkeit der Erde," *Sitzungsber. der Münchener Akad.,* 1904, p. 5; "Über absolute und relative Bewegung," *ibid.,* 1904, p. 383.

be freed from an indeterminateness and arbitrariness. If we imagine the principle applied to the central impact of elastic masses m_1 and m_2 in the form of points, of initial velocities u_1 and u_2 and final velocities v_1 and v_2, we have

$$m_1 u_1{}^2 + m_2 u_2{}^2 = m_1 v_1{}^2 + m_2 v_2{}^2.$$

We can calculate v_1 and v_2 from u_1 and u_2, if we suppose that the principle of energy holds for any velocity of translation c directed in the same sense as u and v. We then have

$$m_1(u_1+c)^2 + m_2(u_2+c)^2 = m_1(v_1+c)^2 + m_2(v_2+c)^2.$$

If we subtract the first equation from the second, we get the equation of the principle of reaction:

$$m_1 u_1 + m_2 u_2 = m_1 v_1 + m_2 v_2,$$

in which c has dropped out. From the first and third equation we can calculate v_1 and v_2. By an analogous treatment of the "absolute" principle of energy, we get Newton's equation of force for a mass-point, and finally the law of reaction, with its corollaries of the conservation of the quantity of motion and the conservation of the center of gravity. The study of this paper is very much to be recommended, since even the conception of mass can be derived by the help of the principle of energy. *Cf.* the section VIII on "Retrospect of the Development of Dynamics."

SYNOPTICAL CRITIQUE OF THE NEWTONIAN
ENUNCIATIONS

1. Now that we have discussed the details with sufficient particularity, we may again survey the form and the disposition of the Newtonian enunciations. Newton premises to his work several definitions, following which he gives the laws of motion. We shall take up the former first.

"*Definition I.* The quantity of any matter is the measure of it by its density and volume conjointly. . . . This quantity is what I shall understand by the term *mass* or *body* in the discussions to follow. It is ascertainable from the weight of the body in question. For I have found, by pendulum-experiments of high precision, that the mass of a body is proportional to its weight; as will hereafter be shown.

"*Definition II.* Quantity of motion is the measure of it by the velocity and quantity of matter conjointly.

"*Definition III.* The resident force [*vis insita,* i. e. the inertia] of matter is a power of resisting, by which every body, so far as in it lies, perseveres in its state of rest or of uniform motion in a straight line.

"*Definition IV.* An impressed force is any action upon a body which changes, or tends to change, its state of rest, or of uniform motion in a straight line.

"*Definition V.* A centripetal force is any force by which bodies are drawn or impelled towards, or tend in any way to reach, some point as center.

"*Definition VI.* The absolute quantity of a centripetal force is a measure of it increasing and diminishing with the efficacy of the cause that propagates it from the center through the space round about.

"*Definition VII*. The accelerative quantity of a centripetal force is the measure of it proportional to the velocity which it generates in a given time.

"*Definition VIII*. The moving quantity of a centripetal force is the measure of it proportional to the motion [See Def. ii.] which it generates in a given time.

"The three quantities or measures of force thus distinguished, may, for brevity's sake, be called absolute, accelerative, and moving forces, being, for distinction's sake, respectively referred to the center of force, to the places of the bodies, and to the bodies that tend to the center: that is to say, I refer moving force to the body, as being an endeavor of the whole towards the center, arising from the collective endeavors of the several parts; accelerative force to the place of the body, as being a sort of efficacy originating in the center and diffused throughout all the several places round about, in moving the bodies that are at these places; and absolute force to the center, as invested with some cause, without which moving forces would not be propagated through the space round about; whether this latter cause be some central body, (such as is a loadstone in a center of magnetic force, or the earth in the center of the force of gravity,) or anything else not visible. This, at least, is the mathematical conception of forces; for their physical causes and seats I do not in this place consider.

"Accelerating force, therefore, is to moving force, as velocity is to quantity of motion. For quantity of motion arises from the velocity and the quantity of matter; and moving force arises from the accelerating force and the same quantity of matter; the sum of the effects of the accelerative force on the several particles of the body being the motive force of the whole. Hence, near the surface of the earth, where the accelerative gravity or gravitating force is in all bodies the same, the motive force of gravity or the weight is as the body [mass]. But if we ascend to higher regions, where the accelerative force of gravity is less, the weight will be equally diminished, always remaining proportional conjointly to the mass and the accelerative force of gravity. Thus, in those regions where the accelerative force of gravity is half as great, the weight of a body will be diminished by one-half. Further, I apply the terms accelerative and motive in one and the same sense to attractions and to impulses. I employ the expressions attraction, impulse, or propensity of any kind towards a center, promiscuously and indifferently, the one for the other; considering those forces not in a physical sense, but mathematically. The reader, therefore, must not infer from any expressions of this kind that I may use, that I take upon myself to explain the kind or the mode of an action, or the causes or the physical reason thereof, or that I attribute forces in a true or physical sense, to centers (which are only mathematical points), when at any time I happen to say that centers attract or that central forces are in action."

2. Definition I is, as has already been set forth, a pseudo-definition. The concept of mass is not made clearer by describing mass as the product of the volume into the density, as density itself denotes simply the mass of unit of volume. The true definition of mass can be deduced only from the dynamical relations of bodies.

To Definition II, which simply enunciates a mode of computation, no objection is to be made. Definition III (*inertia*), however, is rendered superfluous by

Definitions IV-VIII of force, inertia being included and given in the fact that forces are accelerative.

Definition IV defines force as the cause of the acceleration, or tendency to acceleration, of a body. The latter part of this is justified by the fact that in the cases also in which accelerations cannot take place, other attractions that answer thereto, as the compression and distension etc. of bodies occur. The cause of an acceleration towards a definite center is defined in Definition V as centripetal force, and is distinguished in VI, VII, and VIII as absolute, accelerative, and motive. It is, we may say, a matter of taste and of form whether we shall embody the explication of the idea of force in one or in several definitions. In point of principle the Newtonian definitions are open to no objections.

3. The Axioms or Laws of Motion then follow, of which Newton enunciates three:

"*Law I*. Every body perseveres in its state of rest or of uniform motion in a straight line, except in so far as it is compelled to change that state by impressed forces."

"*Law II*. Change of motion [i. e. of momentum] is proportional to the moving force impressed, and takes place in the direction of the straight line in which such force is impressed."

"*Law III*. Reaction is always equal and opposite to action; that is to say, the actions of two bodies upon each other are always equal and directly opposite."

Newton appends to these three laws a number of Corollaries. The first and second relate to the principle of the parallelogram of forces; the third to the quantity of motion generated in the mutual action of bodies; the fourth to the fact that the motion of the center of gravity is not changed by the mutual action of bodies; the fifth and sixth to relative motion.

4. We readily perceive that Laws I and II are contained in the definitions of force that precede. According to the latter, without force there is no acceleration, consequently only rest or uniform motion in a straight line. Furthermore, it is wholly unnecessary tautology, after having established acceleration as the measure of force, to say again that change of motion is proportional to the force. It would have been enough to say that the definitions premised were not arbitrary mathematical ones, but correspond to properties of bodies experimentally given. The third law apparently contains something new. But we have seen that it is unintelligible without the correct idea of mass, which idea, being itself obtained only from dynamical experience, renders the law unnecessary.

What is pleonastic and tautological in Newton's propositions is psychologically comprehensible if we imagine an investigator who, setting out from his familiar ideas of statics, is in the act of establishing the fundamental propositions of dynamics. At one time force is in the focus of consideration as a pull or a pressure, and at another time as determinative of accelerations. When, on the one hand, he recognizes, by the idea of a pressure which is common to all forces, that all forces also determine accelerations, then this twofold notion leads him, on the other hand, to a divided and far from unitary representation of the new fundamental propositions. *Cf. Erkenntnis und Irrtum,* 2nd ed., pp. 140, 315.

The first corollary really does contain something new. But it regards the accelerations determined in a body K by different bodies M, N, P as *self-evidently* independent of each other, whereas this is precisely what should have been explicitly recognized as a *fact of experience*. Corollary Second is a simple application of the law enunciated in Corollary First. The remaining corollaries, likewise, are simple deductions, that is, mathematical consequences, from the conceptions and laws that precede.

5. Even if we adhere absolutely to the Newtonian points of view, and disregard the complications and indefinite features mentioned, which are not removed but merely concealed by the abbreviated designations "Time" and "Space," it is possible to replace Newton's enunciations by much more simple, methodically better arranged, and more satisfactory propositions. Such, in our estimation, would be the following:

a. Experimental Proposition. Bodies set opposite each other induce in each other, under certain circumstances to be specified by experimental physics, contrary *accelerations* in the direction of their line of junction. (The principle of inertia is included in this.)

b. Definition. The mass-ratio of any two bodies is the negative inverse ratio of the mutually induced accelerations of those bodies.

c. Experimental Proposition. The mass-ratios of bodies are independent of the character of the physical states (of the bodies) that condition the mutual accelerations produced, be those states electrical, magnetic, or what not; and they remain, moreover, the same, whether they are mediately or immediately arrived at.

d. Experimental Proposition. The accelerations which any number of bodies $A, B, C \ldots$ induce in a body K, are independent of each other. (The principle of the parallelogram of forces follows immediately from this.)

e. Definition. Moving force is the product of the

mass-value of a body into the acceleration induced in that body.

The theorems *a* to *e* were given in my note "Über die Definition der Masse" in Carl's *Repertorium der Experimentalphysik*, IV, 1868; reprinted in *Erhaltung der Arbeit*, 1872, 2nd ed., Leipzig, 1909. *Cf.* also Poincaré, *La Science et l'hypothèse*, Paris, pp. 110 *et seq.*

Then the remaining arbitrary definitions of the algebraical expressions "momentum," "vis viva," and the like, might follow. But these are by no means indispensable. The propositions above set forth satisfy the requirements of simplicity and parsimony which, on economico-scientific grounds, must be exacted of them. They are, moreover, obvious and clear; for no doubt can exist with respect to any one of them either concerning its meaning or its source; and we always know whether it asserts an experience or an arbitrary convention.

6. Upon the whole, we may say, that Newton discerned in an admirable manner the concepts and principles that were *sufficiently assured* to allow of being further built upon. It is possible that to some extent he was forced by the difficulty and novelty of his subject, in the minds of the contemporary world, to great amplitude, and, therefore, to a certain disconnectedness of presentation, in consequence of which one and the same property of mechanical processes appears several times formulated. To some extent, however, he was, as it is possible to prove, not perfectly clear himself concerning the import and especially concerning the source of his principles. This cannot, however, obscure in the slightest his intellectual greatness. He that has to acquire a new point of view naturally can-

not possess it so securely from the beginning as they that receive it unlaboriously from him. He has done enough if he has discovered truths on which future generations can build. For every new inference therefrom affords at once a new insight, a new control, an extension of our prospect, and a clarification of our field of view. Like the commander of an army, a great discoverer cannot stop to institute petty inquiries regarding the right by which he holds each post of vantage he has won. The magnitude of the problem to be solved leaves no time for this. But at a later period, the case is different. Newton might well have expected of the two centuries to follow that they should further examine and confirm the foundations of his work, and that, when times of greater scientific tranquillity should come, the principles of the subject might acquire an even higher philosophical interest than all that is deducible from them. Then problems arise like those just treated, to the solution of which, perhaps, a small contribution has been made here. We join with the eminent physicists Thomson (Lord Kelvin) in our reverence and admiration of Newton. But we can only comprehend with difficulty his opinion that the Newtonian doctrines still remain the best and most philosophical foundation that can be given.

VIII.

RETROSPECT OF THE DEVELOPMENT OF DYNAMICS.

1. Dynamics has developed in an analogous way to statics. Different special cases of motions of bodies were observed, and people tried to put these observations in the form of rules. But just as little as, from

the observation of a case of equilibrium of the inclined plane or the lever, can be derived a mathematically exact and generally valid rule for equilibrium—on account of the inaccuracy of measurement—so little can the corresponding thing be done for cases of motion. Observation only leads, in the first place, to the conjecturing of laws of motion, which, in their special simplicity and accuracy, are presupposed as *hypotheses* in order to try whether the behavior of bodies can be logically derived from these hypotheses. Only if these hypotheses have shown themselves to hold good in many simple and complicated cases, do we agree to keep them. Poincaré, in his *La science et l'hypothèse,* is, then, right in calling the fundamental propositions of mechanics *conventions* which might very well have proven otherwise.

If we pass in review the period in which the development of dynamics fell—a period inaugurated by Galileo, continued by Huygens, and brought to a close by Newton—its main result will be found to be the perception, that bodies mutually determine in each other *accelerations* dependent on definite spatial and material circumstances, and that there are *masses.* The reason the perception of these facts was embodied in so great a number of principles is wholly an historical one; the perception was not reached at once, but slowly and by degrees. In reality only *one* great fact was established. Different pairs of bodies determine, independently of each other, and mutually, in themselves, pairs of accelerations, whose terms exhibit a constant ratio, the criterion and characteristic of each pair.

Not even men of the caliber of Galileo, Huygens, and Newton were able to perceive this fact at once. Even they could only discover it piece by piece, as it is expressed in the law of falling bodies, in the special law of inertia, in the principle of the parallelogram of forces, in the concept of mass, and so forth. Today, no difficulty exists any longer in apprehending the unity of the whole fact. The practical demands of communication alone can justify its piecemeal presentation in several distinct principles, the number of which is really only determined by scientific taste. What is more, a reference to the reflections set forth above respecting the ideas of time, inertia, and the like, will surely convince us that, accurately viewed, the entire fact has, in all its aspects, not yet been perfectly apprehended.

The point of view reached, as Newton expressly states, has nothing to do with the "unknown causes" of natural phenomena. That which in the mechanics of the present day is called *force* is not a something that lies latent in the natural processes, but is a measurable, actual circumstance of motion, the product of the mass into the acceleration. Also when we speak of the attractions or repulsions of bodies, it is not necessary to think of any hidden causes of the motions produced. We signalize by the term attraction merely an actually existing *resemblance* between events determined by conditions of motion and the results of our volitional impulses. In both cases either actual motion occurs or, when the motion is counteracted by some other circumstance of motion, distortion, compression of bodies, and so forth, are produced.

2. The work which devolved on genius here, was the noting of the connection of certain determinative elements of the mechanical processes. The precise establishment of the form of this connection was rather a task for plodding research, which created the different concepts and principles of mechanics. We can de-

termine the true value and significance of these principles and concepts only by the investigation of their historical origin. In this it appears unmistakable at times, that accidental circumstances have given to the course of their development a peculiar direction, which under other conditions might have been very different. Of this an example shall be given.

Before Galileo assumed the familiar fact of the dependence of the final velocity on the time, and put it to the test of experiment, he essayed, as we have already seen, a different hypothesis, and made the final velocity proportional to the *space* described. He believed that he could conclude from this the proportionality of the spaces fallen through with the squares of the times of falling (*Ediz. Nazionale,* VIII, pp. 373, 374).

Later, by a course of fallacious reasoning, he imagined that this assumption involved a self-contradiction (*Dialogo* 3). His reasoning was, that twice any given distance of descent must, by virtue of the double final velocity acquired, necessarily be traversed in the same time as the simple distance of descent. But since the first half is necessarily traversed first, the remaining half will have to be traversed instantaneously, that is in an interval of time not measurable. Whence, it readily follows, that the descent of bodies generally is instantaneous.

The fallacies involved in this reasoning are manifest. Galileo was, of course, not versed in mental integrations, and having at his command no adequate methods for the solution of problems whose facts were in any degree complicated, he could not but fall into mistakes whenever such cases were presented. If we call s the distance and t the time, the Galilean assumption reads in the language of today $ds/dt = as,$ from which follows $s = A\, \varepsilon^{at}$, where a is a constant of experience and A a constant of integration. This is an entirely different conclusion from that drawn by Galileo. It does not conform, it is true, to experience, and Galileo would probably have taken exception to a result that, as a condition of motion generally, made s different from 0 when t equalled 0. But in itself the assumption is by no means *self*-contradictory.

Let us suppose that Kepler had put to himself the same question. Whereas Galileo always sought after the very simplest solutions of things, and at once rejected hypotheses that did not fit, Kepler's mode of procedure was entirely different. He did not quail before the most complicated assumptions, but worked his way, by the constant gradual modification of his original hypothesis, successfully to his goal, as the history of his discovery of the laws of planetary motion fully shows. Most likely, Kepler, on finding the assumption $ds/dt = a\,s$ would not work, would have tried a number of others, and among them probably the correct one $ds/dt = a\sqrt{s}$. But from this would have resulted an essentially different course of development for the science of dynamics.

In the second infinitesimal supposition of Galileo —of proportionality of the velocity to the time of falling—the triangular surfaces of Galileo's construction (fig. 87) represent, in a beautiful and intuitive way, the paths that are described. With the first supposition, on the other hand, the analogous triangles have no phoronomical signification, and on this account the integration was not successful.

It was only gradually and with great difficulty that the concept of "work" attained its present position of importance; and in our judgment it is to the above-mentioned trifling historical circumstance that the difficulties and obstacles it had to encounter are to be ascribed. As the interdependence of the velocity and the time was, as it chanced, first ascertained, it could not be otherwise than that the relation $v = g\,t$ should appear as the original one, the equation $s = gt^2/2$ as the next immediate, and $g\,s = v^2/2$ as a remoter inference. Introducing the concepts mass (m) and force (p), where $p = mg$, we obtain, by multiplying the three equations by m, the expressions $m\,v = pt$, $ms = pt^2/2$, $ps = mv^2/2$—the fundamental equations of mechanics. Of necessity, therefore, the concepts force and momentum (mv) appear more primitive than the concepts work (ps) and *vis viva* (mv^2). No wonder, then, that, wherever the idea of work appeared, one always sought to replace it by the historically older concepts. The entire dispute of the Leibnizians and Cartesians, which was first composed in a manner by D'Alembert, finds its complete explanation in this fact.

From an unbiassed point of view, we have exactly the same right to inquire after the interdependence of the final velocity and the time as after the interdependence of the final velocity and the distance, and to answer the question by experiment. The first inquiry leads us to the experiential truth, that given bodies in contraposition impart to each other in given *times* definite increments of velocity. The second informs us, that given bodies in contraposition impart to each other for given mutual *displacements* definite increments of velocities. Both propositions are equally justified, and both may be regarded as equally original.

The correctness of this view has been substantiated in our own day by the example of J. R. Mayer. Mayer, a modern mind of the Galilean stamp, a mind wholly

free from the influences of the schools, of his own independent accord actually pursued the last-named
method, and produced by it an extension of science
which the schools did not accomplish until later in a
much less complete and less simple form. For Mayer,
work was the original concept. That which is called
work in the mechanics of the schools, he calls force.
Mayer's error was, that he regarded his method as the
only correct one.

3. We may, therefore, as it suits us, regard the *time*
of descent or the *distance* of descent as the factor determinative of velocity. If we fix our attention on
the first circumstance, the concept of force appears as
the original notion, the concept of work as the derived
one. If we investigate the influence of the second fact
first, the concept of work is the original notion. In
the transference of the ideas reached in the observation
of the motion of descent to more complicated relations,
force is recognized as dependent on the *distance* between the bodies—that is, as a function of the distance,
$f(r)$. The work done through the element of distance
dr is then $f(r)dr$. By the second method of investigation work is also obtained as a function of the distance,
$F(r)$; but in this case we know force only in the form
$d \cdot F(r)/dr$—that is to say, as the limiting value of the
ratio: (increment of work)/(increment of distance).

Galileo cultivated by preference the first of these
two methods. Newton likewise preferred it. Huygens
pursued the second method, without at all restricting
himself to it. Descartes elaborated Galileo's ideas after
a fashion of his own. But his performances are insignificant compared with those of Newton and Huygens, and their influence was soon totally effaced. After
Huygens and Newton, the mingling of the two spheres

of thought, the independence and equivalence of which are not always noticed, led to various blunders and confusions, especially in the dispute between the Cartesians and Leibnizians, already referred to, concerning the measure of force. In recent times, however, inquirers turn by preference now to the one and now to the other. Thus the Galileo-Newtonian ideas are cultivated with preference by the school of Poinsot, the Galileo-Huygenian by the school of Poncelet.

4. Newton operates almost exclusively with the notions of force, mass, and momentum. His sense of the value of the concept of mass places him above his predecessors and contemporaries. It did not occur to Galileo that mass and weight were different things. Huygens, too, in all his considerations, puts weights for masses; as for example in his investigations concerning the center of oscillation. Even in the treatise *De Percussione* (*On Impact*), Huygens always says "corpus majus," the larger body, and "corpus minus," the smaller body, when he means the larger or the smaller mass. Physicists were not led to form the concept mass till they made the discovery that the *same* body can by the action of gravity receive different accelerations. The first occasion of this discovery was the pendulum-observations of Richer (1671-1673), from which Huygens at once drew the proper inferences, and the second was the extension of the dynamical laws to the heavenly bodies. The importance of the first point may be inferred from the fact that Newton, to prove the proportionality of mass and weight on the same spot of the earth, personally instituted accurate observations on pendulums of different materials (*Principia*. Lib. II, Sect. VI, *De Motu et Resistentia Corporum Funependulorum*). In the case of John Bernoulli, also, the first

distinction between mass and weight (in the *Meditatio de Natura Centri Oscillationis. Opera Omnia*, Lausanne and Geneva, T. II, p. 168) was made on the ground of the fact that the same body can receive different gravitational accelerations. Newton, accordingly, disposes of all dynamical questions involving the relations of several bodies to each other, by the help of the ideas of force, mass, and momentum.

5. Huygens pursued a different method for the solution of these problems. Galileo had previously discovered that a body rises by virtue of the velocity acquired in its descent to exactly the same height as that from which it fell. Huygens, generalizing the principle (in his *Horologium Oscillatorium*) to the effect that the center of gravity of any system of bodies will rise by virtue of the velocities acquired in its descent to exactly the same height as that from which it fell, reached the principle of the equivalence of work and *vis viva*. The names of the formulæ which he obtained, were, of course, not supplied until long afterwards.

The Huygenian principle of work was received by the contemporary world with almost universal distrust. People contented themselves with making use of its brilliant consequences. It was always their endeavor to replace its deductions by others. Even after John and Daniel Bernoulli had extended the principle, it was its fruitfulness rather than its evidence that was valued.

We observe, that the Galileo-Newtonian principles were, on account of their greater simplicity and apparently greater evidence, invariably preferred to the Galileo-Huygenian. The employment of the latter is exacted only by necessity in cases in which the employment of the former, owing to the laborious attention to details demanded, is impossible; as in the case of John and Daniel Bernoulli's investigations of the motion of fluids.

If we look at the matter closely, however, the same simplicity and evidency will be found to belong to the Huygenian principles as to the Newtonian propositions. That the velocity of a body is determined by the *time* of descent or determined by the *distance* of descent, are assumptions equally natural and equally simple. The *form* of the law must in both cases be supplied by experience. As a starting-point, therefore, $p\,t = m\,v$ and $p\,s = m\,v^2/2$ are equally well fitted.

6. When we pass to the investigation of the motion of several bodies, we are again compelled, in both cases, to take a second step of an equal degree of certainty. The Newtonian idea of mass is justified by the fact, that, if relinquished, all rules governing events would have an end; that we should forthwith have to expect contradictions of our commonest and crudest experiences; and that the physiognomy of our mechanical environment would become unintelligible. The same thing must be said of the Huygenian principle of work. If we surrender the theorem $\Sigma\,p\,s = \Sigma\,m\,v^2/2$, heavy bodies will, by virtue of their own weights, be able to ascend higher; all known rules of mechanical occurrences will have an end. The *instinctive* factors which entered alike into the discovery of the one view and of the other have been already discussed.

The two spheres of ideas could, of course, have grown up much more independently of each other. But in view of the fact that the two were constantly in contact, it is no wonder that they have become partially merged in each other, and that the Huygenian appears the less complete. Newton is all-sufficient with his

forces, masses, and momenta. Huygens would likewise suffice with work, mass, and *vis viva*. But since
he did not in his time completely possess the idea of
mass, that idea had in subsequent applications to be
borrowed from the other sphere. Yet this also could
have been avoided. If with Newton the mass-ratio of
two bodies can be defined as the inverse ratio of the
velocities generated by the same force, with Huygens
it would be logically and consistently definable as the
inverse ratio of the squares of the velocities generated
by the same work.

The two spheres of ideas consider the mutual dependence on each other of entirely different factors of
the same phenomenon. The Newtonian view is in so
far more complete as it gives us information regarding
the motion of each mass. But to do this it is necessary
to go into great detail. The Huygenian view furnishes
a rule for the whole system. It is only a convenience,
but it is then a mighty convenience, when the *relative
velocities* of the masses are previously and independently known.

7. Thus we are led to see, that in the development of dynamics, just as in the development of statics,
the connection of widely different features of mechanical phenomena engrossed at different times the attention
of inquirers. We may regard the momentum of a system as determined by the forces; or, on the other
hand, we may regard its *vis viva* as determined by the
work. In the selection of the criteria in question the
individuality of the inquirers has great scope. It will
be conceived possible, from the arguments above presented, that our system of mechanical ideas might,
perhaps, have been different, had Kepler instituted
the first investigations concerning the motions of fall-

ing bodies, or had Galileo not committed an error in his first speculations. We shall recognize also that not only a knowledge of the ideas that have been accepted and cultivated by subsequent teachers is necessary for the historical understanding of a science, but also that the rejected and transient thoughts of the inquirers, nay even apparently erroneous notions, may be very important and very instructive. The historical investigation of the development of a science is most needful, lest the principles treasured up in it become a system of half-understood prescripts, or worse, a system of *prejudices*. Historical investigation not only promotes the understanding of that which now is, but also brings new possibilities before us, by showing that which exists to be in great measure *conventional* and *accidental*. From the higher point of view at which different paths of thought converge we may look about us with freer vision and discover routes before unknown.

It has been shown that the present form of our science of mechanics rests on an historical accident. This is also shown in a very instructive way by the remarks of Lieut.-Col. Hartmann in his paper on the "Définition physique de la force" (*Congrès international de philosophie,* Geneva, 1905, p. 728), and in *L'enseignement mathématique* (Paris and Geneva, 1904, p. 425). The author shows the utility of the usual interpretations of different conceptions.

In all the dynamical propositions that we have discussed, *velocity* plays a prominent rôle. The reason for this, in our view, is, that, accurately considered, every single body of the universe stands in some definite relation with every other body in the universe; that any one body, and consequently also any several bodies, cannot be regarded as wholly isolated. Our inability to take in all things at a glance alone compels us to consider a few bodies and for the time being to *neglect* in certain aspects the others, a step accomplished by the introduction of velocity, and therefore of time. We cannot regard it as impossible that *integral* laws, to use an expression of C. Neumann, will some day take the place of the laws of mathematical elements, or differential laws, that now make up the science of mechanics, and that we shall have direct knowledge of the dependence on one another of the *positions* of bodies. In such an event, the concept of force will have become superfluous.

IX.

HERTZ'S MECHANICS.

1. The preceding section (VIII), "Retrospect of the Development of Dynamics," was written in the year 1883. It contains, especially in paragraph 7, an extremely general program of a future system of mechanics, and it is to be remarked that the *Mechanics* of Hertz, which appeared in the year 1894,* marks a distinct advance in the direction indicated. It is impossible in the limited space at our disposal to give any adequate conception of the copious material contained in this book, and besides it is not our purpose to expound new systems of mechanics, but merely to trace the development of ideas relating to mechanics. Hertz's book must, in fact, be read by every one interested in mechanical problems.

* H. Hertz, *Die Principien der Mechanik in neuem Zusammenhange dargestellt.* Leipzig, 1894.

2. Hertz's criticisms of prior systems of mechanics, with which he opens his work, contains some very noteworthy epistemological considerations, which from our point of view (not to be confounded either with the Kantian or with the atomistic mechanical concepts of the majority of physicists), stand in need of certain modifications. The constructive images† (or better, perhaps, the concepts), which we consciously and purposely form of objects, are to be so chosen that the "consequences which necessarily follow from them in thought" agree with the "consequences which necessarily follow from them in nature." It is demanded of these images or concepts that they shall be logically admissible, that is to say, free from all self-contradictions; that they shall be correct, that is, shall conform to the relations obtaining between objects; and finally that they shall be appropriate, and contain the least possible superfluous features. Our concepts, it is true, are formed consciously and purposely by us, but they are nevertheless not formed altogether arbitrarily, but are the outcome of an endeavor on our part to adapt our ideas to our sensuous environment. The agreement of the concepts with one another is a requirement which is logically necessary, and this logical necessity, furthermore, is the only necessity that *we* have knowledge of. The belief in a necessity obtaining in nature arises only in cases where our concepts are closely enough adapted to nature to ensure a correspondence between the logical inference and the fact. But the assumption of an adequate adaptation of our ideas can be refuted at any moment by experience. Hertz's criterion of appropriateness coincides with our criterion of economy.

† Hertz uses the term *Bild* (image or picture) in the sense of the old English philosophical use of *idea,* and applies it to systems of ideas or concepts relating to any province.

Hertz's criticism that the Galileo-Newtonian system of mechanics, particularly the notion of force, lacks clearness (pages 7, 14, 15) appears to us justified only in the case of logically defective expositions, such as Hertz doubtless had in mind from his student days. He himself partly retracts his criticism in another place (pages 9, 47); or at any rate, he qualifies it. But the logical defects of some individual interpretation cannot be imputed to systems as such. To be sure, it is not permissible today (page 7) "to speak of a force acting in one aspect only, or, in the case of centripetal force, to take account of the action of inertia twice, once as a mass and again as a force." But neither is this necessary, since Huygens and Newton were perfectly clear on this point. To characterize forces as being frequently "empty-running wheels," as being frequently not demonstrable to the senses, can scarcely be permissible. In any event, "forces" are decidedly in the advantage on this score, as compared with "hidden masses" and "hidden motions." In the case of a piece of iron lying at rest on a table, both the forces in equilibrium, the weight of the iron and the elasticity of the table, are very easily demonstrable.

Neither is the case with energic mechanics so bad as Hertz would have it, and as to his criticism against the employment of minimum principles, that it involves the assumption of purpose and presupposes tendencies directed to the future, the present work shows in another passage quite distinctly that the simple import of minimum principles is contained in an entirely different property from that of purpose. Every system of mechanics contains references to the future, since all must employ the concepts of time, velocity, etc.

3. Nevertheless, though Hertz's criticism of exist-

ing systems of mechanics cannot be accepted in all their severity, his own novel views must be regarded as a great step in advance. Hertz, after eliminating the concept of force, starts from the concepts of time, space, and mass alone, with the idea in view of giving expression only to that which can actually be observed. The sole principle which he employs may be conceived as a combination of the law of inertia and Gauss's principle of least constraint. Free masses move uniformly in straight lines. If they are put in connection in any manner they deviate, in accordance with Gauss's principle, as little as possible from this motion; their *actual* motion is more nearly that of *free* motion than any other *conceivable* motion. Hertz says the masses move as a result of their connection in a *straightest* path. Every deviation of the motion of a mass from uniformity and rectilinearity is due, in his system, not to a force but to rigid connection with other masses. And where such matters are not visible, he conceives hidden masses with hidden motions. All physical forces are conceived as the effect of such actions. Force, force-function, energy, in his system, are secondary and auxiliary concepts only.

Let us now look at the most important points singly, and ask to what extent was the way prepared for them. The notion of eliminating force may be reached in the following manner. It is part of the general idea of the Galileo-Newtonian system of mechanics to conceive of all connections as replaced by forces which determine the motions required by the connections; conversely, everything that appears as force may be conceived to be due to a connection. If the first idea frequently appears in the older systems, as being historically simpler and more immediate, in the case of

Hertz the latter is the more prominent. If we reflect that in both cases, whether forces or connections be presupposed, the actual dependence of the motions of the masses on one another is given for every instantaneous conformation of the system by linear differential equations between the coördinates of the masses, then the existence of these equations may be considered the essential thing,—the thing established by experience. Physics indeed gradually accustoms itself to look upon the description of the facts by differential equations as its proper aim,—a point of view which was taken also in Chapter V. of the present work (1883). But with these the general applicability of Hertz's mathematical formulations is recognized without our being obliged to enter upon any further interpretation of the forces or connections.

Hertz's fundamental law may be described as a sort of generalized law of inertia, modified by connections of the masses. For the simpler cases, this view was a natural one, and doubtless often forced itself upon the attention. In fact, the principle of the conservation of the center of gravity and of the conservation of areas was actually described in the present work (Chapter III.), as a generalized law of inertia. If we reflect that by Gauss's principle the *connection* of the masses determines a minimum of deviation from those motions which it would describe for itself, we shall arrive at Hertz's fundamental law the moment we consider all the forces as due to the connections. For on severing all connections, only isolated masses moving by the law of inertia are left as ultimate elements. The connection, then, supplies the smallest possible deviation from uniform motion in a straight line.

Gauss very distinctly asserted that no substantially

new principle of mechanics could ever be discovered. And Hertz's principle also is only new in form, for it is identical with Lagrange's equations. The minimum condition which the principle involves does not refer to any enigmatic purpose, but its import is the same as that of all minimum laws. That alone takes place which is dynamically determined (Chapter III.). The deviation from the actual motion is dynamically not determined; this deviation is not present; the actual motion is therefore unique, or, according to the striking designation of Petzoldt, it is uniquely determined.*

It is hardly necessary to remark that the physical side of mechanical problems is not only not disposed of, but is not even so much as touched, by the elaboration of such a formal mathematical system of mechanics. Free masses move uniformly in straight lines. Masses having different velocities and directions if connected mutually affect each other as to velocity, that is, determine in each other accelerations. These physical experiences enter along with purely geometrical and arithmetical theorems into the formulation, for which the latter alone would in no wise be adequate; for that which is uniquely determined mathematically and geometrically only, is for that reason not also uniquely determined mechanically. But we have here discussed at considerable length in Chap. II, that the physical principles in question were not at all self-evident, and that even their exact significance was by no means easy to establish.

4. In the beautiful ideal form which Hertz has given to mechanics, its physical contents have shrunk to

---

* See Petzoldt's excellent article "Das Gesetz der Eindeutigkeit" (*Vierteljahrsschrift für wissenschaftliche Philosophie*, XIX., page 146, especially page 186.)   R. Henke is also mentioned in this article as having approached Hertz's view in his tract (*Ueber die Methode der kleinsten Quadrate* (Leipzig, 1894).

an apparently almost imperceptible residue. It is scarcely to be doubted that Descartes if he lived today would have seen in Hertz's mechanics, far more than in Lagrange's "analytic geometry of four dimensions," his own ideal. For Descartes, who in his opposition to the occult qualities of Scholasticism would grant no other properties to matter than extension and motion, sought to reduce all mechanics and physics to a geometry of motions, on the assumption of a motion indestructible at the start.

5. It is not difficult to analyze the psychological circumstances which led Hertz to his system. After inquirers had succeeded in representing electric and magnetic forces that act at a distance as the results of motions in a medium, the desire must again have awakened to accomplish the same result with respect to the forces of gravitation, and if possible for all forces whatsoever. The idea was therefore very natural to discover whether the concept of force generally could not be eliminated. It cannot be denied that when we can command all the phenomena taking place in a medium, together with the large masses contained in it, by means of a single complete picture, our concepts are on an entirely different plane from what they are when only the relations of these isolated masses as regards acceleration are known. This will be willingly granted even by those who are convinced that the interaction of parts in contact is not more intelligible than action at a distance. The present tendencies in the development of physics are entirely in this direction.

If we are not content to leave the assumption of occult masses and motions in its general form, but should endeavor to investigate them singly and in detail, we should be obliged, at least in the present state of our physical knowledge, to resort, even in the simplest cases, to fantastic and even frequently questionable fictions, to which the *given* accelerations would be far preferable. For example, if a mass m is moving uniformly in a circle of radius r, with a velocity v, which we are accustomed to refer to a centripetal force $\dfrac{mv^2}{r}$ proceeding from the center of the circle, we might instead of this conceive the mass to be rigidly connected at the distance $2r$ with one of the same size having a contrary velocity. Huygens's centripetal impulsion would be another example of a force replaced by a connection. As an ideal program Hertz's mechanics is simpler and more beautiful, but for practical purposes our present system of mechanics is preferable, as Hertz himself (page 47), with his characteristic candor, admits.*

x.

IN ANSWER TO CRITICISMS OF THE VIEWS EXPRESSED
BY THE AUTHOR IN CHAPTERS I AND II

1. The views put forward in the first two chapters of this book were worked out by me a long time ago. At the start they were almost without exception coolly rejected, and only gradually gained friends. All the essential features of my *Mechanics* I stated originally in a brief communication of five octavo pages entitled *On the Definition of Mass.* These were the theorems now given at page 304 of the present book. The com-

* Compare J. Classen, "Die Principien der Mechanik bei Hertz und Boltzmann" (*Jahrbuch der Hamburgischen wissenschaftlichen Anstalten*, XV., p. 1, Hamburg, 1898).

munication was rejected by Poggendorf's *Annalen,* and it did not appear until a year later (1868), in Carl's *Repertorium.* In a lecture delivered in 1871, I outlined my epistemological point of view in natural science generally, and with special exactness for physics. The concept of cause is replaced there by the concept of function; the determining of the dependence of phenomena on one another, the economic exposition of actual facts, is proclaimed as the object, and physical concepts as a means to an end solely. I no longer cared to impose upon any editor the responsibility for the publication of the contents of this lecture, and the same was published as a separate tract in 1872.† In 1874, when Kirchhoff in his *Mechanics* came out with his theory of "description" and other doctrines, which were analogous in part only to my views, and still aroused the "universal astonishment" of his colleagues, I became resigned to my fate. But the great authority of Kirchhoff gradually made itself felt, and the consequence of this also doubtless was that on its appearance in 1883 my *Mechanics* did not evoke so much surprise. In view of the great assistance afforded by Kirchhoff, it is altogether a matter of indifference with me that the public should have regarded, and partly still regards, my interpretation of the principles of physics as a continuation and elaboration of Kirchhoff's views; whilst in fact mine were not only older as to date of publication, but also more radical.*

The agreement with my point of view appears upon the whole to be increasing, and gradually to extend

---

† *Erhaltung der Arbeit,* Prague, 1872: "It should be added that a second edition of *Die Geschichte und die Wurzel des Satzes von der Erhaltung der Arbeit* appeared at Leipzig in 1909, and, as already mentioned, an English translation, under the title *History and Root of the Principle of the Conservation of Energy,* was published at Chicago and London in 1911."

* See the preface to the first edition.

over more extensive portions of my work. It would be more in accord with my aversion for polemical discussions to wait quietly and merely observe what part of the ideas enunciated may be found acceptable. But I cannot suffer my readers to remain in obscurity with regard to the existing disagreements, and I have also to point out to them the way in which they can find their intellectual bearings outside of this book, quite apart from the fact that esteem for my opponents also demands a consideration of their criticisms. These opponents are numerous and of all kinds: historians, philosophers, metaphysicians, logicians, educators, mathematicians, and physicists. I can make no pretence to any of these qualifications in any superior degree. I can only select here the most important criticisms, and answer them in the capacity of a man who has a most lively and ingenious interest in understanding the growth of physical ideas. I hope that this will also make it easy for others to find their way in this field and to form their own judgment.

P. Volkmann in his writings on the epistemology* of physics appears as my opponent only in certain criticisms on individual points, and particularly by his adherence to the old systems and by his predilection for them. It is the latter trait, in fact, that separates us; for otherwise Volkmann's views have much affinity with my own. He accepts my adaptation of ideas, the principle of economy and of comparison, even though his expositions differ from mine in individual features and vary in terminology. I, for my part, find in his writings the important principle of isolation and super-

---

* *Erkenntnisstheoretische Grundzüge der Naturwissenschaft*, Leipzig, 1896,
—*Ueber Newton's Philosophia Naturalis*, Königsberg, 1898.—*Einführung in das Studium der theoretischen Physik*, Leipzig, 1900. Our references are to the last-named work.

position, appropriately emphasized and admirably described, and I willingly accept them. I am also willing to admit that concepts which at the start are not very definite must acquire their "retroactive consolidation" by a "circulation of knowledge," by an "oscillation" of attention. I also agree with Volkmann that from this last point of view Newton accomplished in his day nearly the best that it was possible to do; but I cannot agree with Volkmann when he shares the opinion of Thomson and Tait, that even in the face of the substantially different epistemological needs of the present day, Newton's achievement is definitive and exemplary. On the contrary, it appears to me that if Volkmann's process of "consolidation" be allowed complete sway, it must necessarily lead to enunciations not differing in any essential point from my own. I follow with genuine pleasure the clear and objective discussions of G. Heymans,* yet my antimetaphysical position separates us whether it now is recognized as justified or not. The differences which I have with Höfler† and Poske‡ relate in the main to individual points. So far as principles are concerned, I take precisely the same point of view as Petzoldt,§ and we differ only on questions of minor importance. The numerous criticisms of others, which either refer to the arguments of the writers just mentioned, or are supported by analogous grounds, cannot out of regard for the reader be treated at length. It will be sufficient to describe the character of these

* *Die Gesetze und Elemente des wissenschaftlichen Denkens*, II., Leipzig 1894.

† *Studien zur gegenwärtigen Philosophie der mathematischen Mechanik*, Leipzig, 1900.

‡ *Vierteljahrsschrift für wissenschaftliche Philosophie*, Leipzig, 1884, page 385.

§ "Das Gesetz der Eindeutigkeit (*Vierteljahrsschrift für wissenschaftliche Philosophie, XIX.*, page 146).

differences by selecting a few individual, but important points.

2. A special difficulty seems to be still found in accepting my definition of mass. Streintz (compare p. 267) has remarked in criticism of it that it is based solely upon gravity, although this was expressly excluded in my first formulation of the definition (1868). Nevertheless, this criticism is again and again put forward, and quite recently even by Volkmann (*loc. cit.*, p. 18). My definition simply takes note of the fact that bodies in mutual relationship, whether it be that of action at a distance, so called, or whether rigid or elastic connections be considered, determine in one another changes of velocity (accelerations). More than this, one does not need to know in order to be able to form a definition with perfect assurance and without the fear of building on sand. It is not correct as Höfler asserts (*loc. cit.*, p. 77), that this definition tacitly assumes *one and the same force* acting on both masses. It does not assume even the notion of force, since the latter is built up subsequently upon the notion of mass, and gives then the principle of action and reaction quite independently and without falling into Newton's logical error. In this arrangement one concept is not misplaced and made to rest on another which threatens to give way under it. This is, as I take it, the only really serviceable aim of Volkmann's "circulation" and "oscillation." After we have defined mass by means of accelerations, it is not difficult to obtain from our definition apparently new variant concepts like "capacity for acceleration," "capacity for energy of motion" (Höfler, *loc. cit.*, page 70). To accomplish anything dynamically with the concept of mass, the concept in question must, as I most emphatically insist, be a

dynamical concept. Dynamics cannot be constructed with quantity of matter by itself, but the same can at most be artificially and arbitrarily attached to it (*loc. cit.,* pages 71, 72). Quantity of matter by itself is never mass, neither is it thermal capacity, nor heat of combustion, nor nutritive value, nor anything of the kind. Neither does "mass" play a thermal, but only a dynamical rôle (compare Höfler, *loc. cit.,* pages 71, 72). On the other hand, the different physical quantities are proportional to one another, and two or three bodies of unit mass form, by virtue of the dynamic definition, a body of twice or three times the mass, as is analogously the case also with thermal capacity by virtue of the thermal definition. Our instinctive craving for concepts involving quantities of things, to which Höfler (*loc. cit.,* page 72) is doubtless seeking to give expression, and which amply suffices for every-day purposes, is something that no one will think of denying. But a scientific concept of "quantity of matter" should properly be deduced from the proportionality of the single physical quantities mentioned, instead of, contrariwise, building up the concept of mass upon "quantity of matter." The measurement of mass by means of weight results from my definition quite naturally, whereas in the ordinary conception the measurability of quantity of matter by one and the same dynamic measure is either taken for granted outright, or proof must be given beforehand by special experiments, that equal weights act under all circumstances as equal masses. In my opinion, the concept of mass has here been subjected to·thorough analysis for the first time since Newton. For historians, mathematicians, and physicists appear to have all treated the question as an easy and almost self-evident one. It is, on the contrary, of fundamental significance and is deserving of the attention of my opponents.

3. Many criticisms have been made of my treatment of the law of inertia. I believe I have shown (1868), somewhat as Poske has done (1884), that any deduction of this law from a general principle, like the law of causality, is inadmissible, and this view has now won some support (compare Heymans, *loc. cit.*, page 432). Certainly, a principle that has been universally recognized for so short a time only cannot be regarded as *a priori* self-evident. Heymans (*loc. cit.*, p. 427) correctly remarks that axiomatic certainty was ascribed a few centuries ago to a diametrically opposite form of the law. Heymans sees a supra-empirical element only in the fact that the law of inertia is referred to absolute space, and in the further fact that both in the law of inertia and in its ancient diametrically opposite form something *constant* is assumed in the condition of the body that is left to itself (*loc. cit.*, page 433). We shall have something to say further on regarding the first point, and as for the latter it is psychologically intelligible without the aid of metaphysics, because constant features alone have the power to satisfy us either intellectually or practically—which is the reason that we are constantly seeking for them. Now, looking at the matter from an entirely unprejudiced point of view, the case of these axiomatic certainties will be found to be a very peculiar one. One will strive in vain with Aristotle to convince the common man that a stone hurled from the hand would be necessarily brought to rest at once after its release, were it not for the air which rushed in behind and forced it forwards. But he would put just as little credence in Galileo's theory of infinite uniform motion. On the other hand,

Benedetti's theory of the gradual diminution of the *vis impressa,* which belongs to the period of unprejudiced thought and of liberation from ancient preconceptions, will be accepted by the common man without contradiction. This theory, in fact, is an immediate reflection of experience, while the first-mentioned theories, which idealize experience in contrary directions, are a product of technical professional reasoning. They exercise the illusion of axiomatic certainty only upon the mind of the scholar whose entire customary train of thought would be thrown out of gear by a disturbance of these elements of his thinking. The behavior of inquirers toward the law of inertia seems to me from a psychological point of view to be adequately explained by this circumstance, and I am inclined to allow the question of whether the principle is to be called an axiom, a postulate, or a maxim, to rest in abeyance for the time being. Heymans, Poske, and Petzoldt concur in finding an empirical and a supra-empirical element in the law of inertia. According to Heymans (*loc. cit.,* p. 438) experience simply afforded the opportunity for applying an *a priori* valid principle. Poske thinks that the empirical origin of the principle does not exclude its *a priori* validity (*loc. cit.,* pp. 401 and 402). Petzoldt also deduces the law of inertia in part only from experience, and regards it in its remaining part as given by the law of unique determination. I believe I am not at variance with Petzoldt in formulating the issue here ct stake as follows: It first devolves on experience to inform us what particular dependence of phenomena on one another actually exists, what the determining thing is, and only experience can instruct us on this point. If we are convinced that we have been sufficiently instructed in this regard, then when adequate data are at hand we regard it as unnecessary to keep on waiting for further experiences; the phenomenon is determined for us, and since this alone is determination, it is uniquely determined. In other words, if I have discovered by experience that bodies determine accelerations in one another, then in all circumstances where such determinative bodies are lacking I shall expect with unique determination uniform motion in a straight line. The law of inertia thus results immediately in all its generality, without our being obliged to specialize with Petzoldt; for every deviation from uniformity and rectilinearity takes acceleration for granted. I believe I am right in saying that the same fact is twice formulated in the law of inertia and in the statement that forces determine accelerations (p. 172). If this be granted, then an end is also put to the discussion as to whether a vicious circle is or is not contained in the application of the law of inertia (Poske, Höfler).

My inference as to the probable manner in which Galileo achieved clearness regarding the law of inertia was drawn from a passage in his third Dialogue,* which was literally transcribed from the Paduan edition of 1744, III., page 124, in my tract on *The Con-*

---

* "Constat jam, quod mobile ex quiete in *A* descendens per *AB*, gradus acquirit velocitatis juxta temporis ipsius incrementum: gradum vero in *B* esse maximum acquisitorum, et suapte natura immutabiliter impressum, sublatis scilicet causis accelerationis novae, aut retardationis: accelerationis inquam, si adhuc super extenso plano ulterius progrederetur; retardationis vero, dum super planum acclive *BC* fit reflexio: in horizontali autem *GH* aequabilis motus juxta gradum velocitatis ex *A* in *B* acquisitae in infinitum extenderetur."

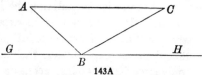

143A

"It is plain now that a movable body, starting from rest at *A* and descending down the inclined plane *AB*, acquires a velocity proportional to the increment of its time: the velocity possessed at *B* is the greatest of the velocities acquired, and by its nature immutably impressed, provided all *servation of Energy.*† Conceiving a body which is rolling down an inclined plane to be conducted upon rising inclined planes of varying slopes, the slight retardation which it suffers on absolutely smooth rising planes of small inclination, and the retardation zero, or unending uniform motion on a horizontal plane, must have occurred to him. Wohlwill was the first to object to this way of looking at the matter (see page 169), and others have since joined him. He asserts that uniform motion in a circle and horizontal motion still occupied distinct places in Galileo's thought, and that Galileo started from the ancient concepts and freed himself only very gradually from them. It is not to be denied that the different *phases* in the intellectual development of the great inquirers have much interest for the historian, and *some one* phase may, in its importance in this respect, be relegated into the background by the others. One must needs be a poor psychologist and have little knowledge of oneself not to know how difficult it is to liberate oneself from traditional views, and how even after that is done the remnants of the old ideas still hover in consciousness and are the cause of occasional backslidings even after the victory has been practically won. Galileo's experience cannot have been different. But with the physicist it is the instant in which a new view flashes forth that is of greatest interest, and it is this instant for which he will always seek. I have sought for it, I believe I have found it; and I am of the opinion that it left its unmistakable

---

causes of new acceleration or retardation are taken away: I say acceleration, having in view its possible further progress along the plane extended; retardation, in view of the possibility of its being reversed and made to mount the ascending plane *BC*. But in the horizontal plane *GH* its uniform motion, with the velocity acquired in the descent from *A* to *B*, will be continued *ad infinitum.*"

† (Eng. Trans., in part, in my *Popular Scientific Lectures*, third edition, Chicago, The Open Court Publishing Co.).

traces in the passage in question. Poske (*loc. cit.*, page 393) and Höfler (*loc. cit.*, pages 111, 112) are unable to give their assent to my interpretation of this passage, for the reason that Galileo does not expressly refer to the limiting case of transition from the inclined to the horizontal plane; although Poske grants that the consideration of limiting cases was frequently employed by Galileo, and although Höfler admits having actually tested the educational efficacy of this device with students. It would indeed be a matter of surprise if Galileo, who may be regarded as the inventor of the principle of continuity, should not in his long intellectual career have applied the principle to this most important case of all for him. It is also to be considered that the passage does not form part of the broad and general discussions of the Italian dialogue, but is tersely couched, in the dogmatic form of a result, in Latin. And in this way also the "velocity immutably impressed" may have crept in.*

The physical instruction which I enjoyed was in all probability just as bad and just as dogmatic as it was the fortune of my older critics and colleagues to enjoy. The principle of inertia was then enunciated as a dogma which accorded perfectly with the system. I could understand very well that disregard of all obstacles to motion led to the principle, or that it must be discovered, as Appelt says, by abstraction; nevertheless, it always remained remote and within the comprehension of supernatural genius only. And where was the guarantee that with the removal of all obstacles the diminution of the velocity also ceased?

* Even granting that Galileo reached his knowledge of the law of inertia only gradually, and that it was presented to him merely as an accidental discovery, nevertheless the following passages which are taken from the Paduan edition of 1744 will show that his limitation of the law to horizontal motion was justified by the inherent nature of the subject treated; and the assumption that Galileo toward the end of his scientific career did not possess a full knowledge of the law, can hardly be maintained.

Poske (*loc. cit.*, p. 395) is of the opinion that Galileo, to use a phrase which I have repeatedly employed, *"discerned"* or *"perceived"* the principle immediately. But what is this discerning? Inquiring man looks here and looks there, and suddenly catches a glimpse of something he has been seeking or even of something quite unexpected, that rivets his attention. Now, I have shown how this "discerning" came about and in what it consisted. Galileo runs his eye over several different uniformly *retarded* motions, and suddenly picks out from among them a uniform, infinitely continued motion, of so peculiar a character that if it occurred by itself alone it would certainly be regarded as something altogether different in kind. But a very

---

"Sagr. Ma quando l'artiglieria si piantasse non a perpendicolo, ma inclinata verso qualche parte, qual dovrebbe esser' il moto della palla? andrebbe ella forse, come nel l'altro tiro, per la linea perpendicolare, e ritornando anco poi per l'istessa?"

"Simpl. Questo non farebbe ella, ma uscita del pezzo seguiterebbe il suo moto per la linea retta, che continua la dirittura della canna, se non in quanto il proprio peso la farebbe declinar da tal dirittura verso terra."

"Sagr. Talche la dirittura della canna è la regolatrice del moto della palla: nè fuori di tal linea si muove, o muoverebbe, se 'l peso proprio non la facesse delinare in giù. . . ."—*Dialogo sopra i due massimi sistemi del mondo.*

"Sagr. But if the gun were not placed in the perpendicular, but were inclined in some direction; what then would be the motion of the ball? Would it follow, perhaps, as in the other case, the perpendicular, and in returning fall also by the same line?"

"Simpl. This it will not do, but having left the cannon it will follow its own motion in the straight line which is a continuation of the axis of the barrel, save in so far as its own weight shall cause it to deviate from that direction toward the earth."

"Sagr. So that the axis of the barrel is the regulator of the motion of the ball: and it neither does nor will move outside of that line unless its own weight causes it to drop downwards. . . ."

"Attendere insuper licet, quod velocitatis gradus, quicunque in mobili reperiatur, est in illo suapte natura indelebiliter impressus, dum externae causae accelerationis, aut retardationis tollantur, quod in solo horizontali plano contingit: nam in planis declivibus adest jam causa accelerationis majoris, in acclivibus vero retardationis. Ex quo pariter sequitur, motum in horizontali esse quoque aeternum: si enim est aequabilis, non debiliatur, aut remittitur, et multo minus tollitur."—*Discorsi e dimostrazioni matematiche. Dialogo terzo.*

"Moreover, it is to be remarked that the degree of velocity a body has is indestructibly impressed in it by its own nature, provided external causes of acceleration or retardation are wanting, which happens only on horizontal planes: for on descending planes there is greater acceleration, and on ascending planes retardation. Whence it follows that motion in a horizontal plane is perpetual: for if it remains the same, it is not diminished, or abated, much less abolished."

minute variation of the inclination transforms this motion into a finite retarded motion, such as we have frequently met with in our lives. And now, no more difficulty is experienced in recognizing the identity between all obstacles to motion and retardation by gravity, wherewith the ideal type of uninfluenced, infinite, uniform motion is gained. As I read this passage of Galileo's while still a young man, a new light concerning the necessity of this ideal link in our mechanics, entirely different from that of the dogmatic exposition, flashed upon me. I believe that every one will have the same experience who will approach this passage without prior bias. I have not the least doubt that Galileo above all others experienced that light. May my critics see to it how their assent is to be avoided.

4. I have now another important point to discuss in opposition to C. Neumann,[*] whose well-known publication on this topic preceded mine[†] shortly. I contended that the direction and velocity which is taken into account in the law of inertia had no comprehensible meaning if the law was referred to "absolute space." As a matter of fact, we can metrically determine direction and velocity only in a space of which the points are marked directly or indirectly by given bodies. Neumann's treatise and my own were successful in directing attention anew to this point, which had already caused Newton and Euler much intellectual discomfort; yet nothing more than partial attempts at solution, like that of Streintz, have resulted. I have remained to the present day the only one who insists upon referring the law of inertia to the earth,

---

[*] *Die Principien der Galilei-Newton'schen Theorie*, Leipzig, 1870.
[†] *Erhaltung der Arbeit*, Prague, 1872. (Translated in part in the article on "The Conservation of Energy," *Popular Scientific Lectures*, third edition, Chicago, 1898.

and in the case of motions of great spatial and temporal extent, to the fixed stars. Any prospect of coming to an understanding with the great number of my critics is, in consideration of the profound differences of our points of view, very slight. But so far as I have been able to understand the criticisms to which my view has been subjected, I shall endeavor to answer them.

Höfler is of the opinion that the existence of "absolute motion" is denied, because it is held to be "inconceivable." But it is a fact of "more painstaking self-observation" that conceptions of absolute motion do exist. Conceivability and knowledge of absolute motion are not to be confounded. Only the latter is wanting here (*loc. cit.,* pages 120, 164). . . . Now, it is precisely with knowledge that the natural inquirer is concerned. A thing that is beyond the ken of knowledge, a thing that cannot be exhibited to the senses, has no meaning in natural science. I have not the remotest desire of setting limits to the imagination of men, but I have a faint suspicion that the persons who imagine they have conceptions of "absolute motions," in the majority of cases have in mind the memory pictures of some actually experienced relative motion; but let that be as it may, for it is in any event of no consequence. I maintain even more than Höfler, viz., that there exist *sensory illusions* of absolute motions, which can subsequently be reproduced at any time. Every one that has repeated my experiments on the sensations of movement has experienced the full sensory power of such illusions. One imagines one is flying off with one's entire environment, which remains at relative rest with respect to the body; or that one is rotating in a space that is distinguished by nothing that is tangible. But no measure can be applied to this space of illusion; its existence cannot be proved to another person, and it cannot be employed for the metrical and conceptual description of the facts of mechanics; it has nothing to do with the space of geometry.* Finally, when Höfler (*loc. cit.,* p. 133) brings forward the argument that "in every relative motion one at least of the bodies moving with reference to each other must be affected with absolute motion,"—I can only say that for the person who considers absolute motion as meaningless in physics, this argument has no force whatever. But I have no further concern here with philosophical questions. To go into details as Höfler has in some places (*loc. cit.,* pp. 124-126) would serve no purpose before an understanding had been reached on the main question.

Heymans (*loc. cit.,* pp. 412, 448) remarks that an inductive, empirical mechanics *could* have arisen, but that as a matter of fact a different mechanics, based on the non-empirical concept of absolute motion, has arisen. The fact that the principle of inertia has always been suffered to hold for absolute motion which is no-where demonstrable, instead of being regarded as hold-ing good for motion with respect to some actually demonstrable system of coördinates, is a problem which is almost beyond power of solution by the empirical theory. Heymans regards this as a problem that can have a metaphysical solution only. In this I cannot agree with Heymans. He admits that relative motions only are given in experience. With this admission, as with that of the possibility of an empirical mechanics,

---

* I flatter myself on being able to resist the temptation to infuse lightness into a serious discussion by showing its ridiculous side, but in reflecting on these problems I was involuntarily forced to think of the question which a very estimable but eccentric man once debated with me as to whether a yard of cloth in one's dreams is as long as a real yard of cloth.—Is the dream-yard to be really introduced into mechanics as a standard of measurement?

I am perfectly content. The rest, I believe, can be explained simply and without the aid of metaphysics. The first dynamic principles were unquestionably built up on empirical foundations. The earth was the body of reference; the transition to the other coördinate systems took place very gradually. Huygens saw that he could refer the motion of impinging bodies just as easily to a boat on which they were placed, as to the earth. The development of astronomy preceded that of mechanics considerably. When motions were observed that were at variance with known mechanical laws when referred to the earth, it was not necessary immediately to abandon these laws again. The fixed stars were present and ready to restore harmony as a new system of reference with the least amount of changes in the concepts. Think only of the oddities and difficulties which would have resulted if in a period of great mechanical and physical advancement the Ptolemaic system had been still in vogue—a thing not at all inconceivable.

But Newton referred all of mechanics to absolute space! Newton is indeed a gigantic personality; little worship of authority is needed to succumb to his influence. Yet even his achievements are not exempt from criticism. It appears to be pretty much one and the same thing whether we refer the laws of motion to *absolute space,* or enunciate them in a perfectly *abstract* form; that is to say, without specific mention of any system of reference. The latter course is unprecarious and even practical; for in treating special cases every student of mechanics looks for some serviceable system of reference. But owing to the fact that the first course, wherever there was any real issue at stake, was nearly always interpreted as having the same meaning as the latter, Newton's error was fraught with much less danger than it would otherwise have been, and has for that reason maintained itself so long. It is psychologically and historically intelligible that in an age deficient in epistemological critique empirical laws should at times have been elaborated to a point where they had no meaning. It cannot therefore be deemed advisable to make metaphysical problems out of the errors and oversights of our scientific fore-fathers, but it is rather our duty to correct them, be they small people or great. I would not be understood as saying that this has never happened.

Let us again call special attention to the fact that Newton in his oft-mentioned Corollary V, which alone has scientific value, does not refer to absolute space.

The most captivating reasons for the assumption of absolute motion were given thirty years ago by C. Neumann (*loc. cit.*, p. 27). If a heavenly body be conceived rotating about its axis and consequently subject to centrifugal forces and therefore oblate, nothing, so far as we can judge, can possibly be altered in its condition by the removal of all the remaining heavenly bodies. The body in question will continue to rotate and will continue to remain oblate. But if the motion be relative only, then the case of rotation will not be distinguishable from that of rest. All the parts of the heavenly body are at rest with respect to one another, and the oblateness would necessarily also disappear with the disappearance of the rest of the universe. I have two objections to make here. Nothing appears to me to be gained by making a meaningless assumption for the purpose of eliminating a contradiction. Secondly, the celebrated mathematician appears to me to have made here too free a use of intellectual experiment, the fruitfulness and value of which cannot be denied. When experimenting in thought, it is permissible to modify *unimportant* circumstances in order to bring out new features in a given case; but it is not to be antecedently assumed that the universe is without influence on the phenomenon here in question. In fact the provoking paradoxes of Newman only disappear with the elimination of absolute space, without carrying out Corollary **V.**

Volkmann (*loc. cit.*, p. 53) advocates an *absolute* orientation by means of the ether. I have already spoken on this point (in the previous editions), but I am extremely curious to know how one ether particle is to be distinguished from another. Until some means of distinguishing these particles is found, it will be preferable to abide by the fixed stars, and where these forsake us to confess that the true means of orientation is still to be found.

5. Taking everything together, I can only say that I cannot well see what is to be altered in my expositions. The various points stand in necessary connection. After it has been discovered that the behavior of bodies toward one another is one in which accelerations are determined—a discovery which was twice formulated by Galileo and Newton, once in a general and again in a special form as a law of inertia—it is possible to give only *one* rational definition of mass, and that a purely dynamical definition. It is not at all, in my judgment, a matter of taste.* The concept of force and the principle of action and reaction follow of themselves. And the elimination of absolute motion is equivalent to the elimination of what is physically meaningless.

---

* My definition of mass takes a more organic and more natural place in Hertz's mechanics than his own, for it contains implicitly the germ of his "fundamental law."

It would not only be taking a very subjective and short-sighted view of science, but it would also be foolhardy in the extreme, were I to expect that my views in their precise individual form should be incorporated without opposition into the intellectual systems of my contemporaries. The history of science teaches that the subjective, scientific philosophies of individuals are constantly being corrected and obscured, and in the philosophy or constructive image of the universe which humanity gradually adopts, only the very strongest features of the thoughts of the greatest men are, after some lapse of time, recognizable. It is merely incumbent on the individual to outline as distinctly as possible the main features of his own view of the world.

ISAAC NEWTON
1642-1727

CHAPTER III.

THE EXTENDED APPLICATION OF THE PRINCI-
PLES OF MECHANICS AND THE DEDUCTIVE
DEVELOPMENT OF THE SCIENCE.

I.

SCOPE OF THE NEWTONIAN PRINCIPLES.

1. The principles of Newton suffice by themselves, without the introduction of any new laws, to explore thoroughly every mechanical phenomenon practically occurring, whether it belongs to statics or to dynamics. If difficulties arise in any such consideration, they are invariably of a mathematical (formal) character, and in no respect concerned with questions of principle. We have given, let us suppose, a number of masses m_1, m_2, m_3. . . . in space, with definite initial velocities v_1, v_2, v_3. . . . We imagine, further, lines of junction drawn between every two masses. In the directions of

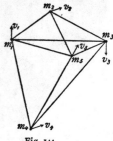

Fig. 144.

these lines of junction are set up the accelerations and counter-accelerations, the dependence of which on the distance it is the business of physics to determine. In a small element of time τ the mass m_5, for example, will traverse in the direction of its initial velocity the distance $v_5\tau$, and in the directions of the lines joining it with the masses m_1, m_2, m_3. . . ., being affected in

such directions with the accelerations φ_1^5, φ_2^5, φ_3^5....,
the distances $(\bar{\varphi}_1^5/2)\tau^2$, $(\varphi_2^5/2)\tau^2$, $(\varphi_3^5/2)\tau^2$.... If
we imagine all these motions to be performed independently of each other, we shall obtain the new position
of the mass m_5 after lapse of time τ. The composition
of the velocities v_5 and $\varphi_1^5\tau$, $\varphi_2^5\tau$, $\varphi_3^5\tau$.... gives the
new initial velocity at the end of time τ. We then
allow a second small interval of time τ to elapse, and,
making allowance for the new spatial relations of the
masses, continue in the same way the investigation of
the motion. In like manner we may proceed with
every other mass. It will be seen, therefore, that, in
point of principle, no embarrassment can arise; the
difficulties which occur are solely of a mathematical
character, where an exact solution in concise symbols,
and not a clear insight into the momentary workings
of the phenomenon, is demanded. If the accelerations
of the mass m_5, or of several masses, collectively neutralize each other, the mass m_5 or the other masses
mentioned are in equilibrium and will move uniformly
onwards with their initial velocities. If, in addition,
the initial velocities in question are $= 0$, both *equilibrium* and *rest* subsist for these masses.

Nor, where a number of the masses m_1, m_2 ....
have considerable extension, so that it is impossible to
speak of a *single* line joining every two masses, is the
difficulty, in point of principle, any greater. We divide
the masses into portions sufficiently small for our purpose, and draw the lines of junction mentioned between
every two such portions. We, furthermore, take into
account the reciprocal relation of the parts of the
same large mass; which relation, in the case of rigid
masses for instance, consists in the parts resisting
every alteration of their distances from one another.

On the alteration of the distance between any two parts of such a mass an acceleration is observed proportional to that alteration. Increased distances diminish, and diminished distances increase in consequence of this acceleration. By the displacement of the parts with respect to one another, the familiar forces of elasticity are aroused. When masses meet in impact, their forces of elasticity do not come into play until contact and an incipient alteration of form take place.

2. If we imagine a heavy perpendicular column resting on the earth, any particle m in the interior of the column which we may choose to isolate in thought, is in equilibrium and at rest. A vertical downward acceleration g is produced by the earth in the particle, which acceleration the particle obeys. But in so doing it approaches nearer to the particles lying beneath it, and the elastic forces thus awakened generate in m a vertical acceleration upwards, which ultimately, when the particle has approached near enough, becomes equal to g. The particles lying above m likewise approach m with the acceleration g. Here, again, acceleration and counter-acceleration are produced, whereby the particles situated above are brought to rest, but whereby m continues to be forced nearer and nearer to the particles beneath it until the acceleration downwards, which it receives from the particles above it, increased by g, is equal to the acceleration it receives in the upward direction from the particles beneath it. We may apply the same reasoning to every portion of the çolumn and the earth beneath it, readily perceiving that the lower portions lie nearer each other and are more violently pressed together than the parts above. Every portion lies between a less closely pressed upper portion and a more closely pressed lower por-

tion; its downward acceleration g is neutralized by a surplus of acceleration upwards, which it experiences from the parts beneath. We comprehend the equilibrium and rest of the parts of the column by imagining *all* the accelerated motions which the reciprocal relation of the earth and the parts of the column determine, as in fact simultaneously performed. The apparent mathematical sterility of this conception vanishes, and it immediately becomes very vivid, when we reflect that in reality no body is completely at rest, but that in all, slight tremors and disturbances are constantly taking place which now give to the accelerations of descent and now to the accelerations of elasticity a slight preponderance. Rest, therefore, is a case of motion, very infrequent, and, indeed, never completely realized. The tremors mentioned are by no means an unfamiliar phenomenon. When, however, we occupy ourselves with cases of equilibrium, we are concerned simply with a *schematic* reproduction in thought of the mechanical facts. We then *purposely* neglect these disturbances, displacements, bendings, and tremors, as here they have no interest for us. All cases of this class, which have a scientific or practical importance, fall within the province of the so-called *theory of elasticity*. The whole outcome of Newton's achievements is that we reach our goal everywhere with one and the same idea, and by means of it are able to reproduce and construct beforehand all cases of equilibrium and motion. All phenomena of a mechanical kind now appear to us as uniform throughout and as made up of the same elements.

3. Let us consider another example. Two masses m, m are situated at a distance a from each

other (Fig. 145). When displaced with respect to each other, elastic forces proportional to the change of distance are supposed to be awakened. Let the masses be movable in the X-direction parallel to a, and their coördinates

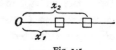

Fig. 145.

be x_1, x_2. If a force f is applied at the point x_2, the following equations obtain:

$$m \frac{d^2 x_1}{dt^2} = p[(x_2 - x_1) - a] \quad \ldots \ldots \ldots \quad (1)$$

$$m \frac{d^2 x_2}{dt^2} = - p[(x_2 - x_1) - a] + f \quad \ldots \ldots \quad (2)$$

where p stands for the force that one mass exerts on the other when their mutual distance is altered by the value 1. All the quantitative properties of the mechanical process are determined by these equations. But we obtain these properties in a more comprehensible form by the integration of the equations. The ordinary procedure is, to find by the repeated differentiation of the equations before us new equations in sufficient number to obtain by elimination equations in x_1 alone or x_2 alone, which are afterwards integrated. We shall here pursue a different method. By subtracting the first equation from the second, we get

$$m \frac{d^2 (x_2 - x_1)}{dt^2} = - 2p[(x_2 - x_1) - a] + f, \text{ or}$$

putting $x_2 - x_1 = u$,

$$m \frac{d^2 u}{dt^2} = - 2p[u - a] + f \quad \ldots \ldots \ldots \quad (3)$$

and by the addition of the first and the second equations

$$m\frac{d^2\,(x_2 + x_1)}{dt^2} = f, \text{ or, putting } x_2 + x_1 = v,$$

$$m\frac{d^2 v}{dt^2} = f \quad . . . . \quad (4)$$

The integrals of (3) and (4) are respectively

$$u = A\sin\sqrt{\frac{2p}{m}}\cdot t + B\cos\sqrt{\frac{2p}{m}}\cdot t + a + \frac{f}{2p} \text{ and}$$

$$v = \frac{f}{m}\cdot\frac{t^2}{2} + Ct + D; \text{ whence}$$

$$x_1 = -\frac{A}{2}\sin\sqrt{\frac{2p}{m}}\cdot t - \frac{B}{2}\cos\sqrt{\frac{2p}{m}}\cdot t + \frac{f}{2m}\cdot\frac{t^2}{2}$$
$$+ Ct - \frac{a}{2} - \frac{f}{4p} + \frac{D}{2},$$

$$x_2 = \frac{A}{2}\sin\sqrt{\frac{2p}{m}}\cdot t + \frac{B}{2}\cos\sqrt{\frac{2p}{m}}\cdot t + \frac{f}{2m}\cdot\frac{t^2}{2}$$
$$+ Ct + \frac{a}{2} + \frac{f}{4p} + \frac{D}{2}.$$

To take a particular case, we will assume that the action of the force f begins at $t = 0$, and that at this time

$$x_1 = 0, \frac{dx_1}{dt} = 0$$

$$x_2 = a, \frac{dx_2}{dt} = 0,$$

that is, the initial positions are given and the initial velocities are $= 0$. The constants A, B, C, D being eliminated by these conditions, we get

$$(5) \quad x_1 = \frac{f}{4p}\cos\sqrt{\frac{2p}{m}}\cdot t + \frac{f}{2m}\cdot\frac{t^2}{2} - \frac{f}{4p},$$

(6) $\quad x_2 = -\dfrac{f}{4p} \cos \sqrt{\dfrac{2p}{m}} \cdot t + \dfrac{f}{2m}\dfrac{t^2}{2} + a + \dfrac{f}{4p}$, and

(7) $\quad x_2 - x_1 = -\dfrac{f}{2p} \cos \sqrt{\dfrac{2p}{m}} \cdot t + a + \dfrac{f}{2p}.$

We see from (5) and (6) that the two masses, in addition to a uniformly accelerated motion with half the acceleration that the force f would impart to one of these masses alone, execute an oscillatory motion symmetrical with respect to their center of gravity. The duration of this oscillatory motion, $T = 2\pi \sqrt{m/2p}$, is smaller in proportion as the force that is awakened in the same mass-displacement is greater (if our attention is directed to two particles of the same body, in proportion as the body is harder). The amplitude of oscillation of the oscillatory motion $f/2p$ likewise decreases with the magnitude p of the force of displacement generated. Equation (7) exhibits the periodic change of distance of the two masses during their progressive motion. The motion of an elastic body might in such case be characterized as vermicular. With hard bodies, however, the number of the oscillations is so great and their excursion so small that they remain unnoticed, and may be left out of account. The oscillatory motion, furthermore, vanishes, either gradually through the effect of some resistance, or when the two masses, at the moment the force f begins to act, are a distance $a + f/2p$ apart and have *equal* initial velocities. The distance $a + f/2p$ that the masses are apart after the vanishing of their vibratory motion, is $f/2p$ greater than the distance of equilibrium a. A tension y, namely, is set up by the action of f, by which the acceleration of the foremost mass is reduced to one-half whilst that of the mass following is increased by the same amount. In this, then, agreeably to our assumption, $p\ y/m = f/2\ m$ or $y = f/2p$. As we see, it is in our power to determine the minutest details of a phenomenon of this character by the Newtonian principles. The investigation becomes (mathematically, yet not in point of principle) more complicated when we conceive a body divided up into a great number of small parts that cohere by elasticity. Here also in the case of sufficient hardness the vibrations may be neglected. Bodies in which we purposely regard the mutual displacement of the parts as evanescent, are called *rigid* bodies.

4. We will now consider a case that exhibits the *schema of a lever*. We imagine the masses M, m_1, m_2 arranged in a triangle and joined by elastic connections. Every alteration of the sides, and consequently also every alteration of the angles, gives rise to accelerations, as the result of which the triangle endeavors to assume its previous form and size. By the aid of the Newtonian principles we can deduce from such a schema the laws of the lever, and at the same time feel that the *form* of the deduction, although it may be

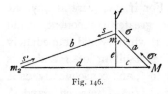

Fig. 146.

more complicated, still remains admissible when we pass from a *schematic* lever composed of three masses to the case of a *real* lever. The mass M we assume either to be in itself very large or conceive it joined by powerful elastic forces to other very large masses (the earth for instance). M then represents an immovable fulcrum.

Let m_1, now, receive from the action of some external force an acceleration f perpendicular to the line of junction $M m_2 = c + d$. Immediately a stretching of the lines $m_1 m_2 = b$ and $m_1 M = a$ is produced, and in the directions in question there are respectively set up the accelerations, as yet undetermined, s and σ, of which the components $s(e/b)$ and $\sigma(e/a)$ are directed oppositely to the acceleration f. Here e is the altitude of the triangle m_1m_2M. The mass m_2 receives the acceleration s', which resolves itself into the two components $s'(d/b)$ in the direction of M and $s'(e/b)$ parallel to f. The former of these determines a slight approach of m_2 to M. The accelerations produced in M by the reactions of m_1 and m_2, owing to its great mass, are imperceptible. We purposely neglect, therefore, the motion of M.

The mass m_1, accordingly, receives the acceleration $f - s(e/b) - \sigma(e/a)$, whilst the mass m_2 suffers the parallel acceleration $s'(e/b)$. Between s and σ a simple relation obtains. If, by supposition, we have a *very rigid* connection, the triangle is only imperceptibly distorted. The components of s and σ *perpendicular* to f destroy each other. For if this were at any one moment not the case, the greater component would produce a further distortion, which would immediately counteract its excess. The resultant of s and σ is therefore directly contrary to f, and consequently, as is readily obvious, $\sigma(c/a) = s(d/b)$. Between s and s', further, subsists the familiar relation $m_1 s = m_2 s'$ or $s = s'(m_2/m_1)$. Altogether m_2 and m_1 receive respectively the accelerations $s'(e/b)$ and $f - s'(e/b)$ $(m_2/m_1)\,(\overline{c+d}/c)$, or, introducing in the place of the variable value $s'(e/b)$ the designation φ, the accelerations φ and $f - \varphi(m_2/m_1)\,(\overline{c+d}/c)$.

At the commencement of the distortion, the acceleration of m_1, owing to the increase of φ, diminishes, whilst that of m_2 increases. If we make the altitude e of the triangle very small, our reasoning still remains applicable. In this case, however, a becomes $= c = r_1$, and $a + b = c + d = r_2$. We see, moreover, that the distortion must continue, φ increase, and the acceleration of m_1 diminish until the stage is reached at which the accelerations of m_1 and m_2 bear to each other the proportion of r_1 to r_2. This is equivalent to a *rotation* of the whole triangle (without further distortion) about M, which mass by reason of the vanishing accelerations is at rest. As soon as rotation sets in, the reason for further alterations of φ ceases. In such a case, consequently,

$$\varphi = \frac{r_2}{r_1}\Big\{ f - \varphi\,\frac{m_2}{m_1}\,\frac{r_2}{r_1} \Big\} \text{ or } \varphi = r_2\,\frac{r_1 m_1 f}{m_1 r_1{}^2 + m_2 r_2{}^2}.$$

For the angular acceleration ψ of the lever we get

$$\psi = \frac{\varphi}{r_2} = \frac{r_1 m_1 f}{m_1 r_1{}^2 + m_2 r_2{}^2}.$$

Nothing prevents us from entering still more into the details of this case and determining the distortions and vibrations of the parts with respect to each other. With sufficiently rigid connections, however, these details may be neglected. It will be perceived that we have arrived, by the employment of the Newtonian principles, at the same result to which the Huygenian view also would have led us. This will not appear strange to us if we bear in mind that the two views are in every re-

spect *equivalent*, and merely start from different aspects
of the same subject-matter. If we had pursued the
Huygenian method, we should have arrived more
speedily at our goal but with less insight into the details of the phenomenon. We should have employed
the work done in some displacement of m_1 to determine the *vires vivæ* of m_1 and m_2, wherein we should
have assumed that the velocities in question v_1, v_2
maintained the ratio $v_1/v_2 = r_1/r_2$. The example
here treated is very well adapted to illustrate what
such an equation of condition means. The equation
simply asserts, that on the slightest deviations of v_1/v_2
from r_1/r_2 powerful forces are set in action which *in
point of fact* prevent all further deviation. The bodies
obey of course, not the *equations*, but the *forces*.

5. We obtain a very obvious case if we put in the
example just treated $m_1 = m_2 = m$ and $a = b$ (Fig.
147). The dynamical state of the system ceases to
change when $\varphi = 2(f - 2\varphi)$, that is, when the accelerations of the masses
at the base and the ver-
tex are given by $2f/5$
and $f/5$. At the com-
mencement of the dis-

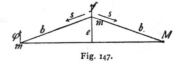

Fig. 147.

tortion φ increases, and simultaneously the acceleration of the mass at the vertex is decreased by double
that amount, until the proportion subsists between the
two of $2:1$.

We have yet to consider the case of *equilibrium* of
a schematic lever, consisting (Fig. 148) of three masses
m_1, m_2, and M, of which the last is again supposed
to be very large or to be elastically connected with
very large masses. We imagine two equal and opposite forces $s, -s$ applied to m_1 and m_2 in the direction
m_1m_2, or, what is the same thing, accelerations im-

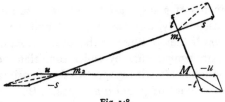

Fig. 148.

pressed inversely proportional to the masses m_1, m_2.
The stretching of the connection $m_1 m_2$ also generates
accelerations inversely proportional to the masses m_1,
m_2, which neutralize the first ones and produce equilibrium. Similarly, along $m_1 M$ imagine the equal and
contrary forces t, — t operative; and along $m_2 M$ the
forces u, — u. In this case also equilibrium obtains.
If M be elastically connected with masses sufficiently
large, — u and — t need not be applied, inasmuch
as the last-named forces are spontaneously evoked the
moment the distortion begins, and always balance the
forces opposed to them. Equilibrium subsists, accordingly, for the two equal and opposite forces s, — s as
well as for the wholly arbitrary forces t, u. As a matter
of fact s, — s destroy each other and t, u pass through
the fixed mass M, that is, are destroyed on distortion
setting in.

The condition of equilibrium readily reduces itself
to the common form when we reflect that the moments of t and u, forces passing through M, are with
respect to M zero, while the moments of s and — s are
equal and opposite. If we compound t and s to p, and
u and — s to q, then, by Varignon's *geometrical* principle of the parallelogram, the moment of p is equal to
the sum of the moments of s and t, and the moment of
q is equal to the sum of the moments of u and — s.

The moments of p and q are therefore equal and opposite. Consequently, *any* two forces p and q will be in *equilibrium* if they produce in the direction $m_1 m_2$ equal and opposite components, by which condition the equality of the moments with respect to M is posited. That then the resultant of p and q also passes through M, is likewise obvious, for s and $-s$ destroy each other and t and u pass through M.

6. The Newtonian point of view, as the example just developed shows us, includes that of Varignon. We were right, therefore, when we characterized the statics of Varignon as a *dynamical* statics, which, starting from the fundamental ideas of modern dynamics, voluntarily restricts itself to the investigation of cases of equilibrium. Only in the statics of Varignon, owing to its abstract form, the significance of many operations, as for example that of the translation of the forces in their own directions, is not so distinctly exhibited as in the instance just treated.

The considerations here developed will convince us that we can dispose by the Newtonian principles of every phenomenon of a mechanical kind which may arise, provided we only take the pains to enter far enough into details. We literally *see through* the cases of equilibrium and motion which here occur, and behold the masses actually impressed with the accelerations they determine in one another. It is the same grand fact, which we recognize in the most various phenomena, or at least can recognize there if we make a point of so doing. Thus a unity, homogeneity, and economy of thought were produced, and a new and wide domain of physical conception opened which before Newton's time was unattainable.

Mechanics, however, is not altogether an end in itself; it has also *problems to solve* that touch the needs of practical life and affect the furtherance of other sciences. Those problems are now for the most part advantageously solved by other methods than the Newtonian—methods whose equivalence to that has already been demonstrated. It would, therefore, be mere impractical pedantry to contemn all other advantages and insist upon always going back to the elementary Newtonian ideas. It is sufficient to have once convinced ourselves that this is always possible. Yet the Newtonian conceptions are certainly the most *satisfactory* and the most lucid; and Poinsot shows a noble sense of scientific clearness and simplicity in making these conceptions the sole foundation of the science.

II.

THE FORMULÆ AND UNITS OF MECHANICS.

1. All the important formulæ of modern mechanics were discovered and employed in the period of Galileo and Newton. The particular designations, which, owing to the frequency of their use, it was found convenient to give them, were for the most part not fixed upon until long afterwards. The systematical mechanical units were not introduced until later still. Indeed, the last named improvement, cannot be regarded as having yet reached its completion.

2. Let s denote the distance, t the time, v the instantaneous velocity, and φ the acceleration of a uniformly accelerated motion. From the researches of Galileo and Huygens, we derive the following equations:

$$v = \varphi t$$
$$s = \frac{\varphi}{2} t^2$$
$$\varphi s = \frac{v^2}{2}$$

$$\left. \right\} \quad \cdots \cdots \cdots \cdots \cdots \quad (1)$$

Multiplying throughout by the mass m, these equations give the following:

$$m v = m \varphi t$$
$$m s = \frac{m \varphi}{2} t^2$$
$$m \varphi s = \frac{m v^2}{2},$$

and, denoting the moving force $m \varphi$ by the letter p, we obtain

$$m v = p t$$
$$m s = \frac{p t^2}{2}$$
$$p s = \frac{m v^2}{2}$$

$$\left. \right\} \quad \cdots \cdots \cdots \cdots \cdots \quad (2)$$

Equations (1) all contain the quantity φ; and each contains in addition two of the quantities s, t, v, as exhibited in the following table:

$$\varphi \left\{ \begin{array}{l} v, t \\ s, t \\ s, v \end{array} \right.$$

Equations (2) contain the quantities m, p, s, t, v; each containing m, p and in addition to m, p two of the three quantities s, t, v, according to the following table:

$$m, p \left\{ \begin{array}{l} v, t \\ s, t \\ s, v \end{array} \right.$$

Questions concerning motions due to constant forces are answered by equations (2) in great variety. If, for example, we want to know the velocity v that a mass m acquires in the time t through the action of a force p, the first equation gives $v = p\,t/m$. If, on the other hand, the *time* be sought during which a mass m with the velocity v can move in opposition to a force p, the same equation gives us $t = m\,v/p$. Again, if we inquire after the *distance* through which m will move with velocity v in opposition to the force p, the third equation gives $s = m\,v^2/2p$. The two last questions illustrate, also, the futility of the Descartes-Leibnizian dispute concerning the measure of force of a body in motion. The use of these equations greatly contributes to confidence in dealing with mechanical ideas. Suppose, for instance, we put to ourselves, the question, what force p will impart to a given mass m the velocity v; we readily see that between m, p, and v *alone*, no equation exists, so that either s or t must be supplied, and consequently the question is an *indeterminate* one. We soon learn to recognize and avoid indeterminate cases of this kind. The distance that a mass m acted on by the force p describes in the time t, if moving with the initial velocity 0, is found by the second equation $s = p\,t^2/2m$.

3. Several of the formulæ in the above-discussed equations have received particular names. The force of a moving body was spoken of by Galileo, who alternately calls it "momentum," "impulse," and "energy." He regards this momentum as proportional to the product of the mass (or rather the weight, for Galileo had no clear idea of *mass*, and for that matter no more had Descartes, nor even Leibniz) into the velocity of the body. Descartes accepted this view. He put

the force of a moving body $= m\,v$, called it *quantity of motion*, and maintained that the sum-total of the quantity of motion in the universe remained constant, so that when one body lost momentum the loss was compensated for by an increase of momentum in other bodies. Newton also employed the designation "quantity of motion" for $m\,v$, and this name has been retained to the present day. [But *momentum* is the more usual term.] For the second member of the first equation, viz. $p\,t$, Belanger, proposed, as late as 1847, the name *impulse*.* The expressions of the second equation have received no particular designations. Leibniz (1695) called the expression $m\,v^2$ of the third equation *vis viva* or *living force*, and he regarded it, in opposition to Descartes, as the true measure of the force of a body in motion, calling the pressure of a body at rest *vis mortua*, or dead force. Coriolis found it more appropriate to give the term $\frac{1}{2}m\,v^2$ the name *vis viva*. To avoid confusion, Belanger proposed to call $m\,v^2$ living force and $\frac{1}{2}m\,v^2$ *living power* [now commonly called in English *kinetic energy*]. For $p\,s$ Coriolis employed the name *work*. Poncelet confirmed this usage, and adopted the *kilogram-meter* (that is, a force equal to the weight of a kilogram acting through the distance of a meter) as the *unit of work*.

4. Concerning the historical details of the origin of these notions "quantity of motion" and "vis viva," a glance may now be cast at the ideas which led Descartes and Leibniz to their opinions. In his *Principia Philosophiæ*, published in 1644, II, 36, DESCARTES expressed himself as follows:

* See, also, Maxwell, *Matter and Motion*, American edition, page 72. But this word is commonly used in a different sense, namely, as "the limit of a force which is infinitely great but acts only during an infinitely short time." See Routh, *Rigid Dynamics*, Part I, pages 65-66.—*Trans.*

"Now that the nature of motion has been examined, we must consider its cause, which may be conceived in two senses: first, as a universal, original cause— the general cause of all the motion in the world; and second, as a special cause, from which the individual parts of matter receive motion which before they did not have. As to the universal cause, it can manifestly be none other than God, who in the beginning created matter with its motion and rest, and who now preserves, by his simple ordinary concurrence, on the whole, the same amount of motion and rest as he originally created. For though motion is only a condition of moving matter, there yet exists in matter a definite quantity of it, which in the world at large never increases or diminishes, although in single portions it changes; namely, in this way, that we must assume, in the case of the motion of a piece of matter which is moving twice as fast as another piece, but in quantity is only one half of it, that there is the same amount of motion in both, and that in the proportion as the motion of one part grows less, in the same proportion must the motion of another, equally large part grow greater. We recognize it, moreover, as a perfection of God, that He is not only in Himself unchangeable, but that also his modes of operation are most rigorous and constant; so that, with the exception of the changes which indubitable experience or divine revelation offer, and which happen, as our faith or judgment show, without any change in the Creator, we are not permitted to assume any others in his works—lest inconstancy be in any way predicated of Him. Therefore, it is wholly rational to assume that God, since in the creation of matter he imparted different motions to its parts, and preserves all matter in the same way and conditions in which he created it, so he similarly *preserves* in it *the same quantity of motion."*

Although signal individual performances in science cannot be gainsaid to Descartes, as his studies on the rainbow and his enunciation of the law of refraction show, his importance nevertheless is contained rather in the great general and revolutionary ideas which he promulgated in philosophy, mathematics, and the natural sciences. The maxim of doubting everything that has hitherto passed for established truth cannot be rated too high; although it was more observed and exploited by his followers than by himself. Analytical geometry with its modern methods is the outcome of his idea to dispense with the consideration of all the details of geometrical figures by the application of algebra, and to reduce everything to the consideration of distances. He was a pronounced enemy of occult qualities in physics, and strove to base all physics on mechanics, which he conceived as a pure geometry of motion. He has shown by his experiments that he regarded no physical problem as insoluble by this method. He took too little note of the fact that mechanics is possible only on the condition that the positions of the bodies are determined in their dependence on one another by a relation of force, by a function of time; and Leibniz frequently referred to this deficiency. The mechanical concepts which Descartes developed with scanty and vague materials could not possibly pass as copies of nature, and were pronounced to be phantasies even by Pascal, Huygens, and Leibniz. It has been remarked, however, in a former place, how strongly Descartes's ideas, in spite of these facts, have persisted to the present day. He also exercises a powerful influence upon physiology by his theory of vision, and by his contention that animals were machines—a theory which he naturally had not the courage to extend to human beings, but by which he anticipated the idea of reflex motion (compare Duhem, *L'évolution des théories physiques,* Louvain, 1896).

The merit of having first *sought after* a more universal and more fruitful point of view in mechanics, cannot be denied Descartes. This is the peculiar task of the philosopher, and it is an activity which constantly exerts a fruitful and stimulating influence on physical science.

Descartes, however, was infected with all the usual errors of the philosopher. He places absolute confidence in his own ideas. He never troubles himself to put them to experiential test. On the contrary, a minimum of experience always suffices him for a maximum of inference. Added to this, is the indistinctness of his conceptions. Descartes did not possess a clear idea of mass. It is hardly allowable to say that Descartes defined $m\,v$ as momentum, although Descartes's scientific successors, feeling the need of more definite notions, adopted this conception. Descartes's greatest error, however—and the one that vitiates all his physical inquiries—is this, that many propositions appear to him self-evident *à priori* concerning the truth of which experience alone can decide. Thus, in the two paragraphs following that cited above (§§37-39) it is asserted as a self-evident proposition that a body preserves unchanged its velocity and direction. The experiences cited in §38 should have been employed, not as a confirmation of an *à priori* law of inertia, but as a foundation on which this law in an empirical sense should be based.

Descartes's view was attacked by LEIBNIZ (1686) in the *Acta Eruditorum,* in a little treatise bearing the title: "A short Demonstration of a Remarkable Error of Descartes and Others, Concerning the Natural Law by which they think that the Creator always preserves the same Quantity of Motion; by which, however, the Science of Mechanics is totally perverted."

In machines in equilibrium, Leibniz remarks, the *loads* are inversely proportional to the velocities of displacement; and in this way the idea arose that the product of a *body* ("corpus," "moles") into its *velocity* is the measure of force. This product Descartes regarded as a constant quantity. Leibniz's opinion, however, is, that this measure of force is only accidentally the correct measure, in the case of the machines. The true measure of force is different, and must be determined by the method which Galileo and Huygens pursued. Every body rises by virtue of the velocity acquired in its descent to a height exactly equal to that from which it fell. If, therefore, we assume, that the same "force" is requisite to raise a body m a height $4h$ as to raise a body $4m$ a height h, we must, since we know that in the first case the velocity acquired in descent is but twice as great as in the second, regarded the product of a "body" into the *square* of its velocity as *the measure of force*.

In a subsequent treatise (1695), Leibniz reverts to this subject. He here makes a distinction between simple pressure (*vis mortua*) and the force of a moving body (*vis viva*), which latter is made up of the sum of the pressure-impulses. These impulses produce, indeed, an "impetus" ($m\,v$), but the impetus produced is not the true measure of force; this, since the cause must be equivalent to the effect, is (in conformity with the preceding considerations) determined by $m\,v^2$. Leibniz remarks further that the possibility of perpetual motion is excluded only by the acceptance of his measure of force.

Leibniz, no more than Descartes, possessed a genuine concept of mass. Where the necessity of such an idea occurs, he speaks of a body (*corpus*), of a load (*moles*), of different-sized bodies of the same specific gravity, and so forth. Only in the second treatise, and there only once, does the expression "massa" occur, in all probability borrowed from Newton. Still, to derive any definite results from Leibniz's theory, we must associate with his expressions the notion of mass, as his successors actually did. As to the rest, Leibniz's procedure is much more in accordance with the methods of science than Descartes's. Two things, however, are confounded: the question of the *measure of force* and the question of the *constancy* of the sums $\Sigma\,m\,v$ and $\Sigma\,m\,v^2$. The two have in reality nothing to do with each other. With regard to the first question, we now know that both the Cartesian and the Leibnizian measure of force, or, rather, the measure of the effectiveness of a body in motion, have, each in a different sense, their justification. Neither measure, however, as Leibniz himself correctly remarked, is to be confounded with the common Newtonian, measure of force.

With regard to the second question, the latter investigations of Newton really proved that for *free* material systems not acted on by external forces the Cartesian sum $\Sigma\,m\,v$ is a constant; and the investigations of Huygens showed that also the sum $\Sigma\,m\,v^2$ is a constant, provided *work* performed by forces does not alter it. The dispute raised by Leibniz rested, therefore, on various *misunderstandings*. It lasted fifty-seven years, till the appearance of D'Alembert's *Traitè de dynamique,* in 1743. To the theological ideas of Descartes and Leibniz, we shall revert in another place.

5. The three equations above discussed, though they are only applicable to *rectilinear* motions produced by *constant* forces, may yet be considered the *fundamental* equations of mechanics. If the motion be rectilinear but the force variable, these equations pass by a slight, almost self-evident, modification into others, which we shall here only briefly indicate, since mathematical developments in the present treatise are wholly subsidiary.

From the first equation we get for variable forces $m\,v = \int p\,d\,t + C$, where p is the variable force, $d\,t$ the time-element of the action, $\int p\,d\,t$ the sum of all the products $p\,.\,d\,t$ from the beginning to the end of the action, and C a constant quantity denoting the value of $m\,v$ before the force begins to act. The second equation passes in like manner into the forms $s = \int dt \int \frac{p}{m} dt + C\,t + D$, with two so-called constants of integration. The third equation must be replaced by

$$\frac{m\,v^2}{2} = \int p\,ds + C.$$

Curvilinear motion may always be conceived as the product of the simultaneous combination of three rectilinear motions, best taken in three mutually perpendicular directions. Also for the components of the motion of this very general case, the above-given equations retain their significance.

6. The mathematical processes of addition, subtraction, and equating possess intelligible meaning only when applied to quantities of the same kind. We cannot add or equate masses and times, or masses and velocities, but only masses and masses, and so on. When, therefore, we have a mechanical equation, the question immediately presents itself whether the mem-

bers of the equation are quantities of *the same kind,* that is, whether they can be measured by *the same* unit, or whether, as we usually say, the equation is *homogeneous.* The units of the quantities of mechanics will form, therefore, the next subject of our investigations.

The choice of units, which are, as we know, quantities of the same kind as those they serve to measure, is in many cases arbitrary. Thus, an arbitrary mass is employed as the unit of mass, an arbitrary length is employed as the unit of length, an arbitrary time as the unit of time. The mass and the length employed as units can be preserved; the time can be reproduced by pendulum-experiments and astronomical observations. But units like a unit of velocity, or a unit of acceleration, cannot be preserved, and are much more difficult to reproduce. These quantities are consequently so connected with the arbitrary fundamental units, mass, length, and time, that they can be easily and at once derived from them. Units of this class are called *derived* or *absolute* units. This latter designation is due to GAUSS, who first derived the magnetic units from the mechanical, and thus created the possibility of a universal comparison of magnetic measurements. The name, therefore, is of historical origin.

As unit of velocity we might choose the velocity with which, say, q units of length are travelled over in unit of time. But if we did this, we could not express the relation between the time t, the distance s, and the velocity v by the usual simple formula $s = v t$, but should have to substitute for it $s = q . v t$. If, however, we define the unit of velocity as the velocity with which the unit of length is travelled over in unit of time, we may retain the form $s = v t$. Among the derived units the simplest possible relations are made to obtain. Thus, as the unit of area and the unit of volume, the square and cube of the unit of length are always employed.

According to this, we assume then, that by unit velocity unit length is described in unit time, that by unit acceleration unit velocity is gained in unit time, that by unit force unit acceleration is imparted to unit mass, and so on.

The derived units depend on the arbitrary fundamental units; they are functions of them. The function which corresponds to a given derived unit is called its *dimension*. The theory of dimensions was laid down by FOURIER, in 1822, in his *Theory of Heat*. Thus, if l denote a length, t a time, and m a mass, the dimensions of a velocity, for instance, are l/t or lt^{-1}. After this explanation, the following table will be readily understood:

NAMES	SYMBOLS	DIMENSIONS
Velocity	v	$l t^{-1}$
Acceleration	φ	$l t^{-2}$
Force	p	$m l t^{-2}$
Momentum	$m v$	$m l t^{-1}$
Impulse	$p t$	$m l t^{-1}$
Work	$p s$	$m l^2 t^{-2}$
Vis viva	$\dfrac{m v^2}{2}$	$m l^2 t^{-2}$
Moment of inertia	Θ	$m l^2$
Statical moment	D	$m l^2 t^{-2}$

This table shows at once that the above-discussed equations are *homogeneous,* that is, contain only members of *the same kind*. Every new expression in mechanics might be investigated in the same manner.

7. The knowledge of the dimensions of a quantity is also important for another reason. Namely, if the

value of a quantity is known for one set of fundamental units and we wish to pass to another set, the value of the quantity in the new units can be easily found from the dimensions. The dimensions of an acceleration, which has, say, the numerical value φ, are $l\,t^{-2}$. If we pass to a unit of length λ times greater and to a unit of time τ times greater, then a number λ times smaller must take the place of l in the expression $l\,t^{-2}$, and a number τ times smaller the place of t. The numerical value of the same acceleration referred to the new units will consequently be $(\tau^2/\lambda)\varphi$. If we take the meter as our unit of length, and the second as our unit of time, the acceleration of a falling body for example is 9.81, or as it is customary to write it, indicating at once the dimensions and the fundamental measures: 9.81 (meter/second^2). If we pass now to the kilometer as our unit of length ($\lambda = 1000$), and to the minute as our unit of time ($\tau = 60$), the value of the same acceleration of descent is $(60 \times 60/1000)9.81$, or 35.316 (kilometer/minute^2).

8. As a unit of length the meter (the length of a platinum-iridium bar at 0°C, kept at Paris and approximately $1/10^7$ of a quadrant of a terrestrial meridian) is commonly employed; as a unit of time, the second (mean sun time, sometimes also sidereal time). In consideration of the above remarks, the velocity of one meter per second is chosen as the unit (of velocity) and as a unit of acceleration that which corresponds to an increase in velocity of one meter per second.

Complications arise in the choice of the unit of mass and unit of force. If one chooses as unit of mass, the mass of the Parisian platinum-iridium kilogram (approximately the mass of a cubic decimeter of water at 4° C) then the force with which this standard is attracted to the earth is not 1, but because $p = m \cdot g$ has the value g; thus in Paris it has the value of 9.808, in other parts of the earth a somewhat different value. The unit of force, then, is that force which in one second transmits to the mass of the kilogram an acceleration of one meter per second. The unit of work is the effect of this unit of force upon one meter distance, and so forth. The consistent metric system of mass in which the mass of the kilogram standard $= 1$, usually is called the absolute system.

The so-called gravitational system of measurement* of mechanical quantities is based upon the fact that the force with which the Parisian kilogram is attracted by the earth, is set at $= 1$. If we then keep the simple relationship $p = mg$, then the mass of the kilogram is not $= 1$, but $1/g$. Accordingly only g such kilograms or

* [The statement written by Mr. C. S. Peirce which was substituted for Sec. 8 in the earlier editions of the *Mechanics* has the following to say in regard to the systems of measurement of mechanical quantities: "Whether the International or the British units are employed, there are two methods of measurement of mechanical quantities, the absolute and the gravitational. The absolute is so called because it is not relative to the acceleration of gravity at any station. This method was introduced by Gauss.

"The special absolute system, widely used by physicists in the United States and Great Britain, is called the Centimeter-Gram-Second system. In this system, writing C for centimeter, G for gram mass, and S for second,

the unit of length is...C;
the unit of mass is..G;
the unit of time is..S;
the unit of velocity is......................................C/S;
the unit of acceleration (which might be called a "galileo," because Galileo Galilei first measured an acceleration) is ....C/S²;
the unit of density is.......................................G/C³;
the unit of momentum is....................................GC/S;
the unit of force (called a *dyne*) is......................GC/S²;
the unit of pressure (called one millionth of an absolute atmosphere) is..............................G/CS²;
the unit of energy (*vis viva*, or work, called an *erg*) is ....½ GC²/S²;
etc.

9.808 kilograms together have the mass 1. The same kilogram in another part of the earth, A, with the acceleration g', will be attracted to the earth not with the force 1, but with the force g'/g. Accordingly g/g' Parisian kilograms at this place correspond to a force of one kilogram. Let us take, then, g' pieces, which at A, exert one kilogram pressure, we will have again g times the mass of the kilogram or the mass 1. However, if we had a body at A, which we knew would be attracted by the force of one kilogram in Paris, we should have to figure naturally, not with g' but with g such bodies in the unit of mass.

A body which weighs p kilograms in Paris (in a vacuum) has the mass p/g. A body which at A exercises p kilograms pressure, has the mass p/g'. The difference between g and g' can be disregarded in many cases, but must be considered when exactness is required.

"The gravitational system of measurement of mechanical quantities, takes the kilogram or pound, or rather the attraction of these towards the earth, compounded with the centrifugal force, which is the acceleration called gravity, and is denoted by g, and is different at different places, as the unit of force, and the foot-pound or kilogram-meter, being the amount of gravitational energy transformed in the descent of a pound through a foot or of a kilogram through a meter, as the unit of energy. Two ways of reconciling these convenient units with the adherence to the usual standard of length naturally suggest themselves, namely, first, to use the pound weight or the kilogram weight divided by g as the unit of mass, and, second, to adopt such a unit of time as will make the acceleration of g, at an initial station, unity. Thus, at Washington, the acceleration of gravity is 980.05 galileos. If, then, we take the centimeter as the unit of length, and the 0.031943 second as the unit of time, the acceleration of gravity will be 1 centimeter for such unit of time squared. The latter system would be for most purposes the more convenient; but the former is the more familiar.

"In either system, the formula $p = m\,g$ is retained; but in the former g retains its absolute value, while in the latter it becomes unity for the initial station. In Paris, g is 980.96 galileos; in Washington it is 980.05 galileos. Adopting the more familiar system, and taking Paris for the initial station, if the unit of force is a kilogram's weight, the unit of length a centimeter, and the unit of time a second, then the unit of mass will be 1/981.0 kilogram, and the unit of energy will be a kilogram-centimeter, or $(1/2)$ $(1000/981.0)$ G C^2/S^2. Then, at Washington the gravity of a kilogram will be, not 1, as at Paris, but $980.1/981.0 = 0.99907$ units or Paris kilogram-weights. Consequently, to produce a force of one Paris kilogram-weight we must allow Washington gravity to act upon $981.0/980.1 = 1.00092$ kilograms." [For some critical remarks on the preceding method of exposition, see *Nature*, in the issue for November 15, 1894.] Trans.

The remaining units in the gravitational system are naturally determined by the choice of the unit of force. Thus the unit of work is that by which the force acts on the distance 1—that is the kilogram-meter. The unit of kinetic energy (*vis viva*) is that which is produced by one unit of work, etc.

If we allow a body to fall, which in Paris (in vacuum) weighs p kilograms at 45° Lat. at sea level (with the acceleration 9.806), then we have, according to absolute measurement, the mass p upon which 9.806 p units of force act; according to gravitational measurements, however, we have the mass $\frac{p}{9.808}$ upon which $p \cdot \frac{9.806}{9.808}$ units of force act. If one meter is the distance fallen, then the work accomplished and the kinetic energy achieved according to absolute measurement is 9.806 p, according to gravitational measure, however, it is $\frac{9.806}{9.808} \cdot p$. The unit of force of the gravitational measure is in round numbers about ten times greater than that of the absolute measure; for the unit of mass the same holds true. For a given quantity of work or kinetic energy in the gravitational system corresponds about a ten times smaller numerical value than in the absolute system.

It must still be noted that instead of the kilogram as unit of mass and the meter as unit of length, in England the gram and centimeter,* in Germany milligram and millimeter are used. Conversion between these units offers no difficulties according to given practices. In mechanics, as in some other branches of

---

* The National Bureau of Standards at Washington defines standards of length, mass, and time (September 19, 1935) as follows:

LENGTH

The primary standard of length in the United States is the United

physics closely allied to it, our calculations involve but three fundamental quantities, quantities of space, quantities of time, and quantities of mass. This circumstance is a source of simplification and power in the science which should not be underestimated.

States Prototype Meter 27, a platinum-iridium line standard having an X-shaped cross section. The length of this bar which is deposited at the National Bureau of Standards in Washington, is known in terms of the International Prototype meter which is deposited at the International Bureau of Weights and Measures at Sèvres, near Paris, France.

A supplementary definition of the meter in terms of wave length of light was adopted provisionally by the Seventh General (International) Conference on Weights and Measures in 1927. According to this definition the relation for red cadmium light-waves under specified conditions of temperature, pressure, and humidity, is

$$1 \text{ meter} = 1\ 553\ 164.13 \text{ wave lengths.}$$

From this relation the wave length of the red radiation from cadmium, under standard conditions of temperature, pressure, and humidity, is found to be 6438.4696×10^{-7} millimeters. (Benoit, Fabry, and Perot. Trav. et Mem. du Bu. Int. des Poids et Mesures, vol. 15, p. 131.)

The United States yard is defined by the relation

$$1 \text{ yard} = \frac{3600}{3937} \text{ meter (exactly)}$$

From this relation it follows that

$$1 \text{ yard} = 0.9144018 \text{ meter (approx.)}$$

and $1 \text{ inch} = 25.4000508$ millimeter (approx.)

For industrial purposes a relation between the yard and the meter has been adopted by the American Standards Association (A.S.A. B48. 1-1933), and by similar organizations in 15 other countries. This relation is

$$1 \text{ inch} = 25.4 \text{ millimeters (exactly)}$$

from which $1 \text{ yard} = 0.9144$ meter (exactly)

The adoption of this relation by industry, for use in making conversions between inches and millimeters, did not change the official definition of the yard or of the meter. Its legal adoption in the United States and in Great Britain would be a very desirable step in the direction of international uniformity in precision length measurements.

MASS

The primary standard of mass for this country is United States Prototype Kilogram 20, which is a platinum-iridium standard kept at the National Bureau of Standards. The value of this mass standard is known in terms of the International Prototype Kilogram, a platinum-iridium standard which is kept at the International Bureau of Weights and Measures.

For many years the British standards were considered to be the primary standards of the United States. Later, for over 50 years, the avoirdupois pound was defined in terms of the Troy Pound of the Mint, which is a brass standard kept at the United States Mint in Philadelphia. In 1911 the Troy Pound of the Mint was superseded, for coinage purposes, by the Troy Pound of the National Bureau of Standards. Since 1893 the avoirdupois pound has been defined in terms of the United States Prototype Kilogram 20 by the relation:

$$1 \text{ avoirdupois pound} = 0.4535924277 \text{ kilogram}$$

Insofar as can be determined, these changes in definition have not made any change in the actual value of the pound.

Distinction between Mass and Weight

The *mass* of a body, as used herein, is the quantity of material in the body. The *weight* of a body is defined as the force with which that body is attracted toward the earth. Confusion sometimes arises from the practice of referring to standards of mass as "weights" and from the fact that such standards are compared by "weighing" one against another by means of a balance. Standard "weights" are, in reality, standards of mass.

Another practice which tends to confusion is that of using the terms kilogram, gram, pound, etc., in two distinct senses; first, to designate units of mass, and second, to designate units of weight or force. For example, a body having a mass of one kilogram is called a kilogram (mass) and the force with which such a body is attracted toward the earth is also called a kilogram (force).

The International Kilogram and the U. S. Prototype Kilogram are specifically defined by the International Conference on Weights and Measures as standards of mass. The U. S. pound, which is derived from the International Kilogram, is, therefore, a standard of mass.

So long as no material is added to or taken from a body its mass remains constant. Its weight, however, varies with the acceleration of gravity "g". For example, a body would be found to weigh more at the poles of the earth than at the equator, and less at high elevations than at sea level. (Standard acceleration of gravity, adopted by the International Committee on Weights and Measures in 1901 is 980.665 cm/sec^2. This value corresponds nearly to the value at latitude 45° and sea level.)

Since standards of mass (or "weights"), are ordinarily calibrated and used on even-arm balances the effects of variations in the acceleration of gravity are self-eliminating and need not be taken into account. Two objects of equal mass will be affected in the same manner and by the same amount by any change in the value of the acceleration of gravity, and thus if they have the same weight, i.e., if they balance each other on an even-arm balance, under one value of "g" they will also balance each other under any other value of "g".

On a spring balance, however, the weight of the body is not balanced against the weight of another body, but against the elastic force of a spring. Therefore, using a very sensitive spring balance, the weight would be found to vary with the acceleration of gravity.

Effect of Air Buoyancy

Another point that must be taken into account in the calibration and use of standards of mass is the buoyancy or lifting effect of the air. A body immersed in any fluid is buoyed up by a force equal to the weight of the displaced fluid. Two bodies of equal mass, if placed one on each pan of an even-arm balance, will balance each other in a vacuum. If compared in air, however, they will not balance each other unless they are of equal volume. If of unequal volume, the larger body will displace the greater volume of air and will be buoyed up by a greater force than will the smaller body, and the larger body will appear to be lighter in weight than the smaller body. The greater the difference in volume, and the greater the density of the air in which the comparison weighing is made, the greater will be the apparent difference in weight. For that reason, in assigning a precise numerical value of apparent mass to a standard, it is necessary to base this value on definite values for the air density and the density of the mass standard of reference.

At the National Bureau of Standards the corrections to be applied to high precision analytical weights are given on the basis of comparison in vacuum and also on the basis of comparisons in air of standard density, 1.2 mg per cm^3, and against brass weights having a density of 8.4 grams per cubic centimeter.

Commercial weights and weighing scales are usually adjusted on the basis of "apparent weight in air against brass weights". That is, commercial weights, regardless of their material, are so adjusted that they will balance a correct brass standard of mass of 8.4 density and of the same nominal value, when compared in air at standard atmospheric density. Weighing scales are so adjusted that they indicate the correct mass of a mass standard of 8.4 density when the standard is weighed in air of standard density, 1.2 mg per cm³. In commercial weighing no correction need be made for variations in air density.

TIME

There is no physical standard of time corresponding to the standards of length and mass. Time is measured in terms of the motion of the earth; (a) on its axis, and (b) around the sun. The time it takes the earth to make a complete rotation on its axis is called a day, and the time it takes it to make a complete journey around the sun, as indicated by its position with reference to the stars, is called a year. The earth makes about 365¼ rotations on its axis (365.2422, more exactly) while making a complete journey around the sun. In other words, there are almost exactly 365¼ solar days in a tropical or solar year. As it would be inconvenient and confusing to have the year, as used in every-day life, contain a fractional part of a day, fractional days are avoided by making the calendar year contain 365 days in ordinary years and 366 days in leap-years. The frequency of occurrence of leap-years is such as to keep the average length of the calendar year as nearly as practicable equal to that of the tropical year, in order that calendar dates may not drift through the various seasons of the tropical year.

The earth, in its journey around the sun, does not move at a uniform speed, and the sun in its apparent motion does not move along the equator but along the ecliptic. Therefore the apparent solar days are not of exactly equal length. To overcome this difficulty time is measured in terms of the motion of a fictitious or "mean" sun, the position of which, at all times, is the same as would be the apparent position of the real sun if the earth moved on its axis and in its journey around the sun at a uniform rate. Ordinary clocks and watches are designed and regulated to indicate time in terms of the apparent motion of this fictitious or "mean sun". It is "mean noon" when this "mean sun" crosses the meridian, and the time between two successive crossings is a "mean solar day". The length of the mean solar day is equal to the average length of the apparent solar day.

The time used by astronomers is sidereal time. This is defined by the rotation of the earth with respect to the stars. A sidereal day is the interval between two successive passages of a star across a meridian. The sidereal day is subdivided into hours, minutes and seconds, the hours being numbered from 1 to 24.

The mean solar day is divided into 24 hours, each hour into 60 minutes, and each minute into 60 seconds. Thus the mean solar second is 1/86400 of a mean solar day, and this mean solar second is the unit in which short time intervals are measured and expressed.

III.

THE LAWS OF THE CONSERVATION OF MOMENTUM, OF THE CONSERVATION OF THE CENTER OF GRAVITY, AND OF THE CONSERVATION OF AREAS.

1. Although Newton's principles are fully adequate to deal with any mechanical problem that may arise, yet it is convenient to contrive for cases occurring more frequently some particular rules, which will enable us to treat problems of this kind by routine forms and to dispense with the minute discussion of them. Newton and his successors developed several such principles. Our first subject will be NEWTON's doctrines concerning *freely movable* material systems.

2. If two free masses m and m' are subjected in the direction of their line of junction to the action of forces that proceed from *other* masses, then, in the interval of time t, the velocities v, v' will be generated, and the equation $(p + p')t = mv + m'v'$ will subsist. This follows from the equations $pt = mv$ and $p't' = m'v'$. The sum $mv + m'v'$ is called the *momentum* of the system, and in its computation oppositely directed forces and velocities are regarded as having opposite signs. If, now, the masses m, m' in addition to being subjected to the action of the external forces p, p' are also acted upon by *internal* forces, that is by such as are mutually exerted by the masses on *one another,* these forces will, by Newton's third law, be equal and opposite, q, $- q$. The sum of the impressed impulses is, then, $(p + p' + q - q)t = (p + p')t$, the same as before; and, consequently, also, the total momentum of the system will be the same. The momentum of a system is thus determined exclusively by *external*

forces, that is, by forces which masses *outside* of the system exert on its parts.

Imagine a number of free masses m, m', m''. . . . distributed in any manner in space and acted on by external forces p, p', p''. . . . whose lines have any directions. These forces produce in the masses in the interval of time t the velocities v, v', v''. . . . Resolve all the forces in three directions x, y, z at right angles to each other, and do the same with the velocities. The sum of the impulses in the x-direction will be equal to the momentum generated in the x-direction; and so with the rest. If we imagine additionally in action between the masses m, m', m''. . . ., pairs of equal and opposite internal forces q, $-q$, r, $-r$, s, $-s$, etc., these forces, resolved, will also give in every direction pairs of equal and opposite components, and will consequently have no influence on the sum-total of the impulses. Once more the momentum is exclusively determined by external forces. The law which states this fact is called the *law of the conservation of momentum.*

3. Another form of the same principle, which Newton likewise discovered, is called the law of the *conservation of the center of gravity.* Imagine in A and B (Fig. 149) two masses, $2m$ and m, in mutual action, say that of electrical repulsion; their center of gravity is situated at S, where $BS = 2AS$. The accelerations they impart to each other are oppositely directed and in the inverse proportion of the masses. If, then, in consequence of the mutual action, $2m$ describes a distance AD, m will necessarily describe a distance $BC = 2AD$. The point S will still remain the position of the center of gravity, as $CS = 2DS$.

$$\overline{\underset{D}{}\quad\underset{A}{}\quad\underset{S}{}\quad\underset{B}{}\quad\underset{C}{}}$$

$2m$ m

Fig. 149.

Therefore, two masses cannot, by *mutual action,* displace their common center of gravity.

If our considerations involve *several* masses, distributed in any way in space, the same result will also be found to hold good for this case. For as *no two* of the masses can displace their center of gravity by mutual action, the center of gravity of the system as a whole cannot be displaced by the mutual action of its parts.

Imagine freely placed in space a system of masses m, m', m''. . . . acted on by *external* forces of any kind. We refer the forces to a system of rectangular coördinates and call the coördinates respectively x, y, z, x', y', z', and so forth. The coördinates of the center of gravity are then

$$\xi = \frac{\Sigma\, m\, x}{\Sigma\, m}, \; \eta = \frac{\Sigma\, m\, y}{\Sigma\, m}, \; \zeta = \frac{\Sigma\, m\, z}{\Sigma\, m},$$

in which expressions x, y, z may change either by uniform motion or by uniform acceleration or by any other law, according as the mass in question is acted on by no external force, by a constant external force, or by a variable external force. The center of gravity will have in all these cases a different motion, and in the first may even be at rest. If now *internal* forces, acting between every two masses, m' and m'', come into play in the system, opposite displacements w', w'' will thereby be produced in the direction of the lines of junction of the masses, such that, allowing for signs, $m'w' + m''w'' = 0$. Also with respect to the components x_1 and x_2 of these displacements the equation $m'x_1 + m''x_2 = 0$ will hold. The internal forces consequently produce in the expressions ξ, η, ζ, only such additions as mutually destroy each other. Consequently, the *motion of the center of gravity* of a system is determined by *external* forces only.

If we wish to know the *acceleration* of the center of gravity of the system, the accelerations of the system's *parts* must be similarly treated. If φ, φ', φ''. . . . denote the accelerations of m, m', m''. . . . in any direction, and φ the acceleration of the center of gravity in the same direction, $\varphi = \Sigma m \varphi / \Sigma m$, or putting the total mass $\Sigma m = M$, $\varphi = \Sigma m \varphi / M$. Accordingly, we obtain the acceleration of the center of gravity of a system in any direction by taking the sum of all the forces in that direction and dividing the result by the total mass. The center of gravity of a system moves exactly as if all the masses and all the forces of the system were concentrated at that center. Just as a single mass can acquire no acceleration without the action of some external force, so the center of gravity of a system can acquire no acceleration without the action of external forces.

4. A few examples may now be given in illustration of the principle of the conservation of the center of gravity.

Imagine an animal *free* in space. If the animal moves in one direction a portion m of its mass, the remainder of it M will be moved in the opposite direction, always so that its center of gravity retains its original position. If the animal draws back, the mass m, the motion of M also will be reversed. The animal is unable, without external supports or forces, to move itself from the spot which it occupies, or to alter motions impressed upon it from without.

A lightly running vehicle A is placed on rails and loaded with stones. A man stationed in the vehicle casts out the stones one after another, in the same direction. The vehicle, supposing the friction to be sufficiently slight, will at once be set in motion in the opposite direction. The center of gravity of the system as a whole (of the vehicle + the stones) will, so far as its motion is not destroyed by external obstacles, continue to remain in its original spot. If the same man would pick up the stones from without and place them in the vehicle, the vehicle in this case would also be set in motion; but not to the same extent as before, as the following example will render evident.

A projectile of mass m is thrown with a velocity v from a cannon of mass M. In the reaction, M also receives a velocity, V, such that, making allowance for the signs, $MV + mv = 0$. This explains the so-called recoil. The relation here is $V = -(m/M)v$; or, for equal velocities of flight, the recoil is less according as the mass of the cannon is greater than the mass of the projectile. If the work done by the powder be expressed by A, the *vires vivæ* will be determined by the equation $MV^2/2 + mv^2/2 = A$; and, the sum of the momenta being by the first-cited equation $= 0$, we readily obtain $V = \sqrt{2Am/M(M+m)}$. Consequently, neglecting the mass of the exploded powder, the recoil vanishes when the mass of the projectile vanishes. If the mass m were not expelled from the cannon but sucked into it, the recoil would take place in the opposite direction. But it would have no time to make itself visible since before any perceptible distance had been traversed, m would have reached the bottom of the bore. As soon, however, as M and m are in rigid connection with each other, as soon, that is, as they are *relatively* at rest to each other, they must be *absolutely* at rest, for the center of gravity of the system as a whole has no motion. For the same reason no considerable motion can take place when the stones in the preceding example are taken into the vehicle, because on the establishment of rigid connections between the vehicle and the stones the opposite momenta generated are destroyed. A cannon sucking in a projectile would experience a perceptible recoil only if the sucked in projectile could fly through it.

Imagine a locomotive freely suspended in the air, or, what will subserve the same purpose, at rest with insufficient friction on the rails. By the law of the conservation of the center of gravity, as soon as the heavy masses of iron in connection with the piston-rods begin to oscillate, the body of the locomotive will be set in oscillation in a contrary direction—a motion which may greatly disturb its uniform progress. To eliminate this oscillation, the motion of the masses of iron worked by the piston-rods must be so compensated for by the contrary motion of other masses that the center of gravity of the system as a whole will remain in one position. In this way no motion of the body of the locomotive will take place. This is done by affixing masses of iron to the driving-wheels.

The facts of this case may be very prettily shown by Page's electromotor (Fig. 150). When the iron core in the bobbin *AB* is projected by the internal forces acting between bobbin and core to the right, the body of the motor, supposing it to rest on lightly movable wheels *r r*, will move to the left. But if to a spoke of the fly-wheel *R* we affix an appropriate balance-weight *a,* which always moves in the contrary direction to the iron core, the sideward movement of the body of the motor may be made to vanish totally.

Of the motion of the fragments of a bursting bomb we know nothing. But it is plain, by the law of the conservation of the center of gravity, that, making al-

lowance for the resistance of the air and the obstacles
the individual parts may meet, the center of gravity of
the system will continue after the bursting to describe
the parabolic path of its original projection.

5. A law closely allied to the law of the center of
gravity, and similarly applicable to *free* systems, is the
principle of the conservation of areas. Although New-

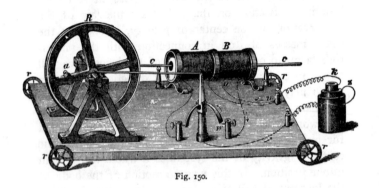

Fig. 150.

ton had, so to say, this principle within his very grasp, it
was nevertheless not enunciated until a long time afterwards by EULER, D'ARCY, and DANIEL BERNOULLI.
Euler and Daniel Bernoulli discovered the law almost
simultaneously (1746), on the occasion of treating a
problem proposed by Euler concerning the motion of
balls in rotatable tubes, being led to it by the consideration of the action and reaction of the balls and the
tubes. D'Arcy (1747) started from Newton's investigations, and generalized the law of sectors which the
latter had employed to explain Kepler's laws.

Two masses m, m' (Fig. 151) are in mutual action.
By virtue of this action the masses describe the distances AB, CD in the direction of their line of junction.

Allowing for the signs, then, $m \cdot AB + m' \cdot CD = 0$.
Drawing *radii vectores* to the moving masses from any

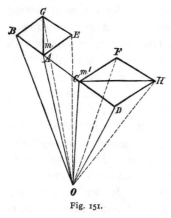

Fig. 151.

point O, and regarding the areas described in opposite senses by the radii as having opposite signs, we further obtain $m \cdot OAB + m' \cdot OCD = 0$. Which is to say, if two masses mutually act on each other, and *radii vectores* be drawn to these masses from any point, the sum of the areas described by the radii multiplied by the respective masses is $= 0$. If the masses are also acted on by external forces and as the effect of these the areas OAE and OCF are described, the joint action of the internal and external forces, during any very small period of time, will produce the areas OAG and OCH. But it follows from Varignon's theorem that

$$mOAG + m'OCH = mOAE + m'OCF +$$
$$mOAB + m'OCD = mOAE + m'OCF;$$

in other words, *the sum of the products of the areas so described into the respective masses which compose a system is unaltered by the action of internal forces.*

If we have several masses, the same thing may be asserted, for every two masses, of the projection on any given plane of the motion. If we draw radii from any point to the several masses, and project on any plane the areas the radii describe, the sum of the products of these areas into the respective masses will be independent of the action of internal forces. This is the *law of the conservation of areas.*

If a single mass not acted on by forces is moving uniformly forward in a straight line and we draw a radius vector to the mass from any point O, the area described by the radius increases proportionally to the time. The same law holds for $\Sigma\, m\, f$, in cases in which several masses not acted on by forces are moving, where we signify by the summation the algebraic sum of all the products of the areas (f) into the moving masses—a sum which we shall hereafter briefly refer to as the sum of the mass-areas. If *internal* forces come into play between the masses of the system, this relation will remain unaltered. It will still subsist, also, if external forces be applied whose lines of action pass through the *fixed* point O, as we know from the researches of Newton.

If the mass be acted on by an external force, the area f described by its radius vector will increase in time by the law $f = a\, t^2/2 + b\, t + c$, where a depends on the accelerative force, b on the initial velocity, and c on the initial position. The sum $\Sigma\, m\, f$ increases by the same law, where several masses are acted upon by external accelerative forces, provided these may be regarded as constant, which for sufficiently small intervals of time is always the case. The law of areas in this case states that the *internal* forces of the system have *no influence* on the increase of the sum of the mass-areas.

A free rigid body may be regarded as a system whose parts are maintained in their relative positions by internal forces. The law of areas is applicable therefore to this case also. A simple instance is afforded by the uniform rotation of a rigid body about an axis passing through its center of gravity. If we call m a portion of its mass, r the distance of the portion from the axis, and α its angular velocity, the sum of the mass-areas produced in unit of time will be Σm $(r/2)r\,\alpha = (\alpha/2)\Sigma m\,r^2$, or, the product of the moment of inertia of the system into half its angular velocity. This product can be altered only by external forces.

6. A few examples may now be cited in illustration of the law.

If two rigid bodies K and K' are connected, and K is brought by the action of internal forces into rotation relatively to K', immediately K' also will be set in rotation, in the opposite direction. The rotation of K generates a sum of mass-areas which, by the law, must be compensated for by the production of an equal, but opposite, sum by K'.

This is very prettily exhibited by the electromotor of Fig. 152. The fly-wheel of the motor is placed in a horizontal plane, and the motor thus attached to a vertical axis, on which it can freely turn. The wires conducting the current dip, in order to prevent their interference with the rotation, into two coaxial gutters of mercury fixed on the axis. The body of the motor (K') is tied by a thread to the stand supporting the axis and the current is turned on. As soon as the fly-wheel (K), viewed from above, begins to rotate in the direction of the hands of a watch, the string is drawn taut and the body of the motor exhibits the tendency to rotate *in the opposite direction*—a rotation which immediately takes place when the thread is burnt away.

The motor is, with respect to rotation about its axis, a free system. The sum of the mass-areas generated, for the case of rest, is $= 0$. But the *wheel* of the motor being set in rotation by the action of the internal electro-magnetic forces, a sum of mass-areas is

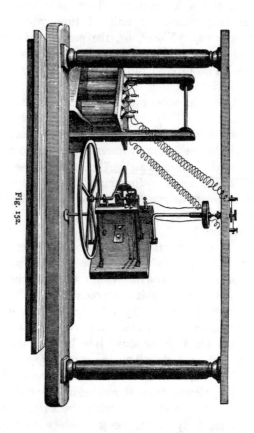

Fig. 152.

produced which, as the total sum must remain $= 0$, is compensated for by the rotation in the opposite direction of the body of the motor. If an index be attached to the body of the motor and kept in a fixed position by an elastic spring, the rotation of the body of the motor cannot take place. Yet every acceleration of the wheel in the direction of the hands of a watch (produced by a deeper immersion of the battery) causes the index to swerve in the opposite direction, and every retardation produces the contrary effect.

A beautiful but curious phenomenon presents itself when the current to the motor is interrupted. Wheel and motor continue at first their movements in opposite directions. But the effect of the friction of the axes soon becomes apparent and the parts gradually assume with respect to each other relative rest. The motion of the body of the motor is seen to diminish; for a moment it ceases; and, finally, when the state of relative rest is reached, it is reversed and assumes the direction of the original motion of the wheel. The *whole* motor now rotates in the direction the wheel did at the start. The explanation of the phenomenon is obvious. The motor is not a *perfectly* free system. It is impeded by the friction of the axes. In a perfectly free system the sum of the mass-areas, the moment the parts re-entered the state of relative rest, would again necessarily be = 0. But in the present instance, an external force is introduced—the friction of the axes. The friction on the axis of the wheel diminishes the mass-areas generated by the wheel and body of the motor alike. But the friction on the axis of the body of the motor only diminishes the sum of the mass-areas generated by the body. The wheel retains, thus, an excess of mass-area, which when the parts are relatively at rest is rendered apparent in the motion of the entire motor. The phenomenon subsequent to the interruption of the current supplies us with a model of what according to the hypothesis of astronomers has taken place on the moon. The tidal wave created by the earth has reduced to such an extent by friction the velocity of rotation of the moon that the lunar day has grown to a month. The fly-wheel represents the fluid mass moved by the tide.

Another example of this law is furnished by *reaction-wheels*. If air or gas be emitted from the wheel (Fig. 153*a*) in the direction of the short arrows, the whole wheel will be set in rotation in the direction of the large arrow. In Fig. 153*b*, another simple reaction-wheel is represented. A brass tube *r r* plugged at both ends and appropriately perforated, is placed on a second brass tube *R*, supplied with a thin steel pivot through which air can be blown; the air escapes at the apertures *O, O'*.

It might be supposed that sucking on the reaction-wheels would produce the opposite motion to that resulting from blowing. Yet this does not usually take place, and the reason is obvious. The air that is sucked into the spokes of the wheel must take part immediately in the motion of the wheel, must enter the condition of relative rest with respect to the wheel; and when the system is completely at rest, the sum of its mass-areas must be = 0. Generally, no perceptible rotation takes place on the sucking in of the air. The circumstances are similar to those of the recoil of a cannon which sucks in a projectile. If, therefore, an elastic ball, which has but one escape-tube, be attached to the reaction-wheel, in the manner represented in Fig. 153*a*, and be alternately squeezed so that the same quantity of air is by turns blown out and sucked in, the wheel will continue to revolve rapidly in the same direction as it did in the case in which we blew into it. This is partly due to the fact that the air

Fig. 153 a.

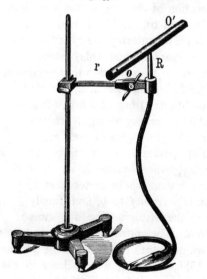

Fig. 153 b.

sucked into the spokes must participate in the motion of the latter and therefore can produce no reactional rotation, but it also results partly from the difference of the motion which the air outside the tube assumes in the two cases. In blowing, the air flows out in jets, and performs rotations. In sucking, the air comes in from all sides, and has no distinct rotation.

Fig. 154.

The correctness of this view is easily demonstrated. If we perforate the bottom of a hollow cylinder, a closed band-box for instance, and place the cylinder on the steel pivot of the tube R, after the side has been slit and bent in the manner indicated in Fig. 154, the box will turn in the direction of the long arrow when blown into and in the direction of the short arrow when sucked on. The air, here, on entering the cylinder, can continue its rotation *unimpeded*, and this motion is accordingly compensated for by a rotation in the opposite direction.

7. The following case also exhibits similar conditions. Imagine a tube (Fig. 155a) which, running straight from a to b, turns at right angles to itself at the latter point, passes to c, describes the circle $c\,d\,e\,f$, whose plane is at right angles to $a\,b$, and whose center is at b, then proceeds from f to g, and, finally continuing the straight line $a\,b$, runs

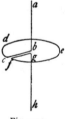

Fig. 155 a.

from g to h. The entire tube is free to turn on an axis $a\,h$. If we pour into this tube, in the manner indicated in Fig. 155b, a liquid, which flows in the direction $c\,d\,e\,f$, the tube will immediately begin to turn in the direction *f e d c*. This impulse, however, ceases, the moment the liquid reaches the point *f*, and flowing out into the radius *fg* is obliged to join in the motion of the latter. By the use of a constant stream of liquid,

Fig. 155 b.

therefore, the rotation of the tube may soon be stopped. But if the stream be interrupted, the fluid, in flowing off through the radius *f g*, will impart to the tube a motional impulse in the direction of its own motion, *c d e f*, and the tube will turn in this direction. All these phenomena are easily explained by the law of areas.

A. Schuster of Manchester has proved in a very beautiful way, in the London *Philosophical Transactions* for 1876 (vol. CLXVI, p. 715), that the forces which set the radiometer of Crookes and Geissler in motion are *inner* forces. If we put the vanes of the radiometer into rotation by means of light, after we have suspended the glass cover bifilarly, this cover immediately shows a tendency to rotate in a sense contrary to the vanes. Schuster was able to measure the magnitude of the forces which here came into action.

V. Dvořák of Agram, the discoverer of the acoustic reaction-wheel, has, at my request, carried out analogous experiments with his reaction-wheel. If we put the resonator-wheel into acoustical rotation, its light cylindrical glass cover, which floated on water, fell at once into rotation in the opposite sense, and this latter rotation, when the wheel only goes on rotating by inertia, also immediately reverses its sense of rotation. My son, Ludwig Mach, has, at my wish, improvised upon the experiment with Dvořák's wheel by replacing the glass cover by a light paraffined paper cover which floated on water. When such a paper cover was suspended bifilarly, every acceleration of the wheel showed an increased tendency to rotation in the opposite sense, and every retardation a diminished tendency of this kind; and this was shown in a very striking manner. Dvořák's experiments are explained by those with the motor represented in Fig. 152 and especially by the experiment of Fig. 153a (Cf. A. Haberditzl, "Über kontinuierliche akustische Rotation und deren Beziehung zum Flächenprinzip," *Sitzungsber. der Wiener Akademie, math.-naturwiss. Klasse,* May 9th, 1878.)

The trade-winds, the deviation of the oceanic currents and of rivers, Foucault's pendulum experiment, and the like, may also be treated as examples of the law of areas. Another pretty illustration is afforded by bodies with variable moments of inertia. Let a body with the moment of inertia Θ rotate with the angular velocity α and, during the motion, let its moment of inertia be transformed by internal forces, say by springs, into Θ', α will then pass into α', where $\alpha\Theta = \alpha'\Theta'$, that is $\alpha' = \alpha(\Theta/\Theta')$. On any considerable diminution of the moment of inertia, a great increase of angular velocity ensues. The principle might conceivably be employed, instead of Foucault's method, to demonstrate the rotation of the earth.

A phenomenon which substantially embodies the conditions last suggested is the following: According to Prof. Tumlirz, a glass funnel, with its axis placed in a vertical position, is rapidly filled with a liquid in such a manner that the stream does not enter in the direction of the axis but strikes the sides. A slow rotatory motion is thereby set up in the liquid which as long as the funnel is full, is not noticed. But when the fluid retreats into the neck of the funnel, its moment of inertia is so diminished and its angular velocity so increased that a violent eddy with considerable axial depression is created. Frequently the entire effluent jet is penetrated by an axial thread of air.

8. If we carefully examine the principles of the center of gravity and of the areas, we shall discover in

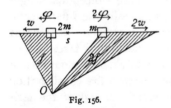

Fig. 156.

both simply convenient modes of expression, for practical purposes, of a well-known property of mechanical phenomena. To the acceleration φ of one mass m there always corresponds a contrary acceleration φ' of a second mass m', where allowing for the signs $m\varphi + m'\varphi' = 0$. To the force $m\varphi$ corresponds the equal and opposite force $m'\varphi'$. When any masses m and $2m$ describe with the contrary accelerations 2φ and φ the distances $2w$ and w (Fig. 156), the position of their center of gravity S remains unchanged, and the sum of their mass-areas with respect to any point O is, allowing for the signs, $2m \cdot f + m \cdot 2f = 0$. This simple exposition shows us, that the principle of the center of gravity expresses the same thing with respect to *parallel coördinates* that the principle of areas expresses with respect to *polar co-ördinates*. Both contain simply the fact of reaction.

The principles in question admit of still another simple construction. Just as a single body cannot, without the influence of external forces, that is, without the aid of a second body, alter its uniform motion of progression or rotation, so also a system of bodies cannot, without the aid of a second system, on which it can, so to speak, brace and support itself, alter what may properly and briefly be called its *mean* velocity of progression or rotation. Both principles contain, thus, a *generalized statement of the law of inertia*, the correctness of which in the present form we not only *see* but *feel*.

This feeling is not unscientific; much less is it detrimental. Where it does not replace conceptual insight but exists by the side of it, it is really the fundamental requisite and sole evidence of a *complete* mastery of mechanical facts. We are ourselves a fragment of mechanics, and this fact profoundly modifies our mental life. No one will convince us that the consideration of mechanico-physiological processes, and of the feelings and instincts here involved, must be excluded from scientific mechanics. If we know principles like those of the center of gravity and of areas only in their abstract mathematical form, without having dealt with the palpable simple facts, which are at once their application and their source, we only half comprehend them, and shall scarcely recognize actual phenomena as examples of the theory. We are in a position like that of a person who is suddenly placed on a high tower but has not previously travelled in the district round about, and who therefore does not know how to interpret the objects he sees.

IV.

THE LAWS OF IMPACT.

1. The laws of impact were the occasion of the enunciation of the most important principles of mechanics, and furnished also the first examples of the application of such principles. As early as 1639, a contemporary of Galileo, the Prague professor, MARCUS MARCI (born in 1595), published in his treatise *De Proportione Motus* (Prague) a few results of his investigations on impact. He knew that a body striking in elastic percussion another of the same size at rest, loses its own motion and communicates an equal quantity to the other. He also enunciates, though not always with the requisite precision, and frequently mingled with what is false, other propositions which still hold good. Marcus Marci was a remarkable man. He possessed for his time very creditable conceptions regarding the composition of motions and "impulses." In the formation of these ideas he pursued a method similar to that which Roberval later employed. He speaks of *partially* equal and opposite motions, and of *wholly* opposite motions, gives parallelogram constructions, and the like, but is unable, although he speaks of an accelerated motion of descent, to reach perfect clearness with regard to the idea of force and consequently also with regard to the composition of forces.

When we add to all this that Marci was on the very verge of anticipating Newton in the discovery of the

IOANNES MARCVS MARCI PHIL: & MEDIC: DOCTOR
et Professor natus Landscronæ Hermundurorum in Boemia
anno 1595. 13 Iunij.

composition of light, however, on account of his incomplete knowledge of the law of refraction here, also,

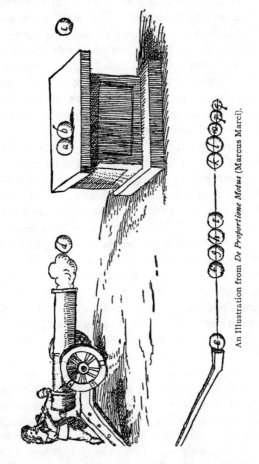

An Illustration from *De Proportione Motus* (Marcus Marci).

he could not reach his goal. According to Wohlwill's
researches (*Zeitschrift für Völkerpsychologie*, 1884,
xv, p. 387) Marci emphatically cannot be considered

as having advanced dynamics in the direction taken by Galileo.

2. GALILEO himself made several experimental attempts to ascertain the laws of impact; but he was not wholly successful in these endeavors. He busied himself principally with the force of a body in motion, or with the "force of percussion," as he expressed it, and endeavored to compare this force with the pressure of a weight at rest, hoping thus to measure it. To this end he instituted an extremely ingenious experiment, which we shall now describe.

A vessel I (Fig. 157) in whose base is a plugged orifice, is filled with water, and a second vessel II is hung beneath it by strings; the whole is fastened to the beam of an equilibrated balance. If the plug is removed from the orifice of vessel I, the fluid will fall in a jet into vessel II. A portion of the pressure due to the resting weight of the water in I is lost and replaced by an action of impact on vessel II. Galileo expected a depression of the whole scale, by which he hoped with the assistance of a counter-weight to determine the effect of the impact. He was to some extent surprised to obtain *no* depression, and he was unable, it appears, perfectly to clear up the matter in his mind.

3. Today, of course, the explanation is not difficult. By the removal of the plug there is produced, first a diminution of the pressure. This consists of two factors: (1) The weight of the jet suspended in the air is lost; and (2) A reaction-pressure upwards is exerted by the effluent jet on vessel I (which acts like a Segner's wheel). Then there is an increase of pressure (Factor 3) produced by the action of the jet on the bottom of vessel II. Before the first drop has reached

the bottom of II, we have only to deal with a diminution of pressure, which, when the apparatus is in full operation, is immediately compensated for. This *initial* depression was, in fact, all that Galileo could observe. Let us imagine the apparatus in operation, and denote the height the fluid reaches in vessel I by h, the corresponding velocity of afflux by v, the distance of the bottom of I from the surface of the fluid in II by k, the

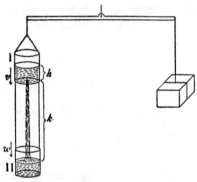

Fig. 157.

velocity of the jet at this surface by w, the area of the basal orifice by a, the acceleration of gravity by g, and the specific gravity of the fluid by s. To determine Factor (1) we may observe that v is the velocity acquired in descent through the distance h. We have, then, simply to picture to ourselves this motion of descent continued through k. The time of descent of the jet from I to II is therefore the time of the descent through $h + k$ less the time of descent through h. During this time a cylinder of base a is discharged with the velocity v. Factor (1), or the weight of the jet suspended in the air, accordingly amounts to

$$\sqrt{2gh}\left[\sqrt{\frac{2(h+k)}{g}} - \sqrt{\frac{2h}{g}}\right] as.$$

To determine Factor (2) we employ the familiar equation $mv = pt$. If we put $t = 1$, then $mv = p$, that is the pressure of reaction upwards on I is equal to the momentum imparted to the fluid jet in unit of time. We will select here the unit of weight as our unit of force, that is, use gravitation measure. We obtain for Factor (2) the expression $[av(s/g)]v = p$, (where the expression in brackets denotes the mass which flows out in unit of time), or

$$a\sqrt{2gh} \cdot \frac{s}{g} \cdot \sqrt{2gh} = 2ahs.$$

Similarly we find the pressure on II to be

$$\left(av \cdot \frac{s}{g}\right)w = q, \text{ or factor 3}:$$

$$a\frac{s}{g}\sqrt{2gh}\sqrt{2g(h+k)}.$$

The total variation of the pressure is accordingly

$$-\sqrt{2gh}\left[\sqrt{\frac{2(h+k)}{g}} - \sqrt{\frac{2h}{g}}\right] as$$
$$-2ahs$$

$$+\frac{as}{g}\sqrt{2gh}\sqrt{2g(h+k)}$$

or, abridged,

$$-2as[\sqrt{h(h+k)} - h] - 2ahs$$
$$+2as\sqrt{h(h+k)},$$

which three factors *completely* destroy each other. In the very necessity of the case, therefore, Galileo could only have obtained a negative result.

We must supply a brief comment respecting Factor

(2). It might be supposed that the pressure on the basal orifice which is lost, is *a h s* and not *2 a h s*. But this *statical* conception would be totally inadmissible in the present, *dynamical* case. The velocity *v* is not generated by gravity instantaneously in the effluent particles, but is the outcome of the mutual pressure between the particles flowing out and the particles left behind; and pressure can only be determined by the momentum generated. The erroneous introduction of the value *a h s* would at once betray itself by self-contradictions.

If Galileo's mode of experimentation had been less elegant, he would have determined without much difficulty the pressure which a *continuous* fluid jet exerts. But he could never, as he soon became convinced, have counteracted by a *pressure* the effect of an instantaneous *impact*. Take—and this is the supposition of Galileo—a freely falling, heavy body. Its final velocity, we know, increases proportionately to the time. The very smallest velocity requires a definite *portion of time* to be produced in (a principle which even Mariotte contested). If we picture to ourselves a body moving vertically upwards with a definite velocity, the body will, according to the amount of this velocity, ascend a definite time, and consequently also a definite distance. The heaviest imaginable body impressed in the vertical upward direction with the smallest imaginable velocity will ascend, be it only a little, in opposition to the force of gravity. If, therefore, a heavy body, be it ever so heavy, receives an instantaneous upward impact from a body in motion, be the mass and velocity of that body ever so small, and such impact imparts to the heavier body the smallest imaginable velocity, that body will, neverthless, yield and move somewhat in the upward direction. The *slightest* impact, therefore, is able to overcome the *greatest* pressure; or, as Galileo says, the force of percussion compared with the force of pressure is *infinitely* great. This result, which is sometimes attributed to intellectual obscurity on Galileo's part, is, on the contrary, a brilliant proof of his intellectual acumen. We should say today, that the force of percussion, the momentum, the impulse, the quantity of motion $m\,v$, is a quantity of different *dimensions* from the pressure p. The dimensions of the former are $m\,l\,t^{-1}$, those of the latter $m\,l\,t^{-2}$. In reality, therefore, pressure is related to momentum of impact as a line is to a surface. Pressure is p, the momentum of impact is $p\,t$. Without employing mathematical terminology it is hardly possible to express the fact better than Galileo did. We now also see why it is possible to measure the impact of a continuous fluid jet by a pressure. We compare the momentum destroyed per second of time with the pressure acting per second of time, that is, homogeneous quantities of the form $p\,t$.

4. The first systematic treatment of the laws of impact was evoked in the year 1668 by a request of the Royal Society of London. Three eminent physicists WALLIS (Nov. 26, 1668), WREN (Dec. 17, 1668), and HUYGENS (Jan. 4, 1669) complied with the invitation of the society, and communicated to it papers in which, independently of each other, they stated, without deductions, the laws of impact. Wallis treated only of the impact of inelastic bodies, Wren and Huygens only of the impact of elastic bodies. Wren, previously to publication, had tested his theorems by experiments, which, in the main, agreed with those of Huygens. These are the experiments to which Newton refers in the *Principia*. The same experiments were, soon after

this, also described, in a more developed form, by Mariotte, in a special treatise, *Sur le Choc des Corps.* Mariotte also described the apparatus now known in physical collections as the percussion-machine.

According to Wallis, the decisive factor in impact is *momentum,* or the product of the mass (*pondus*) into the velocity (*celeritas*). By this momentum the force of percussion is determined. If two inelastic bodies which have equal momenta strike each other, rest will ensue after impact. If their momenta are unequal, the difference of the momenta will be the momentum after impact. If we divide this momentum by the sum of the masses, we shall obtain the velocity of the motion after the impact. Wallis subsequently presented his theory of impact in another treatise, *Mechanica sive de Motu,* London, 1671. All his theorems may be brought together in the formula now in common use, $u = (mv + m'v')/(m + m')$, in which m, m' denote the masses, v, v' the velocities before impact, and u the velocity after impact.

5. The ideas which led Huygens to his results, are to be found in a posthumous treatise of his, *De Motu Corporum ex Percussione,* 1703. We shall examine these in some detail. The assumptions from which Huygens proceeds are: (1) the law of inertia; (2) that elastic bodies of equal mass, colliding with equal and opposite velocities, separate after impact with the same velocities; (3) that all velocities are relatively estimated; (4) that a larger body striking a smaller one at rest imparts to the latter velocity, and loses a part of its own; and finally (5) that when *one* of the colliding bodies preserves its velocity, this also is the case with the *other.*

Huygens, now, imagines two equal elastic masses,

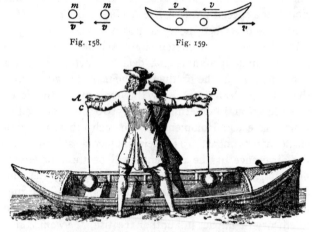

An Illustration from *De Percussione* (Huygens).

which meet with equal and opposite velocities *v*. After the impact they rebound from each other with exactly the same velocities. Huygens is right in *assuming* and not *deducing* this. That elastic bodies exist which recover their form after impact, that in such a transaction no perceptible *vis viva* is lost, are facts which experience alone can teach us. Huygens, now, conceives the occurrence just described, to take place on a boat which is moving with the velocity *v*. For the spectator in the boat the previous case still subsists; but for the spectator on the shore the velocities of the spheres before impact are respectively 2 *v* and 0, and after impact 0 and 2 *v*. An elastic body, therefore, impinging on another of equal mass at rest, communicates to the latter its entire velocity and remains after the impact itself at rest. If we suppose the boat affected with any imaginable velocity, *u*, then for the spectator on the shore the velocities before impact will be respectively

$u + v$ and $u - v$, and after impact $u - v$ and $u + v$. But since $u + v$ and $u - v$ may have *any* values whatsoever, it may be asserted as a principle that equal elastic masses *exchange* in impact their velocities.

A body at rest, however great, is set in motion by a body which strikes it, however small as Galileo pointed out. Huygens, now shows, that the *approach* of the bodies before impact and their recession after impact take place with the *same relative* velocity. A body m impinges

Fig. 160.

on a body of mass M at rest, to which it imparts in impact the velocity, as yet undetermined, w. Huygens, in the demonstration of this proposition, supposes that the event takes place on a boat moving from M towards m with the velocity $w/2$. The initial velocities are, then, $v - w/2$ and $- w/2$; and the final velocities, x and $+ w/2$. But as M has not altered the value, but only the sign, of its velocity, so m, if a loss of *vis viva* is not to be sustained in elastic impact, can only alter the sign of its velocity. Hence, the final velocities are $-(v - w/2)$ and $+ w/2$. As a fact, then, the relative velocity of approach before impact is equal to the relative velocity of separation after impact. Whatever change of velocity a body may suffer, in every case, we can, by the fiction of a boat in motion, and apart from the algebraical signs, keep the value of the velocity the same before and after impact. The proposition holds, therefore, generally.

If two masses M and m collide, with velocities V and v inversely proportional to the masses, M after impact will rebound with the velocity V and m with the velocity v. Let us suppose that the velocities after impact are V_1 and v_1; then by the preceding proposition we must have $V + v = V_1 + v_1$, and by the principle of *vis viva*

$$\frac{MV^2}{2} + \frac{mv^2}{2} = \frac{MV_1^2}{2} + \frac{mv_1^2}{2}.$$

Let us assume, now, that $v_1 = v + w$; then, necessarily, $V_1 = V - w$; but on this supposition

$$\frac{MV_1^2}{2} + \frac{mv_1^2}{2} = \frac{MV^2}{2} + \frac{mv^2}{2} + (M + m)\frac{w^2}{2}.$$

And this equality can, in the conditions of the case, only subsist if $w = 0$; wherewith the proposition above stated is established.

Huygens demonstrates this by a comparison, constructively reached, of the possible heights of ascent of the bodies prior and subsequently to impact. If the velocities of the impinging bodies are not inversely proportional to the masses, they may be made such by the fiction of a boat in motion. The proposition thus includes all imaginable cases.

The conservation of *vis viva* in impact is asserted by Huygens in one of the last theorems (11), which he subsequently also handed in to the London Society. But the principle is unmistakably at the foundation of the previous theorems.

6. In taking up the study of any event or phenomenon A, we may acquire a knowledge of its component elements by approaching it from the point of view of a different phenomenon B, which we already know; in which case our investigation of A will appear as the application of principles before familiar to us. Or, we may begin our investigation with A itself, and, as nature is throughout uniform, reach the same principles originally in the contemplation of A. The investigation of the phenomena of impact was pursued simul-

taneously with that of various other mechanical processes, and both modes of analysis were really presented to the inquirer.

To begin with, we may convince ourselves that the problems of impact can be disposed of by the Newtonian principles, with the help of only a minimum of *new* experiences. The investigation of the laws of impact contributed, it is true, to the discovery of Newton's laws, but the latter do not rest solely on this foundation. The requisite new experiences, not contained in the Newtonian principles, are simply the information that there are *elastic* and *inelastic bodies.* Inelastic bodies subjected to pressure alter their form without recovering it; elastic bodies possess for all their *forms* definite systems of pressures, so that every alteration of form is associated with an alteration of pressure, and *vice versa.* Elastic bodies recover their form; and the forces that induce the form-alterations of bodies do not come into play until the bodies are in contact.

Let us consider two inelastic masses M and m moving respectively with the velocities V and v. If these masses come in contact while possessed of these unequal velocities, internal form-altering forces will be set up in the system M, m. These forces do not alter the quantity of motion of the system, neither do they displace its center of gravity. With the restitution of equal velocities, the form-alterations cease and in inelastic bodies the forces which produce the alterations vanish. Calling the common velocity of motion after impact u, it follows that $M u + m u = M V + m v$, or $u = (MV + mv)/(M + m)$, the rule of Wallis.

Now let us assume that we are investigating the phenomena of impact without a previous knowledge of Newton's principles. We very soon discover, when we so proceed, that velocity is not the *sole* determinative factor of impact; still another physical quality (weight, load, mass, *pondus, moles, massa*) is decisive. The moment we have noted this fact, the simplest case is easily dealt with. If two bodies of equal weight or equal mass collide with equal and opposite velocities (Fig. 161); if, further, the bodies do not separate after impact but retain some common velocity, plainly the sole *uniquely* determined

Fig. 161.

velocity after the collision is the velocity 0. If, further, we make the observation that only the *difference* of the velocities, that is only relative velocity, determines the phenomenon of impact, we shall, by imagining the environment to move, (which experience tells us has no influence on the occurrence,) very readily perceive additional cases. For equal inelastic masses with velocities

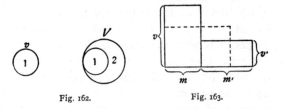

Fig. 162.   Fig. 163.

v and 0 or v and v' the velocity after impact is $v/2$ or $(v + v')/2$. It stands to reason that we can pursue such a line of reflection only after experience has informed us *what* the essential and decisive features of the phenomena are.

If we pass to unequal masses, we must not only know from experience that mass *generally* is of consequence, but also *in what manner* its influence is effective. If, for example, two bodies of masses 1 and 3 with the velocities v and V collide, we might reason thus: We cut out of the mass 3 the mass 1 (Fig. 162), and first make the masses 1 and 1 collide: the resultant velocity is $(v + V)/2$. There are now left, to equalize the velocities $(v + V)/2$ and V, the masses $1 + 1 = 2$ and 2, which applying the same principle gives

$$\frac{\dfrac{v + V}{2} + V}{2} = \frac{v + 3V}{4} = \frac{v + 3V}{1 + 3}.$$

Let us now consider, more generally, the masses m and m', which we represent in Fig. 163 as suitably proportioned horizontal lines. These masses are affected with the velocities v and v', which we represent by ordinates erected on the mass-lines. Assuming that $m < m'$, we cut off from m' a portion m. The offsetting of m and m gives the mass $2m$ with the velocity $(v + v')/2$. The dotted line indicates this relation. We proceed similarly with the remainder $m' - m$. We cut off from $2m$ a portion $m' - m$, and obtain the mass $2m - (m' - m)$ with the velocity $(v + v')/2$ and the mass $2(m' - m)$ with the velocity $[(v + v')/2 + v']/2$. In this manner we may proceed till we have obtained for the whole mass $m + m'$ the *same* velocity u. The constructive method indicated in the figure shows very plainly that here the surface equation $(m + m')u = mv + m'v'$ subsists. We readily perceive, however, that we cannot pursue this line of reasoning except the sum $mv + m'v'$, that is the *form* of the influence of m and v has through some experience or other been previously suggested to us as the determinative and decisive factor. If we renounce the use of the Newtonian principles, then some other specific experiences concerning the import of mv which are equivalent to those principles, are indispensable.

7. The impact of *elastic* masses may also be treated by the Newtonian principles. The sole observation here required is, that a deformation of elastic bodies calls into play *forces of restitution,* which directly depend on the deformation. Furthermore, bodies possess impenetrability; that is to say, when bodies affected with unequal velocities meet in impact, forces which equalize these velocities are produced. If two elastic masses M, m with the velocities C, c collide, a deformation will be effected, and this deformation will not cease until the velocities of the two bodies are equalized. At this instant, inasmuch as only internal forces are involved and therefore the momentum and the motion of the center of gravity of the system remain unchanged, the common equalized velocity will be

$$u = \frac{MC + mc}{M + m}.$$

Consequently, up to this time, M's velocity has suffered a diminution $C - u$; and m's an increase $u - c$.

But elastic bodies being bodies that recover their forms, in *perfectly* elastic bodies the very same forces that produced the deformation, will, only in the inverse order, *again* be brought into play, through the very same elements of time and space. Consequently, on the supposition that m is overtaken by M, M will a second time sustain a diminution of velocity $C - u$, and m will a second time receive an increase of velocity $u - c$. Hence, we obtain for the velocities V, v after impact the expressions $V = 2u - C$ and $v = 2u - c$, or

$$V = \frac{MC + m(2c - C)}{M + m}, \; v = \frac{mc + M(2C - c)}{M + m}.$$

If in these formulæ we put $M = m$, it will follow that $V = c$ and $v = C$; or, if the impinging masses are equal, the velocities which they have will be interchanged. Again, since in the particular case $M/m = -c/C$ or $MC + mc = 0$ also $u = 0$, it follows that $V = 2u - C = -C$ and $v = 2u - c = -c$; that is the masses recede from each other in this case with the same velocities (only oppositely directed) with which they approached. The approach of any two masses M, m affected with the velocities C, c, estimated as positive when in the same direction, takes place with the velocity $C - c$; their separation with the velocity $V - v$. But it follows at once from $V = 2u - C$, $v = 2u - c$, that $V - v = -(C - c)$; that is, the relative velocity of approach and recession is the same. By the use of the expressions $V = 2u - C$ and $v = 2u - c$, we also very readily find the two theorems

$$MV + mv = MC + mc \text{ and}$$

$$MV^2 + mv^2 = MC^2 + mc^2,$$

which assert that the quantity of motion before and after impact, estimated in the same direction, is the same, and that also the *vis viva* of the system before and after impact is the same. We have reached, thus, by the use of the Newtonian principles, all of Huygens's results.

8. If we consider the laws of impact from Huygens's point of view, the following reflections immediately claim our attention. The height of ascent which the center of gravity of any system of masses can reach is given by its *vis viva*, $\frac{1}{2} \Sigma m v^2$. In every case in which work is done by forces, and in such cases the masses follow the forces, this sum is increased by an amount equal to the work done. On the other hand, in every

case in which the system moves in opposition to forces, that is, when work, as we may say, is *done upon* the system, this sum is diminished by the amount of work done. As long, therefore, as the algebraical sum of the work done *on* the system and the work done *by* the system is not changed, whatever other alterations may take place, the sum $\frac{1}{2} \Sigma m v^2$ also remains unchanged. Huygens now, observing that this first property of material systems, discovered by him in his investigations on the pendulum, also obtained in the case of impact, could not help remarking that also the sum of the *vires vivæ* must be the same before and after impact. For in the mutually effected alteration of the forms of the colliding bodies the material system considered has the same amount of work *done on* it as, on the reversal of the alterations, is *done by* it, provided always the bodies develop forces wholly determined by the shapes they assume, and that they regain their original form by means of the same forces employed to effect its alteration. That the latter process takes place, *definite experience* alone can inform us. This law obtains, furthermore, only in the case of so-called *perfectly* elastic bodies.

Contemplated from this point of view, the majority of the Huygenian laws of impact follow at once. Equal masses, which strike each other with equal but opposite velocities, rebound with the same velocities. The velocities are *uniquely* determined only when they are *equal*, and they conform to the principle of *vis viva* only by being the *same* before and after impact. Further it is evident, that if one of the unequal masses in impact changes only the sign and not the magnitude of its velocity, this must also be the case with the other. On this supposition, however, the relative velocity of separation after impact is the same as the velocity of approach before impact. Every imaginable case can be reduced to this one. Let c and c' be the velocities of the mass m before and after impact, and let them be of any value and have any sign. We imagine the *whole* system to receive a velocity u of such magnitude that $u + c = - (u + c')$ or $u = (c - c')/2$. It will be seen thus that it is always possible to discover a velocity of transportation for the system such that the velocity of one of the masses will only change its sign. And so the proposition concerning the velocities of approach and recession holds generally good.

As Huygens's peculiar group of ideas was not fully perfected, he was compelled, in cases in which the velocity-ratios of the impinging masses were not originally known, to draw on the Galileo-Newtonian system for certain conceptions, as was pointed out above. Such an appropriation of the concepts mass and momentum, is contained, although not explicitly expressed, in the proposition according to which the velocity of each impinging mass simply changes its sign when before impact $M/m = - c/C$.

If Huygens had restricted himself wholly to his own point of view, he would scarcely have *discovered* this proposition, although, once discovered, he was able, after his own fashion, to supply its *deduction*. Here, owing to the fact that the momenta produced are equal and opposite, the equalized velocity of the masses on the completion of the change of form will be $u = 0$. When the alteration of form is reversed, and the same amount of work is performed that the system originally suffered, the *same* velocities with *opposite* signs will be *restored*.

If we imagine the entire system affected with a velocity of *translation,* this *particular* case will simultaneously present the *general* case. Let the impinging masses be represented in the figure by $M = BC$ and $m = AC$ (Fig. 164), and their respective velocities by $C = AD$ and $c = BE$. On AB erect the perpendicular CF, and through F draw IK parallel to AB. Then $ID = (m \cdot \overline{C - c})/(M + m)$ and $KE = (M \cdot \overline{C - c})/(M + m)$. On the supposition now

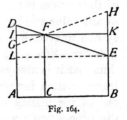

Fig. 164.

that we make the masses M and m collide with the velocities ID and KE, while we simultaneously impart to the system as a whole the velocity

$$u = AI = KB = C - (m \cdot \overline{C - c})/(M + m) =$$
$$c + (M \cdot \overline{C - c})/(M + m) = (MC + mc)/(M + m),$$

the spectator who is moving forwards with the velocity u will see the particular case presented, and the spectator who is at rest will see the general case, be the velocities what they may. The general formulæ of impact, above deduced, follow at once from this conception. We obtain:

$$V = AG = C - 2\frac{m(C - c)}{M + m} = \frac{MC + m(2c - C)}{M + m}$$
$$v = BH = c + 2\frac{M(C - c)}{M + m} = \frac{mc + M(2C - c)}{M + m}.$$

Huygen's successful employment of the fictitious motions is the outcome of the simple perceptions that bodies not affected with *differences* of velocities do not act on one another in impact. All forces of impact are determined by differences of velocity (as all thermal effects are determined by differences of temperature). And since forces generally determine, not velocities, but only changes of velocities, or, again, differences of velocities, consequently, in every aspect of impact the sole decisive factor is *differences* of velocity. With respect to which bodies the velocities are estimated, is indifferent. In fact, many cases of impact which from lack of practice appear to us as different cases, turn out on close examination to be one and the same.

Similarly, the capacity of a moving body for work, whether we measure it with respect to the time of its action by its momentum or with respect to the distance through which it acts by its *vis viva,* has no significance referred to a single body. It is invested with such, only when a second body is introduced, and, in the first case, then, it is the difference of the velocities, and in the second the square of the difference that is decisive. *Velocity* is a physical *level,* like temperature, potential function, and the like.

It remains to be remarked, that Huygens could have reached, originally, in the investigation of the phenomena of impact, the same results that he previously reached by his investigations of the pendulum. In every case there is one thing and one thing only to be done, and that is, *to discover in all the facts the same elements,* or, if we will, to *re*discover in one fact the elements of another which we already know. From which facts the investigation starts, is, however, a matter of historical accident.

9. Let us close our examination of this part of the subject with a few general remarks. The sum of the *momenta* of a system of moving bodies is preserved in impact, both in the case of inelastic and elastic bodies. But this preservation does not take place *precisely* in

the sense of Descartes. The momentum of a body is not diminished in proportion as that of another is increased; a fact which Huygens was the first to note. If, for example, two equal inelastic masses, possessed of equal and opposite velocities, meet in impact, the two bodies lose in the Cartesian sense their entire momentum. If, however, we reckon all velocities *in a given direction* as positive, and all in the opposite as negative, the sum of the momenta *is* preserved. Quantity of motion, conceived in this sense, is always preserved.

The *vis viva* of a system of inelastic masses is altered in impact; that of a system of perfectly elastic masses is preserved. The diminution of *vis viva* produced in the impact of inelastic masses, or produced generally when the impinging bodies move with a common velocity, after impact, is easily determined. Let M, m be the masses, C, c their respective velocities before impact, and u their common velocity after impact; then the loss of *vis viva* is

$$\tfrac{1}{2}MC^2 + \tfrac{1}{2}m\,c^2 - \tfrac{1}{2}(M+m)u^2, \quad \ldots \ldots (1)$$

which in view of the fact that $u = (MC + m\,c)/(M+m)$ may be expressed in the form $\tfrac{1}{2}(Mm/\overline{M+m})(C-c)^2$. Carnot has put this loss in the form

$$\tfrac{1}{2}M(C-u)^2 + \tfrac{1}{2}m(u-c)^2 \quad \ldots \ldots \ldots (2)$$

If we select the latter form, the expressions $\tfrac{1}{2}M(C-u)^2$ and $\tfrac{1}{2}m(u-c)^2$ will be recognized as the *vis viva* generated by the *work of the internal forces*. The loss of *vis viva* in impact is equivalent, therefore, to the work done by the internal or so-called molecular forces. If we equate the two expressions (1) and (2), remembering that $(M + m)u = MC + m\,c,$ we shall obtain an identical equation. Carnot's expression is important for the estimation of losses due to the impact of parts of machines.

In all the preceding expositions we have treated the impinging masses as points which moved only in the direction of the lines joining them. This simplification is admissible when the centers of gravity and the point of contact of the impinging masses lie in one straight line, that is, in the case of so-called direct impact. The investigation of what is called *oblique* impact is somewhat more complicated, but presents no especial interest in point of principle.

A question of a different character was treated by WALLIS. If a body rotates about an axis and its motion is suddenly checked by the retention of one of its points, the force of the percussion will vary with the position (the distance from the axis) of the point arrested. The point at which the intensity of the impact is greatest is called by Wallis the *center of percussion*. If this point be checked, the axis will sustain no pressure. We have no occasion here to enter in detail into these investigations; they were extended and developed by Wallis's contemporaries and successors in many ways.

10. We will now briefly examine, before concluding this section, an interesting application of the laws of impact; namely, the determination of the velocities of projectiles by the *ballistic pendulum*. A mass M is suspended by a weightless and massless string (Fig. 165), so as to oscillate as a pendulum. While in the position of equilibrium it suddenly receives the horizontal velocity V. It ascends by virtue of this velocity to an altitude $h = l\,(1 - \cos \alpha) = V^2/2g$, where l denotes

the length of the pendulum, α the angle of elongation, and g the acceleration of gravity. As the relation

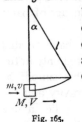

Fig. 165.

$T = \pi \sqrt{l/g}$ subsists between the time of oscillation T and the quantities l, g, we easily obtain $V = (gT/\pi) \sqrt{2(1-\cos \alpha)}$, and by the use of a familiar trigonometrical formula, also

$$V = \frac{2}{\pi} gT \sin \frac{\alpha}{2}.$$

If now the velocity V is produced by a projectile of the mass m which being hurled with a velocity v and sinking in M is arrested in its progress, so that whether the impact is elastic or inelastic, in any case the two masses acquire after impact the *common* velocity V, it follows that $m v = (M + m)V$; or, if m be sufficiently small compared with M, also $v = (M/m)V$; whence finally

$$v = \frac{2}{\pi} \cdot \frac{M}{m} gT \sin \frac{\alpha}{2}.$$

If it is not permissible to regard the ballistic pendulum as a simple pendulum, our reasoning, in conformity with principles before employed, will take the following shape. The projectile m with the velocity v has the momentum $m v$, which is diminished by the pressure p due to impact in a very short interval of time τ to $m V$. Here, then, $m(v - V) = p \tau$. With Poncelet, we reject the assumption of anything like *instantaneous* forces, which generate *instanter* velocities. There are no instantaneous forces. What has been called such are very great forces that produce perceptible velocities in very short intervals of time, but which in other respects do not differ from forces that act continuously. If the force active in impact cannot be regarded as constant during its entire period of action, we have only to put in the place of the expression $p\,\tau$ the expression $\int p\,d\,t$. In other respects the reasoning is the same.

A force equal to that which destroys the momentum of the projectile, acts in reaction on the pendulum. If we take the line of projection of the shot, and consequently also the line of the force, perpendicular to the axis of the pendulum and at the distance b from it, the moment of this force will be $b\,p$, the angular acceleration generated $b\,p/\Sigma\,m\,r^2$, and the angular velocity produced in time τ

$$\varphi = \frac{b \cdot p\,\tau}{\Sigma\,m\,r^2} = \frac{b\,m\,v}{\Sigma\,m\,r^2}.$$

The *vis viva* which the pendulum has at the end of time τ is therefore

$$\tfrac{1}{2}\varphi^2 \Sigma\,m\,r^2 = \tfrac{1}{2}\frac{b^2\,m^2\,v^2}{\Sigma\,m\,r^2}.$$

By virtue of this *vis viva* the pendulum performs the excursion a, and its weight $M\,g$, (a being the distance of the center of gravity from the axis,) is lifted the distance $a(1 - \cos a)$. The work performed here is $M\,g\,a(1 - \cos a)$, which is equal to the above-mentioned *vis viva*. Equating the two expressions we readily obtain

$$v = \frac{\sqrt{2\,M\,g\,a\,\Sigma\,m\,r^2(1 - \cos \alpha)}}{m\,b};$$

and remembering that the time of oscillation is

$$T = \pi \sqrt{\frac{\Sigma\,m\,r^2}{M\,g\,a}},$$

and employing the trigonometrical reduction which was resorted to immediately above, also

$$v = \frac{2}{\pi} \frac{M}{m} \frac{a}{b} \, gT . \sin\frac{\alpha}{2}.$$

This formula is in every respect similar to that obtained for the simple case. The observations requisite for the determination of v, are the mass of the pendulum and the mass of the projectile, the distances of the center of gravity and point of percussion from the axis, and the time and extent of oscillation. The formula also clearly exhibits the dimensions of a velocity. The expressions $2/\pi$ and $\sin (a/2)$ are simple numbers, as are also M/m and a/b, where both numerators and denominators are expressed in units of the same kind. But the factor $g\,T$ has the dimensions $l\,t^{-1}$, and is consequently a velocity. The ballistic pendulum was invented by ROBINS and described by him at length in a treatise entitled *New Principles of Gunnery*, published in 1742.

v.

D'ALEMBERT'S PRINCIPLE.

1. One of the most important principles for the rapid and convenient solution of the problems of mechanics is the *principle of D'Alembert*. The researches concerning the center of oscillation on which almost all prominent contemporaries and successors of Huygens had employed themselves, led directly to a series of simple observations which D'ALEMBERT ultimately generalized and embodied in the principle which goes by his name. We will first cast a glance at these preliminary performances. They were almost without exception evoked by the desire to replace the deduction of

Huygens, which did not appear sufficiently obvious, by one that was more *convincing*. Although this desire was founded, as we have already seen, on a miscomprehension due to historical circumstances, we have, of course, no occasion to regret the new points of view which were thus reached.

2. The first in importance of the founders of the theory of the center of oscillation, after Huygens, is JAMES BERNOULLI, who sought as early as 1686 to explain the compound pendulum by the lever. He arrived, however, at results which not only were obscure but also were at variance with the conceptions of Huygens. The errors of Bernoulli were animadverted on by the Marquis de L'HOPITAL in the *Journal de Rotterdam*, in 1690. The consideration of velocities acquired in *infinitely small* intervals of time in place of velocities acquired in *finite* times—a consideration which the last-named mathematician suggested—led to the removal of the main difficulties that beset this problem; and in 1691, in the *Acta Eruditorum,* and, later, in 1703, in the *Proceedings of the Paris Academy* James Bernoulli corrected his error and presented his results in a final and complete form. We shall here reproduce the essential points of his final deduction.

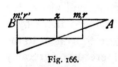

Fig. 166.

A horizontal, massless bar *AB* (Fig. 166) is free to rotate about *A*; and at the distances r, r' from *A* the masses m, m' are attached. The accelerations with which these masses *as thus connected* will fall must be different from the accelerations which they would assume if their connections were severed and they fell freely. There will be one point and one only, at the distance x, as yet unknown, from *A* which will fall with the same acceleration as it would have if it were free, that is, with the acceleration g. This point is termed the center of oscillation.

If m and m' were to be attracted to the earth, not proportionally to their masses, but m so as to fall when free with the acceleration $\varphi = g\,r/x$ and m' with the acceleration $\varphi' = g\,r'/x$, that is to say, if the *natural* accelerations of the masses were proportional to their distances from A, these masses would not interfere with one another when connected. In reality, however, m sustains, in consequence of the connection, an upward component acceleration $g - \varphi$, and m' receives in virtue of the same fact a downward component acceleration $\varphi' - g$; that is to say, the former suffers an upward force of $m(g - \varphi) = g\overline{(x - r/x)}m$ and the latter a downward force of $m'(\varphi' - g) = g\overline{(r' - x/x)}m'$.

Since, however, the masses exert what influence they have on each other solely through the medium of the lever by which they are joined, the upward force upon the one and the downward force upon the other must satisfy the law of the lever. If m in consequence of its being connected with the lever is held back by a force f from the motion which it would take, if free, it will also exert the same force f on the lever-arm r by reaction. It is this reaction pull alone that can be transferred to m' and be balanced there by a pressure $f' = (r/r')f$, and is therefore equivalent to the latter pressure. There subsists, therefore, agreeably to what has been above said, the relation $g\overline{(r' - x/x)}$ $m' = r/r' \cdot g\overline{(x - r/x)}m$ or, $(x - r)m r = (r' - x)$ $m'r'$, from which we obtain $x = (m\,r^2 + m'r'^2)/(mr + m'r')$, exactly as Huygens found it. The generalization of this reasoning, for any number of masses, which need not lie in a single straight line, is obvious.

3. JOHN BERNOULLI (in 1712) attacked the problem of the center of oscillation in a different manner. His performances are consulted in his Collected Works (*Opera,* Lausanne and Geneva, 1762, Vols. II and IV). We shall examine here in detail the main ideas of this physicist. Bernoulli reaches his goal by conceiving the *masses* and *forces* separated.

First, let us consider two simple pendulums of different lengths l, l' whose bobs are affected with gravitational accelerations proportional to the lengths of the pendulums, that is, let us put $l/l' = g/g'$. As the time of oscillation of a pendulum is $T = \pi \sqrt{l/g}$, it follows that the times of oscillation of these pendulums will be the same. Doubling the length of a pendulum, accordingly, while at the same time doubling the acceleration of gravity does not alter the period of oscillation.

Second, though we cannot directly alter the acceleration of gravity at any one spot on the earth, we can do what amounts virtually to this. Thus, imagine a straight massless bar of length $2a$, free to rotate about its middle point; and attach to the one extremity of it the mass m and to the other the mass m'. Then the total mass is $m + m'$ at the distance a from the axis. But the force which acts on it is $(m - m') g$, and the acceleration, consequently, $(m - m'/m + m') g$.

Fig. 167.

Hence, to find the length of the simple pendulum, having the ordinary acceleration of gravity g, which is isochronous with the present pendulum of the length a, we put, employing the preceding theorem,

$$\frac{l}{a} = \frac{g}{\dfrac{m - m'}{m + m'} g}, \text{ or } l = a\, \frac{m + m'}{m - m'},$$

Third, we imagine a simple pendulum of length 1 with the mass m at its extremity. The weight of m produces, by the principle of the lever, the same acceleration as half this force at a distance 2 from the point of suspension. Half the mass m placed at the distance 2, therefore, would suffer by the action of the force impressed at 1 the same acceleration, and a fourth of the mass m would suffer double the acceleration; so that a simple pendulum of the length 2 having the original force at distance 1 from the point of suspension and one-fourth the original mass at its extremity would be isochronous with the original one. Generalizing this reasoning, it is evident that we may transfer any force f acting on a compound pendulum at any distance r, to the distance 1 by making its value rf, and any and every mass placed at the distance r to the distance 1 by making its value r^2m, without changing the time of oscillation of the pendulum. If a force f act on a lever-arm a (Fig. 168) while at the distance r from the axis a mass m is attached, f will be equivalent to a force af/r impressed on m and

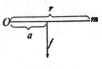

Fig. 168.

will impart to it the linear acceleration af/mr and the angular acceleration af/mr^2. Hence, to find the angular acceleration of a compound pendulum, we divide the sum of the *statical moments* by the sum of the *moments of inertia.*

BROOK TAYLOR, an Englishman,* also developed this idea, on substantially the same principles, but quite independently of John Bernoulli. His solution, however, was not published until some time later, in 1714, in his work, *Methodus Incrementorum.*

* Author of Taylor's theorem, and also of a remarkable work on perspective.—*Trans.*

The above are the most important attempts to solve the problem of the center of oscillation. We shall see that they contain the very same ideas that D'Alembert enunciated in a generalized form.

4. On a system of points M, M', M''.... connected with one another in any way,* the forces P, P', P'' .... are impressed. (Fig. 169.) These forces would impart to the *free* points of the system certain determinate motions. To the *connected* points, however, *different* motions are usually imparted—motions which could be produced by the forces W, W', W''.... These last are the motions which we shall study.

Conceive the force P resolved into W and V, the force P' into W' and V', and the force P'' into W'' and V'', and so on. Since, owing to the connections, only the components W, W', W''.... are effective,

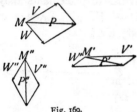

Fig. 169.

therefore, the forces V, V', V''.... must be *equilibrated* by the connections. We will call the forces P, P', P'' the *impressed* forces, the forces W, W', W''...., which produce the actual motions, the *effective* forces, and the forces V, V', V'' .... the forces *gained and lost,* or the *equilibrated* forces. We perceive, thus, that if we resolve the impressed forces into the effective forces and the equilibrated forces, the latter form a system balanced by the connections. This is the principle of D'Alembert. We have allowed ourselves, in its exposition, only the unessential modification of putting forces for the momenta generated by the forces. In

---

* In precise technical language, they are subject to *constraints*, that is, forces regarded as infinite, which compel a certain relation between their motions.—*Trans.*

this form the principle was stated by D'ALEMBERT in his *Traité de dynamique,* published in 1743.

As the system V, V', V''. . . . is in *equilibrium,* the principle of *virtual displacements* is applicable thereto. This gives a second form of D'Alembert's principle. A third form is obtained as follows: The forces P, P' . . . . are the resultants of the components W, W'. . . . and V, V'. . . . If, therefore, we combine with the forces W, W'. . . . and V, V'. . . . the forces $— P$, $— P'$ . . . ., equilibrium will obtain. The force-system $—P$, W, V is in equilibrium. But the system V is independently in equilibrium. Therefore, also the system $— P$, W is in equilibrium, or, what is the same thing, the system P, $— W$ is in equilibrium. Accordingly, if the effective forces with opposite signs be joined to the impressed forces, the two, owing to the connections, will balance. The principle of virtual displacements may also be applied to the system P, $— W$. This LAGRANGE did in his *Mécanique analytique,* 1788.

The fact that equilibrium subsists between the system P and the system $— W$, may be expressed in still another way. We may say that the system W is *equivalent* to

Fig. 170.

the system P. In this form HERMANN (*Phoronomia,* 1716) and EULER (*Comment. Acad. Petrop.,* Old Series, Vol. VII, 1740) employed the principle. It is substantially not different from that of D'Alembert.

5. We will now illustrate D'Alembert's principle by one or two examples.

On a massless wheel and axle with the radii R, r the loads P and Q are hung, which are not in equilibrium. We resolve the force P into (1) W (the force which would produce the actual motion of the mass if this were free) and (2) V, that is, we put $P = W + V$ and also $Q = W' + V'$; it being evident that we may here disregard all motions that are not in the vertical. We have, accordingly, $V = P - W$ and $V' = Q - W'$,

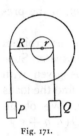

Fig. 171.

and, since the forces V, V' are in equilibrium, also $V \cdot R = V' \cdot r$. Substituting for V, V' in the last equation their values in the former, we get

$$(P - W)R = (Q - W')r \dots (1)$$

which may also be directly obtained by the employment of the second form of D'Alembert's principle. From the conditions of the problem we readily perceive that we have here to deal with a uniformly accelerated motion, and that all that is therefore necessary is to ascertain the acceleration. Adopting gravitation measure, we have the forces W and W', which produce in the masses P/g and Q/g the accelerations γ and γ'; wherefore, $W = (P/g)\gamma$ and $W' = (Q/g)\gamma'$. But we also know that $\gamma' = -\gamma(r/R)$. Accordingly, equation (1) passes into the form

$$\left(P - \frac{P}{g}\gamma\right)R = \left(Q + \frac{Q}{g}\frac{r}{R}\gamma\right)r \dots (2)$$

whence the values of the two accelerations are obtained

$$\gamma = \frac{PR - Qr}{PR^2 + Qr^2} Rg, \text{ and } \gamma' = -\frac{PR - Qr}{PR^2 + Q_r{}^2} rg.$$

These last determine the motion.

It will be seen at a glance that the same result can be obtained by the employment of the ideas of statical moment and moment of inertia. We get by this method for the angular acceleration

$$\varphi = \frac{PR - Qr}{\dfrac{P}{g}R^2 + \dfrac{Q}{g}r^2} = \frac{PR - Qr}{PR^2 + Qr^2} \cdot g;$$

and as $\gamma = R\varphi$ and $\gamma' = -r\varphi$ we re-obtain the preceding expressions.

When the masses and forces are given, the problem of finding the motion of a system is *determinate*. Suppose, however, only the acceleration γ is given with which P moves, and that the problem is to find the loads P and Q that produce this acceleration. We obtain easily from equation (2) the result $P = Q(Rg + r\gamma)$ $r/(g-\gamma)R^2$, that is, a relation between P and Q. One of the two loads therefore is arbitrary. The problem in this form is an *indeterminate* one, and may be solved in an infinite number of different ways.

The following may serve as a second example.

A weight P (Fig. 172) free to move on a vertical straight line AB, is attached to a cord passing over a pulley and carrying a weight Q at the other end. The cord makes with the line AB the variable angle α. The motion of the present case cannot be uniformly accelerated. But if we consider only vertical motions we can easily give for every value of α the momentary acceleration (γ and γ') of

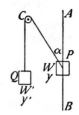

Fig. 172.

P and Q. Proceeding exactly as we did in the last case, we obtain

$$P = W + V,$$
$$Q = W' + V'$$

also

$$V' \cos \alpha = V, \text{ or, since } \gamma' = -\gamma \cos \alpha,$$

$$\left(Q + \frac{Q}{g}\cos\alpha\,\gamma\right)\cos\alpha = P - \frac{P}{g}\,\gamma; \text{ whence}$$

$$\gamma = \frac{P - Q\cos\alpha}{Q\cos^2\alpha + P}\,g$$

$$\gamma' = -\frac{P - Q\cos\alpha}{Q\cos^2\alpha + P}\cos\alpha\,g.$$

Again the same result may be easily reached by the employment of the ideas of statical moment and moment of inertia in a more generalized form. The following reflection will render this clear. The force, or statical moment, that acts on P is $P - Q\cos\alpha$. But the weight Q moves $\cos\alpha$ times as fast as P; consequently its mass is to be taken $\cos^2\alpha$ times. The acceleration which P receives, accordingly is,

$$\gamma = \frac{P - Q\cos\alpha}{\dfrac{Q}{g}\cos^2\alpha + \dfrac{P}{g}} = \frac{P - Q\cos\alpha}{Q\cos^2\alpha + P}\,g.$$

In like manner the corresponding expression for γ' may be found.

The foregoing procedure rests on the simple remark, that not the circular path of the motion of the masses is of consequence, but only the *relative* velocities or *relative* displacements. This extension of the concept moment of inertia may often be employed to advantage.

6. Now that the application of D'Alembert's principle has been sufficiently illustrated, it will not be difficult to obtain a clear idea of its significance. Problems relating to the *motion* of connected points are here disposed of by recourse to experiences concerning the mutual actions of connected bodies reached in the in-

vestigation of problems of *equilibrium*. Where the last mentioned experiences do not suffice, D'Alembert's principle also can accomplish nothing, as the examples adduced will amply indicate. We should, therefore, carefully avoid the notion that D'Alembert's principle is a *general* one which renders special experiences superfluous. Its conciseness and apparent simplicity are wholly due to the fact that it refers us to experiences already in our possession. Detailed knowledge of the subject under consideration founded on exact and minute experience, cannot be dispensed with. This knowledge we must obtain either from the case presented, by a direct investigation, or we must previously have obtained it, in the investigation of some other subject, and carry it with us to the problem in hand. We learn, in fact, from D'Alembert's principle, as our examples show, nothing that we could not also have learned by other methods. The principle fulfils in the solution of problems, the office of a routine-form which, to a certain extent, spares us the trouble of thinking out each new case, by supplying directions for the employment of experiences before known and familiar to us. The principle does not so much promote our *insight* into the processes as it secures us a *practical mastery* of them. The value of the principle is of an economical character.

When we have solved a problem by D'Alembert's principle, we may rest satisfied with the experiences previously made concerning equilibrium, the application of which the principle implies. But if we wish *clearly and thoroughly* to apprehend the phenomenon, that is, to rediscover in it the simplest mechanical elements with which we are familiar, we are obliged to push our researches further, and to replace our experiences concerning equilibrium either by the Newtonian or by the Huygenian conceptions, in some way similar to that pursued on page 353. If we adopt the former alternative, we shall mentally see the accelerated motions enacted which the mutual action of bodies on one another produces; if we adopt the second, we shall directly contemplate the *work* done, on which, in the Huygenian conception, the *vis viva* depends. The latter point of view is particularly convenient if we employ the principle of virtual displacements to express the conditions of equilibrium of the system V or $P - W$. D'Alembert's principle then asserts, that the sum of the virtual moments of the system V, or of the system $P - W$, is equal to zero. The elementary work of the equilibrated forces, if we leave out of account the straining of the connections, is equal to zero. The total work done, then, is performed *solely* by the system P, and the work performed by the system W must, accordingly, be equal to the work done by the system P. All the work that can *possibly* be done is due, neglecting the strains of the connections, to the *impressed* forces. As will be seen, D'Alembert's principle in this form is not essentially different from the principle of *vis viva*.

7. In practical applications of the principle of D'Alembert it is convenient to resolve every force P impressed on a mass m of the system into the mutually perpendicular components X, Y, Z parallel to the axes of a system of rectangular coördinates; every effective force W into corresponding components $m\xi$, $m\eta$, $m\zeta$, where ξ, η, ζ denote accelerations in the directions of the coördinates; and every displacement, in a similar manner, into three displacements δx, δy, δz. As the work done by each component force is effective only in

displacements parallel to the directions in which the components act, the equilibrium of the system $(P,-W)$ is given by the equation

$$\Sigma\{(X-m\xi)\delta x + (Y-m\eta)\delta y + (Z-m\zeta)\delta z\}=0 \quad (1)$$

or

$$\Sigma(X\delta x + Y\delta y + Z\delta z) = \Sigma m(\xi\delta x + \eta\delta y + \zeta\delta z) \quad . \quad . \quad (2)$$

These two equations are the direct expression of the proposition above enunciated respecting the *possible* work of the impressed forces. If this work be $=0$, the particular case of equilibrium results. The principle of virtual displacements flows as a *special* case from this expression of D'Alembert's principle; and this is quite in conformity with reason, since in the general as well as in the particular case the experimental perception of the *import of work* is the sole thing of consequence.

Equation (1) gives the requisite equations of motion; we have simply to express as many as possible of the displacements δx, δy, δz by the others in terms of their relations to the latter, and put the coefficients of the remaining arbitrary displacements $= 0$, as was illustrated in our applications of the principle of virtual displacements.

The solution of a very few problems by D'Alembert's principle will suffice to impress us with a full sense of its convenience. It will also give us the conviction that it is possible, in every case in which it may be found necessary, to solve directly and with perfect insight the very same problem by a consideration of elementary mechanical processes, and to arrive thereby at exactly the same results. Our conviction of the *feasibility* of this operation renders the performance of it, in cases in which purely practical ends are in view, unnecessary.

VI.

THE PRINCIPLE OF VIS VIVA.

1. The principle of *vis viva*, as we know, was first employed by HUYGENS. JOHN and DANIEL BERNOULLI had simply to provide for a greater generality of expression; they added little. If p, p', p''. . . . are weights, m, m', m''. . . . their respective masses, h, h', h''. . . . the distances of descent of the free or connected masses, and v, v', v''. . . . the velocities acquired, the relation obtains

$$\Sigma\, p\, h = \tfrac{1}{2}\, \Sigma\, m\, v^2.$$

If the initial velocities are not $= 0$, but are v_0, v_0', v_0''. . . ., the theorem will refer to the increment of the *vis viva* by the work and read

$$\Sigma\, p\, h = \tfrac{1}{2}\, \Sigma\, m\, (v^2 - v_0^2).$$

The principle still remains applicable when p . . . . are, not weights, but any constant forces, and h . . . not the vertical spaces fallen through, but any paths in the lines of the forces. If the forces considered are variable, the expressions $p\,h$, $p'h'$. . . . must be replaced by the expressions $\int p\, ds$, $\int p'\, ds'$ . . . ., in which p denotes the variable forces and $d\,s$ the elements of distance described in the lines of the forces. Then

or
$$\int p\, ds + \int p'\, ds' + \ldots = \tfrac{1}{2}\, \Sigma\, m\, (v^2 - v_0^2)$$

$$\Sigma \int p\, ds = \tfrac{1}{2}\, \Sigma\, m\, (v^2 - v_0^2) \ldots \ldots \ldots (1)$$

2. In illustration of the principle of *vis viva* we shall first consider the simple problem which we treated by the principle of D'Alembert. On a wheel and axle with the radii R, r hang the weights P, Q (Fig. 173). When this machine is set in motion, work is performed by which the acquired *vis viva* is fully determined. For a rotation of the machine through the angle α, the *work* is

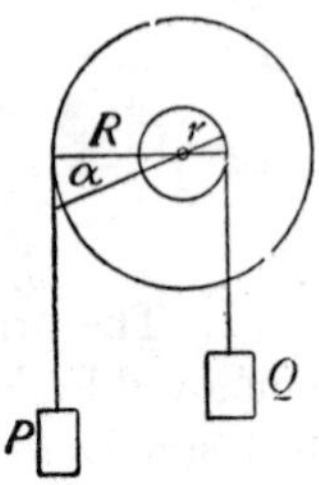

$$P \cdot R \alpha - Q \cdot r \alpha = \alpha (PR - Qr).$$

Fig. 173.

Calling the angular velocity which corresponds to this angle of rotation, φ, the *vis viva* generated will be

$$\frac{P}{g} \frac{(R\varphi)^2}{2} + \frac{Q}{g} \frac{(r\varphi)^2}{2} = \frac{\varphi^2}{2g}(PR^2 + Qr^2).$$

Consequently, the equation obtains

$$\alpha(PR - Qr) = \frac{\varphi^2}{2g}(PR^2 + Qr^2) \quad . \quad . \quad . \quad . \quad (1)$$

Now the motion of this case is a uniformly accelerated motion; consequently, the *same* relation obtains here between the angle α, the angular velocity φ, and the angular acceleration ψ, as obtains in free descent between s, v, g. If in free descent $s = v^2/2g$, then here $\alpha = \varphi^2/2\psi$.

Introducing this value of α in equation (1) we get for the angular acceleration of P,

$$\psi = \frac{PR - Qr}{PR^2 + Qr^2} g,$$

and, consequently, for its absolute acceleration,

$$\gamma = \frac{PR - Qr}{PR^2 + Qr^2} Rg,$$

exactly as in the previous treatment of the problem.

3. As a second example let us consider the case of a

massless cylinder of radius r, in the surface of which, diametrically opposite each other, are fixed two equal masses m, and which in consequence of the weight of these masses rolls without sliding down an inclined plane of the elevation α. First, we must convince ourselves, that in order to represent the total *vis viva* of the system we have simply to sum up the *vis viva* of the motions of rotation and progression. The axis of the cylinder has acquired, we will say, the velocity u in the direction of the length of the inclined plane, and

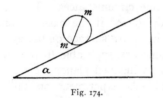

Fig. 174.

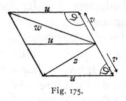

Fig. 175.

we will denote by v the absolute velocity of rotation of the surface of the cylinder. The velocities of rotation v of the two masses m make with the velocity of progression u the angles θ and θ' (Fig. 175), where $\theta + \theta' = 180°$. The compound velocities w and z satisfy therefore the equations

$$w^2 = u^2 + v^2 - 2uv\cos\theta$$
$$z^2 = u^2 + v^2 - 2uv\cos\theta'.$$

But since $\cos\theta = -\cos\theta'$, it follows that

$$w^2 + z^2 = 2u^2 + 2v^2, \text{ or,}$$
$$\tfrac{1}{2}mw^2 + \tfrac{1}{2}mz^2 = \tfrac{1}{2}m2u^2 + \tfrac{1}{2}m2v^2 = mu^2 + mv^2.$$

If the cylinder moves through the angle φ, m describes in consequence of the rotation the space $r\varphi$, and the axis of the cylinder is likewise displaced a distance $r\varphi$. As the spaces traversed are to each other, so also are the velocities v and u, which therefore are equal. The total *vis viva* may accordingly be expressed by $2m\,u^2$. If l is the distance the cylinder travels along the length of the inclined plane, the work done is $2\,m\,g\,.\,l\sin a = 2\,m\,u^2$; whence $u = \sqrt{g\,l\,.\sin a}$. If we compare with this result the velocity acquired by a body in *sliding* down an inclined plane, namely, the velocity $\sqrt{2gl\sin a}$, it will be observed that the contrivance we are here considering moves with only one-half the acceleration of descent that (friction neglected) a sliding body would under the same circumstances. The reasoning of this case is not altered if the mass be uniformly distributed over the entire surface of the cylinder. Similar considerations are applicable to the case of a *sphere* rolling down an inclined plane. It will be seen, therefore, that Galileo's experiment on falling bodies is in need of a quantitative correction.

Next, let us distribute the mass m uniformly over the surface of a cylinder of radius R, which is coaxial with and rigidly joined to a massless cylinder of radius r, and let the latter roll down the inclined plane. Since here $v/u = R/r$, the principle of *vis viva* gives $m\,g\,l\sin a = \frac{1}{2}m\,u^2(1 + R^2/r^2)$, whence

$$u = \sqrt{\dfrac{2\,g\,l\sin\alpha}{1 + \dfrac{R^2}{r^2}}}$$

For $R/r = 1$ the acceleration of descent assumes its previous value $g/2$. For very large values of R/r the acceleration of descent is very small. When $R/r = \infty$ it will be impossible for the machine to roll down the inclined plane at all.

As a third example, we will consider the case of a chain, whose total length is l, and which lies partly on a horizontal plane and partly on a plane having the angle of elevation α. If we imagine the surface on which the chain rests to be very smooth, any very small portion of the chain left hanging over on the in-

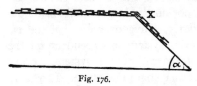

Fig. 176.

clined plane will draw the remainder after it. If μ is the mass of unit of length of the chain and a portion x is hanging over, the principle of *vis viva* will give for the velocity v acquired the equation

$$\frac{\mu l v^2}{2} = \mu x g \frac{x}{2} \sin \alpha = \mu g \frac{x^2}{2} \sin \alpha,$$

or $v = x \sqrt{g \sin \alpha / l}$. In the present case, therefore, the velocity acquired is proportional to the space described. The very law holds that Galileo first conjectured was the law of freely falling bodies. The same reflections, accordingly, are admissible here as at page 308.

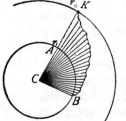

Fig. 177.

4. Equation (1), the equation of *vis viva*, can always be employed, to solve problems of moving bodies, when the *total distance* traversed and the force that acts in each element of the distance are known. It was disclosed, however, by the labors of Euler, Daniel Bernoulli, and Lagrange, that cases occur in which the principle of *vis viva* can be employed without a knowledge of the *actual path* of the motion. We shall see later

on that Clairaut also rendered important services in this field.

Even Galileo knew that the velocity of a heavy falling body depended solely on the *vertical height* descended through, and not on the length or *form* of the path traversed. Similarly, Huygens finds that the *vis viva* of a heavy material system is dependent on the *vertical heights* of the masses of the system. Euler was able to make a further step in advance. If a body K (Fig. 177) is attracted towards a fixed center C in obedience to some given law, the increase of the *vis viva* in the case of rectilinear approach is calculable from the initial and terminal distances $(r_0, r,)$ But the increase is the same, if K passes at all from the position r_0 to the position $r,$, independently of the *form of its path, KB.* For the elements of the work done must be calculated from the projections on the radius of the actual displacements, and are thus ultimately the same as before.

If K is attracted towards several fixed centers C, C', C''. . . ., the increase of its *vis viva* depends on the initial distances r_0, r_0', r_0''. . . . and on the terminal distances $r,$, $r,'$, $r,''$. . . ., that is on the initial and terminal *positions* of K. Daniel Bernoulli extended this idea, and showed further that where movable bodies are in a state of *mutual* attraction the change of *vis viva* is determined solely by their initial and terminal distances from one another. The *analytical* treatment of these problems was perfected by Lagrange. If we join a point having the coördinates a, b, c with a point having the coördinates x, y, z, and denote by r the length of the line of junction and by α, β, γ the angles that line makes with the axes of x, y, z, then, according to Lagrange, because

$$r^2 = (x - a)^2 + (y - b)^2 + (z - c)^2,$$

$$\cos \alpha = \frac{x - a}{r} = \frac{dr}{dx}, \quad \cos \beta = \frac{y - b}{r} = \frac{dr}{dy},$$

$$\cos \gamma = \frac{z - c}{r} = \frac{dr}{dz}.$$

Accordingly, if $f(r) = \dfrac{dF(r)}{dr}$ is the repulsive force, or the negative of the attractive force acting between the two points, the components will be

$$X = f(r) \cos \alpha = \frac{dF(r)}{dr} \frac{dr}{dx} = \frac{dF(r)}{dx},$$

$$Y = f(r) \cos \beta = \frac{dF(r)}{dr} \frac{dr}{dy} = \frac{dF(r)}{dy},$$

$$Z = f(r) \cos \gamma = \frac{dF(r)}{dr} \frac{dr}{dz} = \frac{dF(r)}{dz}.$$

The force-components, therefore, are the partial differential coefficients of *one and the same* function of *r*, or of the coördinates of the repelling or attracting points. Similarly, if several points are in mutual action, the result will be

$$X = \frac{dU}{dx}$$

$$Y = \frac{dU}{dy}$$

$$Z = \frac{dU}{dz},$$

where U is a function of the coördinates of the points. This function was subsequently called by Hamilton[*] the *force-function*.

Transforming, by means of the conceptions here reached, and under the suppositions given, equation

[*] *On a General Method in Dynamics, Phil. Trans.* for 1834. See also C. G. J. Jacobi, *Vorlesungen über Dynamik*, edited by Clebsch, 1866.

(1) into a form applicable to rectangular coördinates, we obtain

$$\Sigma \int (Xdx + Ydy + Zdz) = \Sigma \tfrac{1}{2} m (v^2 - v_o^2) \text{ or,}$$

since the expression to the left is a complete differential,

$$\Sigma \left(\int \frac{dU}{dx} dx + \frac{dU}{dy} dy + \frac{dU}{dz} dz \right) =$$
$$\Sigma \int dU = \Sigma (U_1 - U_o) = \Sigma \tfrac{1}{2} m (v^2 - v_o^2),$$

where U_1 is a function of the terminal values and U_0 the *same* function of the initial values of the coördinates. This equation has received extensive applications, but it simply expresses the knowledge that under the conditions designated the *work done* and therefore also the *vis viva* of a system is *dependent* on the *positions,* or the coördinates, of the bodies constituting it.

If we imagine all masses fixed and only a single one in motion, the work changes only as U changes. The equation $U = constant$ defines a so-called level surface, or surface of equal work. Movement upon such a surface produces no work. U increases in the direction in which the forces tend to move the bodies.

VII.

THE PRINCIPLE OF LEAST CONSTRAINT.

1. GAUSS enunciated (in Crelle's *Journal für Mathematik,* IV, 1829, p. 233) a new law of mechanics, the principle of *least constraint.* He observed, that, in the form which mechanics has historically assumed, dynamics is founded upon statics, (for example, D'Alembert's principle on the principle of virtual displace-

ments,) whereas one naturally would expect that in the highest stage of the science statics would appear as a particular case of dynamics. Now, the principle which Gauss supplied, and which we shall discuss in this section, includes both dynamical and statical cases. It meets, therefore, the requirements of scientific and logical æsthetics. We have already pointed out that this is also true of D'Alembert's principle in its Lagrangian form and the mode of expression above adopted. No *essentially new principle,* Gauss remarks, can now be established in mechanics; but this does not exclude the discovery of *new points of view,* from which mechanical phenomena may be fruitfully contemplated. Such a new point of view is afforded by the principle of Gauss.

2. Let m, m' .... be masses, connected in any manner with one another (Fig. 178). These masses, if *free,* would, under the action of the forces impressed on them, describe in a very short element of time the spaces $a\ b, a'\ b'$ ....; but in consequence of their *connections* they describe in the

Fig. 178.

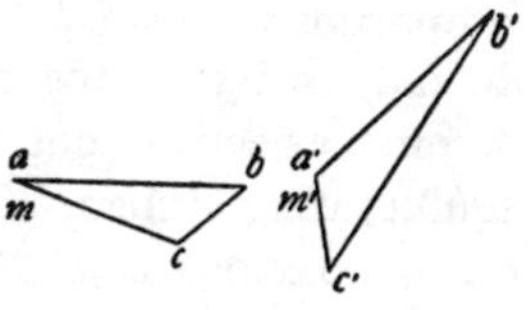

same element of time the spaces $a\ c, a'\ c'$ .... Now, Gauss's principle asserts, that the motion of the connected points is such that, *for the motion actually taken,* the sum of the products of the mass of each material particle into the square of the distance of its deviation from the position it would have reached if free, namely $m(b\ c)^2 + m'\ (b'\ c')^2 + \ .\ .\ .\ . = \Sigma\ m\ (b\ c)^2$, is a *minimum,* that is, is smaller for the actual motion than for any other conceivable motion *in the same connections.* If this sum, $\Sigma\ m(b\ c)^2$, is less for *rest* than for any motion, equilibrium will obtain. The principle includes, thus, both statical and dynamical cases.

The sum $\Sigma\, m(b\, c)^2$ is called the "constraint."* In forming this sum it is plain that the velocities present in the system may be neglected, as the relative positions of *a, b, c* are not altered by them.

3. The new principle is equivalent to that of D'Alembert; it may be used in place of the latter; and, as Gauss has shown, can also be deduced from it. The *impressed* forces carry the free mass *m* in an element of time through the space *a b,* the *effective* forces carry the same mass in the same time in consequence of the connections through the space *a c.* We resolve *a b* into *a c* and *c b* (Fig. 179); and do the same for all the masses. It is thus evident that forces corresponding to the distances *c b, c′ b′* .... and proportional to *m c b, m′ c′ b′* ...,

Fig. 179.

do not, owing to the connections, become effective, but form with the connections an equilibrating system. If, therefore, we erect at the terminal positions *c, c′* .... the virtual displacements *c γ, c′ γ′* ...., forming with *c b, c′ b′* .... the angles θ, θ' .... we may apply, since by D'Alembert's principle forces proportional to *m c b, m′ c′ b′* .... are here in equilibrium, the principle of virtual velocities. Doing so, we shall have

$$\Sigma\, m\, c\, b\,.\, c\, \gamma \cos \theta \gtreqless 0 \quad \ldots \ldots \ldots \ldots \ldots \quad (1)$$

* Professor Mach's term is *Abweichungssumme.* The *Abweichung* is the *declination* or *departure* from free motion, called by Gauss the *Ablenkung.* (See Dühring, *Principien der Mechanik,* §§ 168, 169; Routh, *Rigid Dynamics,* Part I, §§ 390-394.) The quantity $\Sigma\, m\,(b\,c)^2$ is called by Gauss the *Zwang*; and German mathematicians usually follow this practice. In English, the term *constraint* is established in this sense, although it is also used with another, hardly quantitative meaning, for the force which restricts a body absolutely to moving in a certain way.—*Trans.*

But

$$(b\gamma)^2 = (bc)^2 + (c\gamma)^2 - 2bc \cdot c\gamma \cos\theta,$$
$$(b\gamma)^2 - (bc)^2 = (c\gamma)^2 - 2bc \cdot c\gamma \cos\theta, \text{ and}$$
$$\Sigma m(b\gamma)^2 - \Sigma m(bc)^2 = \Sigma m(c\gamma)^2 - 2\Sigma mbc \cdot c\gamma \cos\theta \quad (2)$$

Accordingly, since by (1) the second member of the right-hand side of (2) can only be $= 0$ or *negative*, that is to say, as the sum $\Sigma m(c\gamma)^2$ can never be diminished by the subtraction, but only *increased*, therefore the left-hand side of (2) must also always be positive and consequently $\Sigma m(b\gamma)^2$ always greater than $\Sigma m (bc)^2$, which is to say, every conceivable constraint from unhindered motion is greater than the constraint for the actual motion.

4. The declination, bc, for the very small element of time τ, may, for purposes of practical treatment, be designated by s, and following Scheffler (Schlömilch's *Zeitschrift für Mathematik und Physik*, 1858, III, p. 197), we may remark that $s = \gamma\tau^2/2$, where γ denotes acceleration. Consequently, $\Sigma m s^2$ may also be expressed in the forms

$$\Sigma m \cdot s \cdot s = \frac{\tau^2}{2} \Sigma m\gamma \cdot s = \frac{\tau^2}{2} \Sigma p \cdot s = \frac{\tau^4}{4} \Sigma m\gamma^2,$$

where p denotes the force that produces the declination from free motion. As the constant factor in no wise affects the minimum condition, we may say, the actual motion is always such that

$$\Sigma m s^2 \quad \dotfill \quad (1)$$

or

$$\Sigma p s \quad \dotfill \quad (2)$$

or

$$\Sigma m \gamma^2 \quad \dotfill \quad (3)$$

is a minimum.

5. We will first employ, in our illustrations, the third form. Here again, as our first example, we select the motion of a wheel and axle by the overweight of one of its parts and shall use the designations above frequently employed (Fig. 180). Our problem is, to so determine the actual accelerations γ of P and γ, of Q, that (P/g) $(g-\gamma)^2 + (Q/g)$ $(g-\gamma_r)^2$ shall be a minimum, or, since $\gamma_r = -\gamma$ (r/R), so that $P(g-\gamma)^2 + Q(g + \gamma \cdot r/R)^2 = N$ shall assume its smallest value. Putting, to this end,

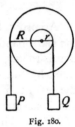

Fig. 180.

$$\frac{dN}{d\gamma} = -P(g-\gamma) + Q\left(g + \gamma\frac{r}{R}\right)\frac{r}{R} = 0,$$

we get $\gamma = (\overline{PR - Qr}/\overline{PR^2 + Qr^2})Rg$, exactly as in the previous treatments of the problem.

As our second example, the motion of descent on an inclined plane may be taken. In this case we shall employ the first form, $\Sigma m\,s^2$. Since we have here only to deal with one mass, our inquiry will be directed to finding that acceleration of descent γ for the plane by which the square of the declination (s^2) is made a minimum. By Fig. 181 we have

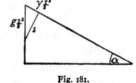

Fig. 181.

$$s^2 = \left(g\frac{\tau^2}{2}\right)^2 + \left(\gamma\frac{\tau^2}{2}\right)^2 - 2\left(g\frac{\tau^2}{2} \cdot \gamma\frac{\tau^2}{2}\right)\sin\alpha,$$

and putting $d(s^2)/d\gamma = 0$, we obtain, omitting all

constant factors, $2\gamma - 2g \sin a = 0$ or $\gamma = g \cdot \sin a$, the familiar result of Galileo's researches.

The following example will show that Gauss's principle also embraces cases of equilibrium. On the arms a, a' of a lever (Fig. 182) are hung the heavy masses m, m'. The principle requires that $m(g - \gamma)^2 + m'(g - \gamma')^2$ shall be a minimum. But $\gamma' = -\gamma(a'/a)$.

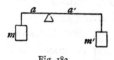

Fig. 182.

Further, if the masses are inversely proportional to the lengths of the lever-arms, that is to say, if $m/m' = a'/a$, then $\gamma' = -\gamma \, (m/m')$. Consequently, $m(g - \gamma)^2 + m'(g + \gamma \cdot m/m')^2 = N$ must be made a minimum. Putting $d N/d\gamma = 0$, we get $m(1 + m/m')\gamma = 0$ or $\gamma = 0$. Accordingly, in this case *equilibrium* presents the least constraint from free motion.

Every *new* cause of constraint, or restriction upon the freedom of motion, increases the quantity of constraint, but the increase is always the least possible. If two or more systems be connected, the motion of least constraint from the motions of the unconnected systems is the actual motion.

Fig. 183.

If, for example, we join together several simple pendulums so as to form a compound linear pendulum, the latter will oscillate with the motion of least constraint from the motion of the single pendulums. The simple pendulum, for any excursion a, receives, in the direction of its path, the acceleration $g \sin a$. Denoting, therefore, by $\gamma \sin a$ the acceleration corresponding to this excursion at the axial distance 1 on the compound pendulum, $\Sigma m(g \sin a - r \gamma \sin a)^2$ or $\Sigma m(g - r\gamma)^2$ will be the quantity to be made a minimum. Consequently, $\Sigma m(g - r\gamma)r = 0$, and $\gamma = g(\Sigma m r / \Sigma m r^2)$. The problem is thus disposed of in the simplest manner. But this simple solution is possible only because the *experiences* that Huygens, the Bernoullis, and others collected long before, are implicitly contained in Gauss's principle.

6. The *increase* of the quantity of constraint, or declination, from free motion by *new* causes of constraint may be exhibited by the following examples:

Over two stationary pulleys A, B, and beneath a movable pulley C (Fig. 184), a cord is passed, each extremity of which is weighted with a load P; and on C a load $2P + p$ is placed. The movable pulley will now descend with the acceleration $(p/\overline{4P + p})g$. But if we make the pulley A fast, we impose upon the system a new cause of constraint, and the quantity of constraint, or declination, from free motion will be in-

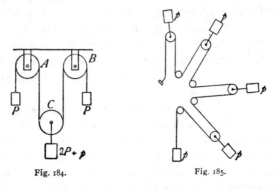

Fig. 184.   Fig. 185.

creased. The load suspended from B, since it now moves with double the velocity, must be reckoned as possessing four times its original mass. The movable pulley accordingly sinks with the acceleration

$(p/\overline{6P + p})g$. A simple calculation will show that the constraint in the latter case is greater than in the former.

A number, n, of equal weights, p, lying on a smooth horizontal surface, are attached to n small movable pulleys through which a cord is drawn in the manner indicated in figure 185 and loaded at its free extremity with p. According as *all* the pulleys are *movable* or *all except one* are *fixed*, we obtain for the motive weight p, allowing for the relative velocities of the masses as referred to p, respectively, the accelerations $(4n/\overline{1+4n})g$ and $(4/5)g$. If all the $n + 1$ masses are movable, the deviation assumes the value $pg/\overline{4n + 1}$, which increases as n, the number of the movable masses, is decreased.

7. Imagine a body of weight Q, movable on rollers on a horizontal surface, and having an inclined plane face. On this inclined face a body of weight P is placed. We now perceive *instinctively* that P will descend with *quicker* acceleration when Q is movable and can give way, than it will when Q is fixed and P's descent more hindered. To any distance of descent h of P a horizontal velocity v and a vertical velocity u of

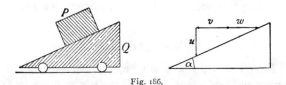

Fig. 186.

P and a horizontal velocity w of Q correspond. Owing to the conservation of the quantity of horizontal motion, (for here only internal forces act,) we have $Pv = Qw$, and for obvious geometrical reasons (Fig. 186) also

$$u = (v + w) \tan \alpha$$

The velocities, consequently, are

$$u = u$$

$$v = \frac{Q}{P + Q} \cot \alpha \cdot u,$$

$$w = \frac{P}{P + Q} \cot \alpha \cdot u.$$

For the work Ph performed, the principle of *vis viva* gives

$$Ph = \frac{P}{g} \frac{u^2}{2} + \frac{P}{g} \left(\frac{Q}{P + Q} \cot \alpha \right)^2 \frac{u^2}{2} + \frac{Q}{g} \left(\frac{P}{P + Q} \cot \alpha \right)^2 \frac{u^2}{2}.$$

Multiplying by $\frac{g}{P}$, we obtain

$$gh = \left(1 + \frac{Q}{P + Q} \frac{\cos^2 \alpha}{\sin^2 \alpha} \right) \frac{u^2}{2}.$$

To find the *vertical* acceleration γ with which the space h is described, be it noted that $h = u^2/2\gamma$. Introducing this value in the last equation, we get

$$\gamma = \frac{(P + Q) \sin^2 \alpha}{P \sin^2 \alpha + Q} \cdot g.$$

For $Q = \infty$, $\gamma = g \sin^2 \alpha$, the same as on a stationary inclined plane. For $Q = 0$, $\gamma = g$, as in free descent. For finite values of $Q = mP$, we get,

since $\dfrac{1 + m}{\sin^2 \alpha + m} > 1$,

$$\gamma = \frac{(1 + m) \sin^2 \alpha}{m + \sin^2 \alpha} \cdot g > g \sin^2 \alpha.$$

The making of Q stationary, being a newly imposed cause of constraint, accordingly *increases* the quantity of constraint, or declination, from free motion.

To obtain γ, in this case, we have employed the principle of the conservation of momentum and the principle of *vis viva*. Employing Gauss's principle, we should proceed as follows: To the velocities denoted as u, v, w the accelerations γ, δ, ε correspond. Remarking that in the free state the only acceleration is the vertical acceleration of P, the others vanishing, the procedure required is, to make

$$\frac{P}{g}(g-\gamma)^2 + \frac{P}{g}\delta^2 + \frac{Q}{g}\varepsilon^2 = N$$

a minimum. As the problem possesses significance only when the bodies P and Q touch, that is only when $\gamma = (\delta + \varepsilon)\tan\alpha$, therefore, also

$$N = \frac{P}{g}[g - (\delta + \varepsilon)\tan\alpha]^2 + \frac{P}{g}\delta^2 + \frac{Q}{g}\varepsilon^2.$$

Forming the differential coefficients of this expression with respect to the two remaining independent variables δ and ε, and putting each equal to zero, we obtain

$$-[g - (\delta + \varepsilon)\tan\alpha]\,P\tan\alpha + P\delta = 0 \text{ and}$$
$$-[g - (\delta + \varepsilon)\tan\alpha]\,P\tan\alpha + Q\varepsilon = 0.$$

From these two equations follows immediately $P\delta - Q\varepsilon = 0$, and, ultimately, the same value for γ that we obtained before.

We will now look at this problem from another point of view. The body P describes at an angle β with the horizon the space s, of which the horizontal and vertical components are v and u, while simultaneously Q describes the horizontal distance w. The force-component that acts in the direction of s is P sin

β, consequently the acceleration in this direction, allowing for the relative velocities of P and Q, is

$$\frac{P \cdot \sin \beta}{\dfrac{P}{g} + \dfrac{Q}{g}\left(\dfrac{w}{s}\right)^2}.$$

Employing the following equations which are directly deducible,

$$Qw = Pv$$
$$v = s \cos \beta$$
$$u = v \tan \beta.$$

the acceleration in the direction of s becomes

$$\frac{Q \sin \beta}{Q + P \cos^2 \beta} g$$

and the vertical acceleration corresponding thereto is

$$\gamma = \frac{Q \sin^2 \beta}{Q + P \cos^2 \beta} \cdot g,$$

an expression, which as soon as we introduce by means
of the equation $u = (v + w) \tan \alpha$, the angle-func-
tions of α for those of β, again assumes the form above
given. By means of our extended conception of mo-
ment of inertia we reach, accordingly, the same result
as before.

Finally we will deal with this problem in a direct
manner. The body P does not descend on the movable inclined plane with the vertical acceleration g, with
which it would fall if free, but with a different vertical
acceleration, γ. It sustains, therefore, a vertical coun-
terforce $(P/g)(g - \gamma)$. But as P and Q, friction
neglected, can only act on each other by means of a
pressure S, *normal* to the inclined plane, therefore

$$\frac{P}{g}(g - \gamma) = S \cos \alpha \text{ and}$$

$$S \sin \alpha = \frac{Q}{g} \varepsilon = \frac{P}{g} \delta.$$

From this is obtained

$$\frac{P}{g}(g - \gamma) = \frac{Q}{g} \varepsilon \cot \alpha,$$

and by means of the equation $\gamma = (\delta + \varepsilon) \tan \alpha$, ultimately, as before,

$$\gamma = \frac{(P + Q)\sin^2 \alpha}{P \sin^2 \alpha + Q} g \cdot \cdot \cdot \cdot \cdot \cdot \cdot \cdot \cdot \cdot (1)$$

$$\delta = \frac{Q \sin \alpha \cos \alpha}{P \sin^2 \alpha + Q} g \cdot \cdot \cdot \cdot \cdot \cdot \cdot \cdot \cdot \cdot (2)$$

$$\varepsilon = \frac{P \sin \alpha \cos \alpha}{P \sin^2 \alpha + Q} g \cdot \cdot \cdot \cdot \cdot \cdot \cdot \cdot \cdot \cdot (3)$$

If we put $P = Q$ and $\alpha = 45°$, we obtain for this particular case $\gamma = \frac{2}{3}g$, $\delta = \frac{1}{3}g$, $\varepsilon = \frac{1}{3}g$. For $P/g = Q/g = 1$ we find the "constraint," or declination from free motion, to be $g^2/3$. If we make the inclined plane stationary, the constraint will be $g^2/2$. If P moved on a stationary inclined plane of elevation β, where $\tan \beta = \gamma/\delta$, that is to say, in the same path in which it moves on the movable inclined plane, the constraint would only be $g^2/5$. And, in that case it would, in reality, be less impeded than if it attained the same acceleration by the displacement of Q.

8. The examples treated will have convinced us that *no substantially new* insight or perception is afforded by Gauss's principle. Employing form (3) of the principle and resolving all the forces and accelerations in the mutually perpendicular coördinate-directions, giving here the letters the same significations as in equa-

tion (1) on page 432, we get in place of the declination, or constraint, $\Sigma m \gamma^2$, the expression

$$N = \Sigma m\left[\left(\frac{X}{m}-\xi\right)^2 + \left(\frac{Y}{m}-\eta\right)^2 + \left(\frac{Z}{m}-\zeta\right)^2\right] \quad (4)$$

and by virtue of the minimum condition

$$dN = 2 \Sigma m\left[\left(\frac{X}{m}-\xi\right)d\xi + \left(\frac{Y}{m}-\eta\right)d\eta + \left(\frac{Z}{m}-\zeta\right)d\zeta\right] = 0.$$

or $\Sigma[(X-m\xi)d\xi + (Y-m\eta)d\eta + (Z-m\zeta)d\zeta] = 0.$

If no connections exist, the coefficients of the (in that case arbitrary) $d\xi$, $d\eta$, $d\zeta$, severally made $= 0$ give the equations of motion. But if connections do exist, we have the same relations between $d\xi$, $d\eta$, $d\zeta$ as above in equation (1), at page 432, between δx, δy, δz. The equations of motion come out the same; as the treatment of the *same* example by D'Alembert's principle and by Gauss's principle fully demonstrates. The first principle, however, gives the equations of motion directly, the second only after differentiation. If we seek an expression that shall give by differentiation D'Alembert's equations, we are led perforce to the principle of Gauss. The principle, therefore, is new only in *form* and not in *matter*. Nor does it possess any advantage over the Lagrangian form of D'Alembert's principle in respect of competency to comprehend both statical *and* dynamical problems, as has been pointed out before (page 441).

There is no need of seeking a mystical or *metaphysical* reason for Gauss's principle. The expression "least constraint" may seem to promise something of the sort; but the name proves nothing. The answer to the question, "*In what* does this constraint consist?" cannot be derived from metaphysics, but must be sought in the facts. The expression (2) of page 443, or (4) of page 452, which is made a minimum, represents the *work* done in an element of time by the deviation of the constrained motion from the free motion. This work, *the work due to the constraint,* is less for the motion actually performed than for any other possible motion.

9. Once we have recognized *work* as the factor determinative of motion, once we have grasped the meaning of the principle of virtual displacements to be, that motion can never take place except where work can be performed, the following converse truth also will involve no difficulty, namely, that *all* the work that *can* be performed in an element of time actually *is* performed. Consequently, the total diminution of work due in an element of time to the connections of the system's parts is restricted to the portion annulled by the *counter-work* of those parts. It is again merely a new aspect of a familiar fact with which we have here to deal.

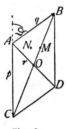

Fig. 187.

This relation is displayed in the very simplest cases. Let there be two masses *m* and *m* at *A*, the one impressed with a force *p*, the other with the force *q* (Fig. 187). If we connect the two, we shall have the mass 2 *m* acted on by a resultant force *r*. Supposing the spaces described in an element of time by the free masses to be represented by *A C*, *A B*, the space described by the conjoint, or double, mass will be $A O = \frac{1}{2} A D$. The deviation, or constraint, is $m(\overline{O B^2} + \overline{OC^2})$. It is less than it would be if the mass arrived at the end of the element of time in M or indeed in any point lying outside of BC, say N, as the simplest geometrical considerations will show. The deviation is proportional to the expression $\overline{p^2 + q^2 + 2pq} \cos \theta/2$, which in the case of equal and opposite forces becomes $2p^2$, and in the case of equal and like-directed forces zero.

Two forces p and q act on the same mass. The force q we resolve parallel and at right angles to the direction of p in r and s. The work done in an element of time is proportional to the squares of the forces, and if there are no connections is expressible by $p^2 + q^2 = p^2 + r^2 + s^2$. If now r acts directly counter to the force p, a diminution of work will be effected and the sum mentioned becomes $(p - r)^2 + s^2$. Even in the principle of the composition of forces, or of the mutual independence of forces, the properties are contained which Gauss's principle makes use of. This will best be perceived by imagining all the accelerations simultaneously performed. If we discard the obscure verbal form in which the principle is clothed, the metaphysical impression which it gives also vanishes. We see the simple fact; we are disillusioned, but also enlightened.

The elucidations of Gauss's principle presented here are in great part derived from the paper of Scheffler cited above. Some of his opinions which I have been unable to share I have modified. We cannot, for example, accept as new the principle which he himself propounds, for both in form and in import it is *identical* with the D'Alembert-Lagrangian.

The paper of Lipschitz ("Bemerkungen zu dem Prinzip des kleinsten Zwanges," *Journal für Math.*, LXXXII, 1877, pp. 316 *et seqq.*) contains profound investigations on the principle of Gauss. Many elementary examples, on the other hand, are to be found in K. Hollefreund's *Anwendungen des Gauss'schen Prinzips vom kleinsten Zwange* (Berlin, 1897). On the principle here spoken of and allied principles, see *Ostwald's Klassiker*, No. 167: *Abhandlungen über die Prinzipien der Mechanik von Lagrange, Rodrigues, Jacobi und Gauss*, edited by Philip E. B. Jourdain (Leipzig, 1908). The notes of Jourdain on pp. 31-68 go beyond the needs of a first orientation, and this orientation is the object of the present elementary book.

What is said in section 9 stands in need of completion. If the masses of the system have no velocity, the actual motions only enter in the sense of possible work, which is consistent with the conditions of the system (C. Neumann, *Ber. der kgl. sächs. Ges. der Wiss.*, XLIV, 1892, p. 184). But if the masses have velocities, which can even be directed against the impressed forces, then the motions which are determined by the velocities and forces are superposed (Boltzmann, *Ann. der Phys. und Chem.*, LVII, 1896, p. 45), and Ostwald's maximum-principle (*Lehrbuch der allgem. Chemie*, II, 1, 1892, p. 37) is, according to Zemplén's excellent and universally comprehensible remark (*Ann. der Phys. und Chem.*, X, 1903, p. 428), unsuitable for the description of *mechanical* events, because it does not take account of the *inertia* of the masses. However, it remains correct that the (virtual) works which are consistent with the conditions become actual. My text, which was drawn up before 1882, could not, of course, take account of the attempts to found an energetical mechanics of two years later. For the rest, I cannot value these attempts so little as some do. Even the old "classical" mechanics has not arrived at its present form without passing through analogous stages of error. In particular, Helm's view (*Die Energetik nach ihrer geschichtlichen Entwickelung,* Leipzig, 1898, pp. 205-252) can hardly be objected to. *Cf.* my exposition of the equal justification of the conceptions of work and force (*Ber. der Wiener Akad.,* December 1873), and also many passages of my *Mechanics,* particularly pp. 307 *et seq.*

VIII.

THE PRINCIPLE OF LEAST ACTION.

1. MAUPERTUIS enunciated, in 1747, a principle which he called *"le principe de la moindre quantité d'action,"* the principle of *least action.* He declared this principle to be one which eminently accorded with the wisdom of the Creator. He took as the measure of the "action" the product of the mass, the velocity, and the space described, or $m v s$. *Why,* it must be confessed, is not clear. By mass and velocity definite quantities may be understood; not so, however, by space, when the time is not stated in which the space is described. If, however, unit of time be meant, the distinction of space and velocity in the examples treated by Maupertuis is, to say the least, peculiar. It appears that Maupertuis reached this obscure expression by an unclear mingling of his ideas of *vis viva* and the principle of virtual velocities. Its indistinctness will be more saliently displayed by the details.

2. Let us see how Maupertuis applies his principle. If M, m be two inelastic masses, C and c their velocities before impact, and u their common velocity after impact, Maupertuis requires, (putting here velocities for

spaces,) that the "action" expended in the change of the velocities in impact shall be a minimum. Hence, $M(C-u)^2 + m(c-u)^2$ is a minimum; that is, $M(C-u) + m(c-u) = 0$; or

$$u = \frac{MC + mc}{M + m}.$$

For the impact of elastic masses, retaining the same designations, only substituting V and v for the two velocities after impact, the expression $M(C-V)^2 + m(c-v)^2$ is a minimum; that is to say,

$$M(C-V)dV + m(c-v)dv = 0 \quad \ldots \ldots \quad (1)$$

In consideration of the fact that the velocity of approach before impact is equal to the velocity of recession after impact, we have

$$C - c = -(V - v) \text{ or}$$
$$C + V - (c + v) = 0 \quad \ldots \ldots \ldots \ldots \quad (2)$$

and

$$dV - dv = 0 \quad \ldots \ldots \ldots \ldots \ldots \quad (3)$$

The combination of equations (1), (2), and (3) readily gives the familiar expressions for V and v. These two cases may, as we see, be viewed as processes in which the least change of *vis viva* by reaction takes place, that is, in which the *least counter-work* is done. They fall, therefore, under the principle of Gauss.

3. Peculiar is Maupertuis's deduction of the *law of the lever*. Two masses M and m (Fig. 188) rest on a bar a, which the fulcrum divides into the portions x and $a - x$. *If the bar be set in rotation,* the velocities and the spaces described will be proportional to the lengths of the lever-arms, and $M x^2 + m(a-x)^2$

is the quantity to be made a minimum, that is $M x - m(a - x) = 0$; whence $x = m\,a/\overline{M + m}$ —a condition that in the case of *equilibrium* is actually fulfilled. In criticism of this, it is to be remarked, first, that masses not subject to gravity or other forces, as Maupertuis here tacitly assumes, are *always* in equilibrium, and,

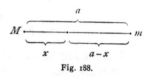

Fig. 188.

secondly, that the inference from Maupertuis's deduction is that the principle of least action is fulfilled *only* in the case of equilibrium, a conclusion which it was certainly not the author's intention to demonstrate.

If it were sought to bring this treatment into approximate accord with the preceding, we should have to assume that the *heavy* masses M and m constantly produced in each other during the process the least possible change of *vis viva*. On that supposition, we should get, designating the arms of the lever briefly by a, b, the velocities acquired in unit of time by u, v, and the acceleration of gravity by g, as our minimum expression, $M(g - u)^2 + m(g - v)^2$; whence $M(g - u)\,du + m(g - v)\,dv = 0$. But in view of the connection of the masses as lever,

$$\frac{u}{a} = -\frac{v}{b}, \text{ and}$$

$$du = -\frac{a}{b}\,dv;$$

whence these equations correctly follow

$$u = a\,\frac{Ma - mb}{Ma^2 + mb^2}\,g, \quad v = -b\,\frac{Ma - mb}{Ma^2 + mb^2}\,g,$$

and for the case of equilibrium, where $u = v = 0$,

$$Ma - mb = 0.$$

Thus, this deduction also, when we come to rectify it, leads to Gauss's principle.

4. Following the precedent of Fermat and Leibniz, Maupertuis also treats by his method the *motion of light*. Here again, however, he employs the notion "least action" in a totally different sense. The expression which for the case of refraction shall be a minimum, is $m \cdot AR + n \cdot RB$, where AR and RB denote the paths described by the light in the first and second media respectively, and m and n the corresponding velocities. True, we really do obtain here, if R be determined in conformity with the minimum condition, the result $\sin \alpha / \sin \beta = n/m = const$. But before, the "action" consisted in the *change* of the expressions mass $\times$ velocity $\times$ distance; now, however, it is constituted of the *sum* of these expressions. Before, the spaces described in unit of time were considered; in the present case the *total* spaces traversed are taken. Should not $m \cdot AR - n \cdot RB$ or $(m-n)(AR-RB)$ be taken as a minimum, and if not, why not? But even if we accept Maupertuis's conception, the reciprocal values of the velocities of the light are obtained, and not the actual values.

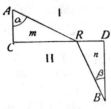

Fig. 189.

It will thus be seen that Maupertuis really had no principle, properly speaking, but only a vague formula, which was forced to do duty as the expression of different familiar phenomena not really brought under one conception. I have found it necessary to enter into some detail in this matter, since Maupertuis's performance, though it has been unfavorably criticized by all mathematicians, is, nevertheless, still invested with a sort of historical halo. It would seem almost as if something of the pious faith of the church had crept into mechanics. However, the mere *endeavor* to gain a more extensive view, although beyond the powers of the author, was not altogether without results. Euler, at least, if not also Gauss, was stimulated by the attempt of Maupertuis.

5. Euler's view is, that the *purposes* of the phenomena of nature afford as good a basis of explanation as their *causes*. If this position be taken, it will be presumed *a priori* that all natural phenomena present a maximum or minimum. Of what character this maximum or minimum is, can hardly be ascertained by metaphysical speculations. But in the solution of mechanical problems by the ordinary methods, it is possible, if the requisite attention be bestowed on the matter, to find the expression which in all cases is made a maximum or a minimum. Euler is thus not led astray by any metaphysical propensities, and proceeds much more scientifically than Maupertuis. He seeks an expression whose variation put $= 0$ gives the ordinary equations of mechanics.

For a *single* body moving under the action of forces Euler finds the requisite expression in the formula $\int v\,ds$, where ds denotes the element of the path and v the corresponding velocity. This expression is smaller for the path *actually* taken than for any other infinitely adjacent neighboring path between the same initial and terminal points, which the body may be *constrained* to take. Conversely, therefore, by *seeking* the path that makes $\int v\,ds$ a minimum, we can also determine the path. The problem of minimizing $\int v\,ds$ is, of course, as Euler assumed, a permissible one, only when v depends on the position of the elements ds, that is to say, when the principle of *vis viva* holds for the forces, or a force-function exists, or what is the same thing, when v is a simple function of coördinates. For a motion in a plane the expression would accordingly assume the form

$$\int \varphi\,(x,y)\,\sqrt{1+\left(\frac{dy}{dx}\right)^2}\,.\,dx$$

In the simplest cases Euler's principle is easily verified. If no forces act, v is constant, and the curve of motion becomes a straight line, for which $\int v\,d\,s = v\int d\,s$ is unquestionably *shorter* than for any other

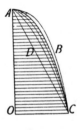

Fig. 190

curve between the same terminal points. Also, a body moving on a curved surface without the action of forces or friction, preserves its velocity, and describes on the surface a *shortest* line.

The consideration of the motion of a projectile in a parabola ABC (Fig. 190) will also show that the quantity $\int v\,d\,s$ is smaller for the parabola than for any other neighboring curve; smaller, even, than for the *straight* line ABC between the same terminal points. The velocity, here, depends solely on the vertical space described by the body, and is therefore the same for all curves whose altitude above OC is the same. If we divide the curves by a system of horizontal straight lines into elements which severally correspond, the elements to be multiplied by the same v's, though in the upper portions smaller for the straight line AD than for AB, are in the lower portions just the reverse; and as it is here that the larger v's come into play, the sum upon the whole is smaller for ABC than for the straight line.

Putting the origin of the coördinates at A, reckoning the abscissas x vertically downwards as positive, and calling the ordinates perpendicular thereto y, we obtain for the expression to be minimized

$$\int_0^x \sqrt{2g(a+x)}\sqrt{1+\left(\frac{dy}{dx}\right)^2}.\,dx,$$

where g denotes the acceleration of gravity and a the distance of descent corresponding to the initial velocity. As the condition of minimum the calculus of variations gives

$$\frac{\sqrt{2g(a+x)}\frac{dy}{dx}}{\sqrt{1+\left(\frac{dy^2}{dx}\right)}} = C \text{ or}$$

$$\frac{dy}{dx} = \frac{C}{\sqrt{2g(a+x)-C^2}} \text{ or}$$

$$y = \int \frac{C\,dx}{\sqrt{2g(a+x)-C^2}},$$

and, ultimately,

$$y = \frac{C}{g}\sqrt{2g(a+x)-C^2} + C',$$

where C and C' denote constants of integration that pass into $C = \sqrt{2ga}$ and $C' = 0$, if for $x = 0$, $dx/dy = 0$ and $y = 0$ be taken. Therefore, $y = 2\sqrt{a\,x}$. By this method, accordingly, the path of a projectile is shown to be of parabolic form.

6. Subsequently, Lagrange drew *express* attention to the fact that Euler's principle is applicable only in

cases in which the principle of *vis viva* holds. Jacobi pointed out that we cannot assert that $\int v\,ds$ for the actual motion is a *minimum,* but simply that the *variation* of this expression, in its passage to an infinitely adjacent neighboring path, is $= 0$. Generally, indeed, this condition coincides with a maximum or minimum, but it is possible that it should occur *without* such; and the minimum property in particular is subject to certain limitations. For example, if a body, constrained to move on a spherical surface, is set in motion by some impulse, it will describe a great circle, generally a shortest line. But if the length of the arc described exceeds 180°, it is easily demonstrated that there exist shorter infinitely adjacent neighboring paths between the terminal points.

7. So far, then, this fact only has been pointed out, that the ordinary equations of motion are obtained by equating the variation of $\int v\,ds$ to zero. But since the properties of the motion of bodies or of their paths may always be defined by differential expressions equated to zero, and since furthermore the condition that the variation of an integral expression shall be equal to zero is likewise given by differential expressions equated to zero, unquestionably *various other* integral expressions may be devised that give by variation the ordinary equations of motion, without its following that the integral expressions in question must possess on that account any particular *physical* significance.

8. The striking fact remains, however, that so *simple* an expression as $\int v\,ds$ does possess the property mentioned, and we will now endeavor to ascertain its physical import. To this end the analogies that exist between the motion of masses and the motion of light, as well as between the motion of masses and the equilib-

rium of strings—analogies noted by John Bernoulli and by Möbius—will stand us in stead.

A body on which no forces act, and which therefore preserves its velocity and direction constant, describes a straight line. A ray of light passing through a homogeneous medium (one having everywhere the same index of refraction) describes a straight line. A string, acted on by forces at its extremities only, assumes the shape of a straight line.

A body that moves in a curved path from a point A to a point B and whose velocity $v = \varphi(x, y, z)$ is a function of coördinates, describes between A and B a curve for which generally $\int v\, ds$ is a minimum. A ray of light passing from A to B describes the same curve, if the refractive index of its medium, $n = \varphi(x, y, z)$, is the same function of coördinates; and in this case $\int n\, ds$ is a minimum. Finally, a string passing from A to B will assume this curve, if its tension $S = \varphi(x, y, z)$ is the same above-mentioned function of coördinates; and for this case, also, $\int S\, ds$ is a minimum.

The *motion of a mass* may be readily deduced from the *equilibrium of a string,* as follows. On an element ds of a string, at its two extremities, the tensions S, S' act, and supposing the force on unit of length to be P, in addition a force $P \cdot ds$. These three forces, which we shall represent in magnitude and direction by BA, BC, BD (Fig. 191), are in equilibrium. If now, a body, with a velocity v represented in magnitude and direction by AB, enter the element of the path

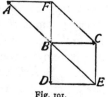

Fig. 191.

ds, and receive within the same the velocity component $BF = -BD$, the body will proceed onward with the

velocity $v' = BC$. Let Q be an accelerating force
whose action is directly opposite to that of P; then for
unit of time the acceleration of this force will be Q,
for unit of length of the string Q/v, and for the element of the string $(Q/v)ds$. The body will move,
therefore, in the *curve of the string*, if we establish between the forces P and the tensions S, in the case of
the string, and the accelerating forces Q and the velocity v in the case of the mass, the relation

$$P : -\frac{Q}{v} = S : v.$$

The minus sign indicates that the directions of P and
Q are opposite.

A closed circular string is in equilibrium when between the tension S of the string, everywhere constant,
and the force P falling radially outwards on unit of
length, the relation $P = S/r$ obtains, where r is the
radius of the circle. A body will move with the constant velocity v in a circle, when between the velocity
and the accelerating force Q acting radially inwards
the relation

$$\frac{Q}{v} = \frac{v}{r} \text{ or } Q = \frac{v^2}{r} \text{ obtains.}$$

A body will move with *constant* velocity v in *any* curve
when an accelerating force $Q = v^2/r$ constantly acts
on it in the direction of the center of curvature of each
element. A string will lie under a constant tension S
in any curve if a force $P = S/r$ acting outwardly from
the center of curvature of the element is impressed on
unit of length of the string.

No concept analogous to that of force is applicable
to the *motion of light*. Consequently, the deduction of
the motion of light from the equilibrium of a string or

the motion of a mass must be differently effected. A mass, let us say, is moving with the velocity $A B = v$ (Fig. 192). A force in the direction $B D$ is impressed on the mass which produces an in-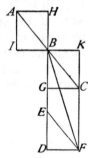crease of velocity $B E$, so that by the composition of the velocities $B C = A B$ and $B E$ the new velocity $B F = v'$ is produced. If we resolve the velocities v, v' into components paral-lel and perpendicular to the force in question, we shall perceive that the *parallel components* alone are *changed* by the action of the force. This being the case, we get, denoting by k the

Fig. 192.

perpendicular component, and by α and α' the angles v and v' make with the direction of the force,

$$k = v \sin \alpha$$
$$k = v' \sin \alpha' \text{ or}$$
$$\frac{\sin \alpha}{\sin \alpha'} = \frac{v'}{v}.$$

If, now, we picture to ourselves a ray of light that penetrates in the direction of v a refracting plane at right angles to the direction of action of the force, and thus passes from a medium having the index of refrac-tion n into a medium having the index of refraction n', where $n/n' = v/v'$, this ray of light will describe the same path as the body in the case above. If, there-fore, we wish to imitate the motion *of a mass* by the *motion of a ray of light* (in the same curve), we must everywhere put the indices of refraction, n, *propor-tional* to the velocities. To deduce the indices of re-fraction from the forces, we obtain for the velocity

$$d\left(\frac{v^2}{2}\right) = P\,dq, \text{ and}$$

for the index of refraction, by analogy,

$$d\left(\frac{n^2}{2}\right) = P\,dq,$$

where P denotes the force and $d\,q$ a distance-element in the direction of the force. If $d\,s$ is the element of the path and α the angle made by it with the direction of the force, we have then

$$d\left(\frac{v^2}{2}\right) = P\cos\alpha\,.\,ds$$

$$d\left(\frac{n^2}{2}\right) = P\cos\alpha\,.\,ds.$$

For the path of a projectile, under the conditions above assumed, we obtained the expression $y = 2\sqrt{a\,x}$. This same parabolic path will be described by a ray of light, if the law $n = \sqrt{2g(a+x)}$ be taken as the index of refraction of the medium in which it travels.

9. We will now investigate more accurately the manner in which this minimum property is related to the *form* of the curve. Let us take, first, (Fig. 193) a broken straight line $A\,B\,C$, which intersects the straight line MN, put $A\,B = s$, $B\,C = s'$, and seek the condition that makes $vs + v'\,s'$ a minimum for the line that passes through the fixed points A and B, where v and v' are supposed to have different, though constant, values above and below MN. If we displace the point B an infinitely small distance to D, the new line through A and C will remain parallel to the original one, as the drawing symbolically shows. The expression $v\,s + v'\,s'$ is increased hereby by an amount

$$-v\,m\sin\alpha + v'\,m\sin\alpha',$$

where $m = D\,B$. The alteration is accordingly proportional to $-v \sin \alpha + v' \sin \alpha'$, and the condition of minimum is that

$$-v \sin \alpha + v' \sin \alpha' = 0, \text{ or } \frac{\sin \alpha}{\sin \alpha'} = \frac{v'}{v}.$$

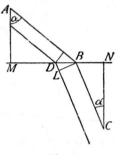
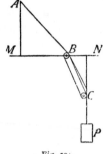

Fig. 193.   Fig. 194.

If the expression $s/v + s'/v'$ is to be made a minimum, we have, in a similar way,

$$\frac{\sin \alpha}{\sin \alpha'} = \frac{v}{v'}.$$

If, next we consider the case of a string stretched in the direction $A\,B\,C$, the tensions of which S and S' are different above and below $M\,N$, in this case it is the minimum of $S\,s + S'\,s'$ that is to be dealt with. To obtain a distinct idea of this case, we may imagine the string stretched once between A and B and thrice between B and C, and finally a weight P attached. Then $S = P$ and $S' = 3\,P$. If we displace the point B a distance m, any diminution of the expression $S\,s + S'\,s'$ thus effected, will express the increase of *work* which the attached weight P performs. If $-S\,m \sin \alpha + S'\,m' \sin \alpha' = 0$, no work is performed. Hence, the *minimum* of $S s + S' s'$ corresponds to a *maximum* of work. In the present case the principle of least action is simply a *different form* of the principle of virtual displacements.

Now suppose that $A\,B\,C$ is a ray of light, whose velocities v and v' above and below $M\,N$ are to each other

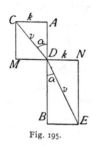

Fig. 195.

as 3 to 1. The motion of light between two points A and B is such that the light reaches B in a minimum of time. The physical reason of this is simple. The light travels from A to B, in the form of elementary waves, by different routes. Owing to the periodicity of the light, the waves generally destroy each other, and only those that reach the designated point in equal times, that is, in equal phases, produce a result. But this is true only of the waves that arrive by the *minimum path* and its adjacent neighboring paths. Hence, for the path actually taken by the light $s/v + s'/v'$ is a minimum. And since the indices of refraction n are inversely proportional to the velocities v of the light, therefore also $n\,s + n'\,s'$ is a minimum.

In the consideration of the *motion of a mass* the condition that $v\,s + v'\,s'$ shall be a minimum, strikes us as something novel. (Fig. 195.) If a mass, in its passage through a plane MN, receive, as the result of the action of a force impressed in the direction $D\,B$, an increase of velocity, by which v, its original velocity, is made v', we have for the path actually taken by the mass the equation $v \sin \alpha = v' \sin \alpha' = k$. *This equation, which is also the condition of minimum, simply states that only the velocity-component parallel to the direction of the force is altered, but that the component k at right angles thereto remains unchanged.* Thus, here also, Euler's principle simply states a familiar fact in a new form.

To the exposition given on the preceding pages, in the year 1883, I have the following remarks to add. It will be seen that the principle of least action, like all other minimum principles in mechanics, is a simple expression of the fact that in the instances in question precisely so much happens as possibly *can* happen under the circumstances, or as is determined, viz., *uniquely* determined, by them. The deduction of cases of equilibrium from unique determination has already been discussed, and the same question will be considered in a later place. With respect to dynamic questions, the import of the principle of unique determination has been better and more perspicuously elucidated than in my case by J. Petzoldt in a work entitled *Maxima, Minima und Oekonomie* (Altenburg, 1891). He says (*loc. cit.,* page 11): "In the case of all motions, the paths actually traversed admit of being interpreted as *signal* instances chosen from an infinite number of conceivable instances. Analytically, this has no other meaning than that expressions may always be found which yield the differential equations of the motion when their variation is equated to zero, —for the variation vanishes only when the integral assumes a unique value."

As a fact, it will be seen that in the instances which were treated above an increment of velocity is uniquely determined only in the direction of the force, while an infinite number of equally legitimate incremental components of velocity at right angles to the force are conceivable, which are, however, for the reason given, excluded by the principle of unique determination. I am in entire accord with Petzoldt when he says: "The theorems of Euler and Hamilton, and not less that of Gauss, are thus nothing more than analytic expressions for the fact of experience that the phenomena of nature are uniquely determined." The uniqueness of the minimum is determinative.

I should like to quote here, from a note which I published in the November number of the Prague *Lotos* for 1873, the following passage: "The static and dynamical principles of mechanics may be expressed as isoperimetrical laws. The anthropomorphic conception is, however, by no means essential, as may be seen, for example, in the principle of virtual velocities. If we have once perceived that the work A determines velocity, it will readily be seen that where work is not done when the system passes into all adjacent positions, no velocity can be acquired, and consequently that equilibrium obtains. The condition of equilibrium will therefore be $\delta A = 0$; where A need not necessarily be exactly a maximum or minimum. These laws are not absolutely restricted to mechanics; they may be of very general scope. If the change in the form of a phenomenon B be dependent on a phenomenon A, the condition that B shall pass over into a certain form will be $\delta A = 0$."

As will be seen, I grant in the foregoing passage that it is possible to discover analogies for the principle of least action in the various departments of physics without reaching them through the circuitous course of mechanics. I look upon mechanics, not as the ultimate explanatory foundation of all the other provinces, but rather, owing to its superior formal development, as an admirable prototype of such an explanation. In this respect, my view differs apparently little from that of the majority of physicists, but the difference is an essential one after all. In further elucidation of my meaning, I should like to refer to the discussions which I have given in my *Principles of Heat* (particularly pages 192, 318, and 356, German edition), and also to my article "On Comparison in Physics" (*Popular Scientific Lectures*, English translation, page 236). Noteworthy articles touching on this point are: C. Neumann, "Das Ostwald'sche Axiom des Energieumsatzes" (*Berichte der k. sächs. Gesellschaft*, 1892, p. 184), and Ostwald, "Ueber das Princip des ausgezeichneten Falles" (*loc. cit.*, 1893, p. 600).

10. The minimum condition $-v \sin \alpha + v' \sin \alpha' = 0$ may also be written, if we pass from a finite broken straight line to the elements of curves, in the form

$$-v \sin \alpha + (v + dv) \sin (\alpha + d\alpha) = 0$$

or

$$d(v \sin \alpha) = 0$$

or, finally,

$$v \sin \alpha = const.$$

In agreement with this, we obtain for the motion of light

$$d(n \sin \alpha) = 0, \; n \sin \alpha = const,$$

$$d\left(\frac{\sin \alpha}{v}\right) = 0, \; \frac{\sin \alpha}{v} = const,$$

and for the equilibrium of a string

$$d(S \sin \alpha) = 0, \; S \sin \alpha = const.$$

To illustrate the preceding remarks by an example, let us take the parabolic path of a projectile, where α

always denotes the angle that the element of the path makes with the perpendicular. Let the velocity be $v = \sqrt{2g(a+x)}$, and let the axis of the y-ordinates be horizontal. The condition $v \cdot \sin \alpha = const$, or $\sqrt{2g(a+x)} \cdot dy/ds = const$, is identical with that which the calculus of variation gives, and we now know its *simple physical* significance. If we picture to ourselves a string whose tension is $S = \sqrt{2g(a+x)}$, an arrangement which might be effected by fixing frictionless pulleys on horizontal parallel rods placed in a vertical plane, then passing the string through these a sufficient number of times, and finally attaching a weight to the extremity of the string, we shall obtain

Fig. 196.

again, for equilibrium (Fig. 196), the preceding condition, the physical significance of which is now obvious. When the distances between the rods are made infinitely small the string assumes the parabolic form. In a medium, the refractive index of which varies in the vertical direction by the law $n = \sqrt{2g(a+x)}$, or in which the velocity of light varies similarly by the law $v = 1/\sqrt{2g(a+x)}$, a ray of light will describe a path which is a parabola. If we should make the velocity in such a medium $v = \sqrt{2g(a+x)}$, the ray would describe a cycloidal path, for which, not $\int \sqrt{2g(a+x)} \cdot ds$, but the expression $\int ds / \sqrt{2g(a+x)}$ would be a minimum.

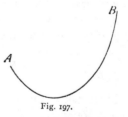

Fig. 197.

11. In comparing the equilibrium of a string with the motion of a mass, we may employ in place of a

string wound round pulleys, a simple, homogeneous cord, provided we subject the cord to an appropriate system of forces. We readily observe that the systems of forces that make the tension, or, as the case may be, the velocity, the *same* function of coördinates, are *different*. If we consider, for example, the force of gravity, $v = \sqrt{2g(a+x)}$. A string, however, subjected to the action of gravity, forms a catenary, the tension of which is given by the formula $S = m - n\,x$, where m and n are constants. The analogy subsisting between the equilibrium of a string and the motion of a mass is substantially conditioned by the fact that for a string subjected to the action of forces possessing a force-function U, there obtains in the case of equilibrium the easily demonstrable equation $U + S = $ const. This *physical* interpretation of the principle of least action is here illustrated only for simple cases; but it may also be applied to cases of greater complexity, by imagining groups of surfaces of equal tension, of equal velocity, or equally refractive indices constructed which divide the string, the path of the motion, or the path of the light into elements, and by making a in such a case represent the angle which these elements make with the respective surface-normals. The principle of least action was extended to systems of masses by Lagrange, who presented it in the form

$$\delta\,\Sigma\,m\int v\,d\,s = 0.$$

If we reflect that the principle of *vis viva*, which is the real foundation of the principle of least action, is not annulled by the connection of the masses, we shall comprehend that the latter principle is in this case also valid and physically intelligible.

IX.

HAMILTON'S PRINCIPLE.

1. It was remarked above that *various* expressions can be devised whose variations equated to zero give the ordinary equations of motion. An expression of this kind is contained in Hamilton's principle

$$\delta \int_{t_0}^{t_1} (U + T)\, dt = 0, \text{ or}$$

$$\int_{t_0}^{t_1} (\delta U + \delta T)\, dt = 0,$$

where δU and δT denote the variations of the work and the *vis viva,* vanishing for the initial and terminal epochs. Hamilton's principle is easily deduced from D'Alembert's, and, conversely, D'Alembert's from Hamilton's; the two are in fact identical, their difference being merely that of form.*

2. We shall not enter here into any extended investigation of this subject, but simply exhibit the identity of the two principles by an *example*—the same that served to illustrate the principle of D'Alembert: the motion of a wheel and axle by the over-weight of one of its parts (Fig. 198). In place of the actual motion, we may imagine, performed in the same interval of time, a *different* motion, varying infinitely little from the actual motion, but coinciding

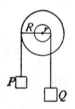

Fig. 198.

exactly with it at the beginning and end. There are thus produced in every element of time $d\,t$, variations

* Compare, for example, Kirchhoff. *Vorlesungen über mathematische Physik, Mechanik,* p. 25 *et seq.,* and Jacobi, *Vorlesungen über Dynamik,* p. 58.

of the work (δU) and of the *vis viva* (δT); variations, that is, of the values U and T realized in the actual motion. But for the actual motion, the integral expression, above stated, is $= 0$, and may be employed, therefore, to determine the actual motion. If the angle of rotation performed varies in the element of time $d\,t$ an amount α from the angle of the actual motion, the variation of the work corresponding to such an alteration will be

$$\delta\,U = (PR - Q\,r)\,\alpha = M\,\alpha.$$

The *vis viva*, for any given angular velocity ω, is

$$T = \frac{1}{g}\,(PR^2 + Q\,r^2)\,\frac{\omega^2}{2},$$

and for a variation $\delta\omega$ of this velocity the variation of the *vis viva* is

$$\delta T = \frac{1}{g}\,(PR^2 + Q\,r^2)\,\omega\,\delta\,\omega.$$

But if the angle of rotation varies in the element $d\,t$ an amount α,

$$\delta\,\omega = \frac{d\,\alpha}{d\,t} \text{ and}$$

$$\delta\,T = \frac{1}{g}\,(PR^2 + Q\,r^2)\,\omega\frac{d\,\alpha}{d\,t} = N\frac{d\,\alpha}{d\,t}.$$

The form of the integral expression, accordingly, is

$$\int_{t_0}^{t_1}\left[M\alpha + N\frac{d\,\alpha}{d\,t}\right]d\,t = 0.$$

But as

$$\frac{d}{d\,t}\,(N\alpha) = \frac{dN}{d\,t}\,\alpha + N\frac{d\,\alpha}{d\,t},$$

therefore,

$$\int_{t_0}^{t_1}\left(M-\frac{dN}{dt}\right)\alpha \cdot dt + (N\alpha)\Big|_{t_0}^{t_1} = 0.$$

The second term of the left-hand member, though, drops out, because, by hypothesis, at the beginning and end of the motion $\alpha = 0$. Accordingly, we have

$$\int_{t_0}^{t_1}\left(M-\frac{dN}{dt}\right)\alpha\,dt = 0,$$

an expression which, since α in every element of time is arbitrary, cannot subsist unless generally

$$M - \frac{dN}{dt} = 0.$$

Substituting for the symbols the values they represent, we obtain the familiar equation

$$\frac{d\omega}{dt} = \frac{PR - Qr}{PR^2 + Qr^2}g.$$

D'Alembert's principle gives the equation

$$\left(M-\frac{dN}{dt}\right)\alpha = 0,$$

which holds for every *possible* displacement. We might, in the converse order, have started from this equation, have thence passed to the expression

$$\int_{t_0}^{t_1}\left(M-\frac{dN}{dt}\right)\alpha\,dt = 0,$$

and, finally, from the latter proceeded to the same result

$$\int_{t_0}^{t_1} \left(M\alpha + N\frac{d\alpha}{dt} \right) dt - (N\alpha)_{t_0}^{t_1} =$$

$$\int_{t_0}^{t_1} \left(M\alpha + N\frac{d\alpha}{dt} \right) dt = 0.$$

3. As a second and more simple example let us consider the motion of vertical descent. For every infinitely small displacement s the equation subsists $[mg - m(dv/dt)]s = 0$, in which the letters retain their conventional significance. Consequently, this equation obtains

$$\int_{t_0}^{t_1} \left(mg - m\frac{dv}{dt} \right) s \cdot dt = 0,$$

which, as the result of the relations

$$d\frac{(mvs)}{dt} = m\frac{dv}{dt}s + mv\frac{ds}{dt} \text{ and}$$

$$\int_{t_0}^{t_1} \frac{d(mvs)}{dt}dt = (mvs)_{t_0}^{t_1} = 0,$$

provided s vanishes at both limits, passes into the form

$$\int_{t_0}^{t_1} \left(mgs + mv\frac{ds}{dt} \right) dt = 0,$$

that is, into the form of Hamilton's principle.

Thus, through all the apparent differences of the mechanical principles a common fundamental sameness is seen. These principles are not the expression of different facts, but, in a measure, are simply views of different *aspects* of the same fact.

X.

SOME APPLICATIONS OF THE PRINCIPLES OF MECHANICS

TO HYDROSTATIC AND HYDRODYNAMIC QUESTIONS.

1. We will now supplement the examples which we have given of the application of the principles of mechanics, as they applied to rigid bodies, by a few hydrostatic and hydrodynamic illustrations. We shall first discuss the laws of equilibrium of a *weightless* liquid subjected exclusively to the action of so-called molecular forces. The forces of gravity we neglect in our considerations. A liquid may, in fact, be placed in circumstances in which it will behave as if no forces of gravity acted. The method of this is due to PLA-TEAU.* It is effected by immersing olive oil in a mixture of water and alcohol of the same density as the oil. By the principle of Archimedes the gravity of the masses of oil in such a mixture is exactly counterbalanced, and the liquid really acts as if it were devoid of weight.

2. First, let us imagine a weightless liquid mass free in space (Fig. 199). Its molecular forces, we know, act only at very small distances. Taking as our radius the distance at which the molecular forces cease to exert a measurable influence, let us describe about a particle *a, b, c* in the interior of the mass a sphere—the so-called sphere of action. This sphere of action is regu-

---

* *Statique expérimentale et théorique des liquides,* 1873.

larly and uniformly filled with other particles. The
resultant force on the central particles *a, b, c* is there-

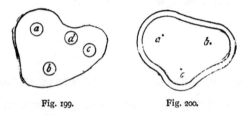

Fig. 199.   Fig. 200.

fore zero. Those parts only that lie at a distance from
the bounding surface less than the radius of the sphere
of action are in different dynamic conditions from the
particles in the interior. If the radii of curvature of
the surface-elements of the liquid mass be all regarded
as very great compared with the radius of the sphere
of action, we may cut off from the mass a superficial
stratum of the thickness of the radius of the sphere of
action in which the particles are in different physical
conditions from those in the interior. If we convey
a particle *a* in the interior of the liquid from the position *a* to the position *b* or *c,* the physical condition
of this particle, as well as that of the particles which
take its place, will remain unchanged. No work can
be done in this way. Work can be done only when a
particle is conveyed from the superficial stratum into
the interior, or, from the interior into the superficial
stratum. That is to say, work can be done only by a
change of size of the surface. The consideration whether
the density of the superficial stratum is the same as
that of the interior, or whether it is constant throughout the entire thickness of the stratum, is not primarily
essential. As will readily be seen, the variation of the
surface-area is equally the condition of the perform-

ance of work when the liquid mass is immersed in a second liquid, as in Plateau's experiments.

We now inquire whether the work which by the transportation of particles into the interior effects a diminution of the surface-area is positive or negative, that is, whether work is performed or work is expended. If we put two fluid drops in contact, they will coalesce *of their own accord;* and as by this action the area of the surface is diminished, it follows that the work that produces a diminution of superficial area in a liquid mass is *positive.* Van der Mensbrugghe has demonstrated this by a very pretty experiment. A square wire frame is dipped into a solution of soap and water, and on the soap-film formed a loop of moistened thread is placed (Fig. 201). If the film within the loop be punctured, the film outside the loop will contract till the thread bounds a circle in the mid-

Fig. 201.

dle of the liquid surface. But the circle, of all plane figures of the same circumference, has the greatest area; consequently, the liquid film has contracted to a minimum. The following will now be clear. A weightless liquid, the forces acting on which are molecular forces, will be in *equilibrium* in all forms in which a system of virtual displacements produces *no* alteration of the liquid's superficial area. But all infinitely small changes of form may be regarded as *virtual* which the liquid admits without alteration of its *volume.* Consequently, equilibrium subsists for all liquid forms for which an infinitely small deformation produces a superficial variation = 0. For a given volume a *minimum* of superficial area gives stable equilibrium; a *maximum* unstable equilibrium.

Among all solids of the same volume, the sphere
has the least superficial area. Hence, the form which
a free liquid mass will assume, the form of stable equilibrium, is the sphere. For this form a maximum of
work is done; for it, no more can be done. If the
liquid adheres to rigid bodies, the form assumed is dependent on various collateral conditions, which render
the problem more complicated.

3. The connection between the *size* and the *form* of
the liquid surface may be investigated as follows. We
imagine the closed outer surface of the liquid to receive without alteration of the liquid's volume an infinitely small variation (Fig. 202). By two sets of
mutually perpendicular lines of curvature, we cut up
the original surface into infinitely small rectangular
elements. At the angles of these elements, on the original surface, we erect normals to the surface, and
determine thus the angles of the corresponding elements of the varied surface. To every element dO of
the original surface there now
corresponds an element dO' of
the varied surface; by an infinitely small displacement,
δn, along the normal, outwards or inwards, dO passes
into dO' and into a corresponding variation of magnitude.

Fig. 202.

Let $d p$, $d q$ be the sides of the element dO. For the
sides $d p'$, $d q'$ of the element dO', then, these relations
obtain

$$dp' = dp\left(1 + \frac{\delta n}{r}\right)$$

$$dq' = dq\left(1 + \frac{\delta n}{r'}\right),$$

where r and r' are the radii of curvature of the principal sections touching the elements of the lines of curvature p, q, or the so-called principal radii of curvature (Fig. 203).* The radius of curvature of an outwardly convex element is reckoned as positive, that of an outwardly concave element as negative, in the usual manner. For the variation of the element we obtain, accordingly,

$$\delta . dO = dO' - dO = dp\,dq\left(1 + \frac{\delta n}{r}\right)\left(1 + \frac{\delta n}{r'}\right) - dp\,dq.$$

Neglecting the higher powers of δn we get

$$\delta . dO = \left(\frac{1}{r} + \frac{1}{r'}\right)\delta n . dO.$$

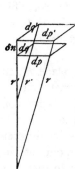

The variation of the whole surface, then, is expressed by

$$\delta O = \int\left(\frac{1}{r} + \frac{1}{r'}\right)\delta n . dO \ \ . \ \ . \ \ . \ \ . \ \ (1)$$

Furthermore, the normal displacements must be so chosen that

$$\int \delta n \cdot d O = 0 \ . \ . \ . \ . \ . \ . \ . \ . \ (2)$$

Fig. 203.

the superficial elements (in the latter case reckoned as negative) shall be equal to zero, or the *volume* remain constant.

Accordingly, expressions (1) and (2) can be put simultaneously $= 0$ only if $1/r + 1/r'$ has the *same value* for all points of the surface. This will be readily seen from the following consideration. Let the ele-

---

* The normal at any point of a surface is cut by normals at infinitely neighboring points that lie in two directions on the surface from the original point, these two directions being at right angles to each other; and the distances from the surface at which these normals cut are the two principal, or extreme, radii of curvature of the surface.—*Trans.*

ments dO of the original surface be symbolically
represented by the ele-
ments of the line $A\,X$
(Fig. 204) and let the
normal displacements
$\delta\,n$ be erected as ordinates thereon in the

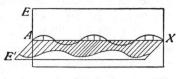

Fig. 204.

plane E, the outward displacements upwards as positive and the inward displacements downwards as negative. Join the extremities of these ordinates so as to
form a curve, and take the quadrature of the curve,
reckoning the surface above AX as positive and that
below it as negative. For all systems of $\delta\,n$ for which
this quadrature $= 0$, the expression (2) also $= 0$, and
all such systems of displacements are admissible, that
is, are virtual displacements.

Now let us erect as ordinates, in the plane E', the
values of $1/r + 1/r'$ that belong to the elements dO. A
case may be easily imagined in which the expressions
(1) and (2) assume coincidently the value zero. Should,
however, $1/r + 1/r'$ have *different* values for different
elements, it will always be possible without altering
the zero-value of the expression (2), so to distribute
the displacements $\delta\,n$ that the expression (1) shall be
different from zero. Only on the condition that $1/r +
1/r'$ has *the same value* for all the elements, is expres-
sion (1) necessarily and universally equated to zero
with expression (2).

Accordingly, from the two conditions (1) and (2)
it follows that $1/r + 1/r' = const$; that is to say, the
sum of the reciprocal values of the principal radii of
curvature, or of the radii of curvature of the principal
normal sections, is, in the case of equilibrium, constant
for the whole surface. By this theorem the dependence

of the *area* of a liquid surface on its superficial *form* is defined. The train of reasoning here pursued was first developed by GAUSS,* in a much fuller and more special form. It is not difficult, however, to present its essential points in the foregoing simple manner.

4. A liquid mass, left wholly to itself, assumes, as we have seen, the spherical form, and presents an absolute minimum of superficial area. The equation $1/r + 1/r' = const$ is here visibly fulfilled in the form $2/R = const$, R being the radius of the sphere. If the free surface of the liquid mass be bounded by two solid circular rings, the planes of which are parallel to each other and perpendicular to the line joining their middle points, the surface of the liquid mass will assume the form of a surface of revolution. The nature of the meridian curve and the volume of the inclosed mass are determined by the radius of the rings R, by the distance between the circular planes, and by the value of the expression $1/r + 1/r'$ for the surface of revolution. When

$$\frac{1}{r} + \frac{1}{r'} = \frac{1}{r} + \frac{1}{\infty} = \frac{1}{R},$$

the surface of revolution becomes a cylindrical surface. For $1/r + 1/r' = 0$, where one normal section is convex and the other concave, the meridian curve assumes the form of the catenary. Plateau visibly demonstrated these cases by pouring oil on two circular rings of wire fixed in the mixture of alcohol and water mentioned above.

Now let us picture to ourselves a liquid mass bounded by surface-parts for which the expression $1/r + 1/r'$ has a positive value, and by other parts

* *Principia Generalia Theoriæ Figuræ Fluidorum in Statu Æquilibrii*, Göttingen, 1830; *Werke*, **V**, 29, Göttingen, 1867.

for which the same expression has a negative value, or, more briefly expressed, by convex and concave surfaces. It will be readily seen that any displacement of the superficial elements outwards along the normal will produce in the concave parts a diminution of the superficial area and in the convex parts an increase. Consequently, *work* is *performed* when *concave surfaces* move outwards and *convex* surfaces *inwards*. Work also is performed when a superficial portion moves outwards for which $1/r + 1/r' = + a$, while simultaneously an equal superficial portion for which $1/r + 1/r' > a$ moves inwards.

Hence, when *differently curved* surfaces bound a liquid mass, the convex parts are forced inwards and the concave outwards till the condition $1/r + 1/r' = const$ is fulfilled for the entire surface. Similarly, when a *connected* liquid mass has *several* isolated surface-parts, bounded by rigid bodies, the value of the expression $1/r + 1/r'$ must, for the state of equilibrium be the same for all free portions of the surface.

For example, if the space between the two circular rings in the mixture of alcohol and water referred to above, be filled with oil, it is possible, by the use of a sufficient quantity of oil, to obtain a cylindrical surface whose two bases are spherical segments. The curvatures of the lateral and basal surfaces will accordingly fulfil the condition $1/R + 1/\infty = 1/\varrho + 1/\varrho$, or $\varrho = 2R$, where ϱ is the radius of the sphere and R that of the circular rings. Plateau verified this conclusion by experiment.

5. Let us now study a weightless liquid mass which encloses a hollow space. The condition that $1/r + 1/r'$ shall have the same value for the interior and exterior surfaces, is here not realizable. On the contrary, as

this sum has always a greater positive value for the closed exterior surface than for the closed interior surface, the liquid will perform work, and, flowing from the outer to the inner surface, cause the hollow space to disappear. If, however, the hollow space be occupied by a fluid or gaseous substance subjected to a determinate pressure, the work done in the last-mentioned process can be *counteracted* by the work expended to produce the compression, and thus equilibrium may be produced.

Fig. 205.

Let us picture to ourselves a liquid mass confined between two similar and similarly situated surfaces very near each other (Fig. 205). A *bubble* is such a system. Its primary condition of equilibrium is the exertion of an excess of pressure by the inclosed gaseous contents. If the sum $1/r + 1/r'$ has the value $+ a$ for the exterior surface, it will have for the interior surface very nearly the value $- a$. A bubble, left wholly to itself, will always assume the spherical form. If we conceive such a spherical bubble, the thickness of which we neglect, the total diminution of its superficial area, on the shortening of the radius r by dr, will be $16\, r\, \pi\, d\, r$. If, therefore, in the diminution of the surface by unit of area the work A is performed, then $A \cdot 16\, r\, \pi\, d\, r$ will be the total amount of work to be compensated for by the work of compression $p \cdot 4r^2\, \pi\, d\, r$ expended by the pressure p on the inclosed contents. From this follows $4\, A/r = p$; from which A may be easily calculated if the measure of r is obtained and p is found by means of a manometer introduced in the bubble.

An *open spherical* bubble cannot subsist. If an

open bubble is to become a figure of equilibrium, the sum $1/r + 1/r'$ must not only be constant for each of the two bounding surfaces, but must also be equal for both. Owing to the opposite curvatures of the surfaces, then, $1/r + 1/r' = 0$. Consequently, $r = -r'$ for all points. Such a surface is called a minimal surface; that is, it has the smallest area consistent with its containing certain closed contours. It is also a surface of zero-sum of principal curvatures; and its elements, as we readily see, are saddle-shaped. Surfaces of this kind are obtained by constructing closed space-curves of wire and dipping the wire into a solution of soap and water.* The soap-film assumes of its own accord the form of the curve mentioned.

6. Liquid figures of equilibrium, made up of thin films, possess a peculiar property. The work of the forces of gravity affects the *entire* mass of a liquid; that of the molecular forces is restricted to its superficial film. Generally, the work of the forces of gravity preponderates. But in thin films the molecular forces come into very favorable conditions, and it is possible to produce the figures in question without difficulty in the open air. Plateau obtained them by dipping wire polyhedrons into solutions of soap and water. Plane liquid films are thus formed, which meet one another at the edges of the framework. When thin plane films are so joined that they meet at a hollow edge, the law $1/r + 1/r' = const$ no longer holds for the liquid surface, as this sum has the value zero for plane surfaces and for the hollow edge a very large negative value. Conformably, therefore, to the views reached above, the liquid should run out of the films,

---

* The mathematical problem of determining such a surface, when the forms of the wires are given, is called *Plateau's Problem.—Trans.*

the thickness of which would constantly decrease, and escape at the edges. This is, in fact, what happens. But when the thickness of the films has decreased to a certain point, then, for *physical* reasons, which are, as it appears, not yet perfectly known, a *state of equilibrium* is effected.

Yet, notwithstanding the fact that the fundamental equation $1/r + 1/r' = const$ is not fulfilled in these figures, because very thin liquid films, especially films of viscous liquids, present physical conditions somewhat different from those on which our original suppositions were based, these figures present, neverthless, in all cases a *minimum* of superficial area. The liquid films, connected with the wire edges and with one another, always meet at the edges by threes at approximately equal angles of 120°, and by fours in corners at approximately equal angles. And it is geometrically demonstrable that these relations correspond to a minimum of superficial area. In the great diversity of phenomena here discussed but one fact is expressed, namely that the molecular forces do work, positive work, when the superficial area is diminished.

7. The figures of equilibrium which Plateau obtained by dipping wire polyhedrons in solutions of soap, form systems of liquid films presenting a remarkable *symmetry*. The question accordingly forces itself upon us: What has equilibrium to do with symmetry and regularity? The explanation is obvious. In every symmetrical system every deformation that tends to destroy the symmetry is complemented by an equal and opposite deformation that tends to restore it. In each deformation positive or negative work is done. One condition, therefore, though not an absolutely sufficient one, that a maximum or minimum of work

corresponds to the form of equilibrium, is thus supplied by symmetry. Regularity is successive symmetry. There is no reason, therefore, to be astonished that the forms of equilibrium are often symmetrical and regular.

8. The science of mathematical hydrostatics arose in connection with a special problem—that of *the shape of the earth*. Physical and astronomical data had led Newton and Huygens to the view that the earth is an oblate ellipsoid of revolution. NEWTON attempted to calculate this oblateness by conceiving the rotating earth as a fluid mass, and assuming that all fluid filaments drawn from the surface to the center exert the same pressure on the center. HUYGENS's assumption was that the directions of the forces are perpendicular to the superficial elements. BOUGUER combined both assumptions. CLAIRAUT, finally (*Théorie de la figure de la terre*, Paris, 1743), pointed out that the fulfilment of *both* conditions does not assure the subsistence of equilibrium.

Clairaut's starting-point is this. If the fluid earth is in equilibrium, we may, without disturbing its equilibrium, imagine any portion of it solidified. Accordingly, let all of it be solidified but a canal *A B*, of any

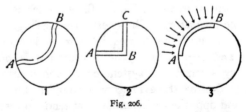

Fig. 206.

form (Fig. 206, 1). The liquid in this canal must also be in equilibrium. But now the conditions which control equilibrium are more easily investigated. If equilibrium exists in *every imaginable canal of this kind,* then the *entire* mass will be in equilibrium. Incidentally Clairaut remarks, that the Newtonian assumption is realized when the canal passes through the center (illustrated in Fig. 206, 2), and the Huygenian when the canal passes along the surface (Fig. 206, 3).

But the kernel of the problem, according to Clairaut, lies in a different view. In *all imaginable* canals, even in one *which returns into itself,* the fluid must be in equilibrium. Hence, if cross-sections be made at any two points M and N of the canal of Fig. 207, the two fluid columns $M\,P\,N$ and $M\,Q\,N$ must exert on the surfaces of section at M and N equal pressures. The terminal pressure of a fluid column of any such canal cannot, therefore, depend on the *length* and the *form* of the fluid column, but must depend solely on the *position* of its terminal points.

Imagine in the fluid in question a canal $M\,N$ of any form (Fig. 208) referred to a system of rectangular co-

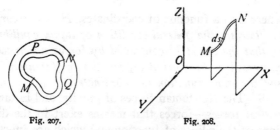

Fig. 207.   Fig. 208.

ördinates. Let the fluid have the *constant* density ϱ and let the force-components $X,\,Y,\,Z$ acting on unit of mass of the fluid in the coördinate directions, be functions of the coördinates $x,\,y,\,z$ of this mass. Let the element of length of the canal be called $d\,s$, and let its projections on the axes be $d\,x,\,d\,y,\,d\,z$. The force-components acting on unit of mass in the direction of the canal are then $X(dx/ds)$, $Y(dy/ds)$, $Z(dz/ds)$. Let q be the cross-section; then, the total force impelling the element of mass $\rho\, q\, d\, s$ in the direction $d\, s$, is

$$\rho q\, ds \left(X\frac{dx}{ds} + Y\frac{dy}{ds} + Z\frac{dz}{ds} \right).$$

This force must be balanced by the increment of pressure through the element of length, and consequently must be put equal to $q \cdot dp$. We obtain, accordingly, $dp = \varrho(Xdx + Ydy + Zdz)$. The difference of pressure (p) between the two extremities M and N is found by integrating this expression from M to N. But as this difference is not dependent on the form of the canal but solely on the position of the extremities M and N, it follows that $\varrho(Xdx + Ydy + Zdz)$, or, the density being constant, $Xdx + Ydy + Zdz$, must be a complete differential. For this it is necessary that

$$X = \frac{dU}{dx}, \; Y = \frac{dU}{dy}, \; Z = \frac{dU}{dz},$$

where U is a function of coördinates. *Hence, according to Clairaut, the general condition of liquid equilibrium is, that the liquid be controlled by forces which can be expressed as the partial differential coefficients of one and the same function of coördinates.*

9. The Newtonian forces of gravity, and in fact all *central* forces—forces that masses exert in the directions of their lines of junction and which are functions of the distances between these masses—possess this property. Under the action of forces of this character the equilibrium of fluids is possible. If we know U, we may replace the first equation by

$$dp = \rho \left(\frac{dU}{dx}\, dx + \frac{dU}{dy}\, dy + \frac{dU}{dz}\, dz \right)$$

or

$$dp = \rho\, dU \text{ and } p = \rho U + const.$$

The totality of all the points for which $U = const$ is a surface, a so-called *level surface*. For this surface also $p = const$. As all the force-relations, and, as we now see, all the pressure-relations, are determined by the nature of the function U, the pressure-relations, accordingly, supply a diagram of the force-relations, as was before remarked on page 119.

In the theory of Clairaut, here presented, is contained, beyond all doubt, the idea that underlies the doctrine of *force-function* or *potential*, which was afterwards developed with such splendid results by Laplace, Poisson, Green, Gauss, and others. As soon as our attention has been directed to this property of certain forces, namely, that they can be expressed as derivatives of the same function U, it is at once recognized as a highly convenient and *economical* course to investigate in the place of the forces themselves the function U.

If the equation

$$dp = \varrho(X\, dx + Y\, dy + Z\, dz) = \varrho\, dU$$

be examined, it will be seen that $X\, dx + Y\, dy + Z\, dz$ is the element of the *work* performed by the forces on unit of mass of the fluid in the displacement ds, whose projections are dx, dy, dz. Consequently, if we transport unit mass from a point for which $U = C_1$ to another point, indifferently chosen, for which $U = C_2$, or, more generally, from the surface $U = C_1$ to the surface $U = C_2$, we perform, no matter by what path the conveyance has been effected, the *same* amount of work. All the points of the first surface present, with respect to those of the second, the same difference of pressure; the relation always being such, that

$$p_2 - p_1 = \varrho(C_2 - C_1),$$

where the quantities designated by the same indices belong to the same surface.

10. Let us picture to ourselves a group of such very closely adjacent surfaces (Fig. 209) of which every two successive ones differ from each other by the same, very small, amount of work required to transfer a mass from one to the other; in other words, imagine the surfaces $U = C$, $U = C + dC$, $U = C + 2\,dC$, and so forth.

A mass moving on a level surface evidently performs no work. Hence, every component force in a direction tangential to the surface is $= 0$; and the direction of the *resultant force* is everywhere normal to the surface. If we call dn the element of the normal intercepted between two consecutive surfaces, and f the force requisite to convey unit mass from the one surface to the other through

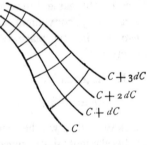

$C + 3dC$
$C + 2\,dC$
$C + dC$
C

Fig. 209.

this element, the work done is $f \cdot dn = dC$. As dC is by hypothesis everywhere constant, the force $f = dC/dn$ is inversely proportional to the distance between the surfaces considered. If, therefore, the surfaces U are known, the *directions of the forces* are given by the elements of a system of curves everywhere at right angles to these surfaces, and the inverse distances between the surfaces measure the *magnitude* of the forces.* These surfaces and curves also confront us in the other departments of physics. We meet them as equipotential surfaces and lines of force in electrostatics and magnetism, as isothermal surfaces and lines of flow in the theory of the conduction of heat, and as equipotential surfaces and lines of flow in the treatment of electrical and liquid contents.

11. We will now illustrate the fundamental idea of Clairaut's doctrine by another, very simple example. Imagine two mutually perpendicular planes to cut the paper at right angles in the straight lines $O X$ and $O Y$ (Fig. 210). We assume that a force-function exists $U = -x y$, where x and y are the distances from the two planes. The force-components parallel to OX and OY are then respectively

$$X = \frac{dU}{dx} = -y$$

and

$$Y = \frac{dU}{dy} = -x.$$

The level surfaces are cylindrical surfaces, whose generating lines are at right angles to the plane of the paper, and whose directrices, $xy = const$, are equilateral hyperbolas. The lines of force are obtained by turning the first mentioned system of curves through an angle of 45° in the plane of the paper about O. If

* The same conclusion may be reached as follows. Imagine a water pipe laid from New York to Key West, with its ends turning up vertically, and of glass. Let a quantity of water be poured into it, and when equilibrium is attained, let its height be marked on the glass at both ends. These two marks will be on one level surface. Now pour in a little more water and again mark the heights at both ends. The additional water in New York balances the additional water in Key West. The gravities of the two are equal. But their quantities are proportional to the vertical distances between the marks. Hence, the force of gravity on a fixed quantity of water is inversely as those vertical distances, that is, inversely as the distances between consecutive level surfaces.—*Trans.*

a unit of mass pass from the point r to O by the route $r\,p\,O$, or $r\,q\,O$, or by any other route, the work done is always $O\,p \times O\,q$. If we imagine a closed canal $O\,p\,r\,q\,O$ filled with a liquid, the liquid in the canal will be in equilibrium. If transverse sections be made at any two points, each section will sustain at both its surfaces the same pressure.

We will now modify the example slightly. Let the forces be $X = -y$, $Y = -a$, where a has a constant value. There exists now no function U so constituted that $X = dU/dx$ and $Y = dU/dy$; for in such a case it would be necessary that $dX/dy = dY/dx$, which is obviously not true. There is therefore no force-function, and consequently no level surfaces. If unit of mass be transported from r to O by

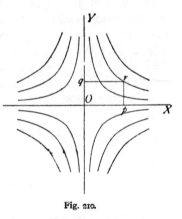

Fig. 210.

the way of p, the work done is $a \times O\,q$. If the transportation be effected by the route $r\,q\,O$, the work done is $a \times O\,q + O\,p \times O\,q$. If the canal $O\,p\,r\,q\,O$ were filled with a liquid, the liquid could not be in equilibrium, but would be forced to rotate constantly in the direction $O\,p\,r\,q\,O$. Currents of this character, which revert into themselves but continue their motion indefinitely, strike us as something quite foreign to our experience. Our attention, however, is directed by this to an important property of the forces of nature, to the property, namely, that the work of such forces may be expressed as a function of coördinates. Whenever exceptions to this principle are observed, we are disposed to regard them as apparent, and seek to clear up the difficulties involved.

12. We shall now examine a few problems of liquid *motion*. The founder of the theory of hydrodynamics is TORRICELLI. Torricelli,† by observations on liquids, discharged through orifices in the bottom of vessels, discovered the following law. If the time occupied in the complete discharge of a vessel be divided into n equal intervals and the quantity discharged in the last, the nth, interval be taken as the unit, there will be discharged in the $(n-1)^{\text{th}}$, the $(n-2)^{\text{th}}$, the $(n-3)^{\text{th}}$. . . . interval, respectively, the quantities 3, 5, 7 . . . . and so forth. An analogy between the motion of falling bodies and the motion of liquids is thus clearly suggested. Further, the perception is an immediate one, that the most curious consequences would ensue if the liquid, by its reversed velocity of efflux, could rise higher than its original level. Torricelli remarked, in fact, that it can rise *at the utmost* to this height, and assumed that it would rise *exactly* as high if all resistances could be removed. Hence, neglecting all resistances, the velocity of efflux, v, of a liquid discharged through an orifice in the bottom of a vessel is connected with the height h of the surface of the liquid by the equation $v = \sqrt{2gh}$; that is to say, the velocity of efflux is the *final* velocity of a body *freely* falling through the height h, or liquid-head; for only with this velocity can the liquid just rise again to the surface.*

Torricelli's theorem consorts excellently with the

† *De Motu Gravium Projectorum*, 1643.

* The early inquirers deduce their propositions in the incomplete form of proportions, and therefore usually put v proportional to $\sqrt{gh}$ or $\sqrt{h}$.

rest of our knowledge of natural processes; but we feel, nevertheless, the need of a more exact insight. VARIGNON attempted to deduce the principle from the relation between force and the *momentum* generated by force. The familiar equation $p\,t = m\,v$ gives, if by α we designate the area of the basal orifice, by h the pressure-head of the liquid, by s its specific gravity, by g the acceleration of a freely falling body, by v the velocity of efflux, and by τ a small interval of time, this result

$$\alpha\,h\,s\,.\,\tau = \frac{\alpha\,v\,\tau\,s}{g}\,.\,v \text{ or } v^2 = g\,h.$$

Here $\alpha\,h\,s$ represents the pressure acting during the time τ on the liquid mass $\alpha\,v\,\tau\,s/g$. Remembering that v is a final velocity, we get, more exactly,

$$\alpha\,h\,s\,.\,\tau = \frac{\alpha\,\dfrac{v}{2}\,.\,\tau\,s}{g}\,.\,v,$$

and thence the correct formula

$$v^2 = 2\,g\,h.$$

13. DANIEL BERNOULLI investigated the motions of fluids by the principle of *vis viva*. We will now treat the preceding case from this point of view, only rendering the idea more modern. The equation which we employ is $p\,s = m\,v^2/2$. In a vessel of transverse section q (Fig. 211), into which a liquid of the specific gravity s is poured till the head h is reached, the surface sinks, say, the small distance $d\,h$, and

Fig. 211.

the liquid mass $q \cdot d\,h \cdot s/g$ is discharged with the velocity v. The work done is the same as though the weight $q \cdot d\,h \cdot s$ had descended the distance h. The path of the motion in the vessel is not of consequence here. It makes no difference whether the stratum $q \cdot d\,h$ is discharged directly through the basal orifice, or passes, say, to a position a, while the liquid at a is displaced to b, that at b displaced to c, and that at c discharged. The work done is in each case $q \cdot d\,h \cdot s \cdot h$. Equating this work to the *vis viva* of the discharged liquid, we get

$$q \cdot dh \cdot s \cdot h = \frac{q \cdot dh \cdot s}{g} \frac{v^2}{2}, \text{ or}$$

$$v = \sqrt{2gh}.$$

The sole assumption of this argument is that *all* the work done in the vessel appears as *vis viva* in the liquid discharged, that is to say, that the velocities within the vessel and the work spent in overcoming friction therein may be *neglected*. This assumption is not very far from the truth if vessels of sufficient width are employed, and no violent rotatory motion is set up.

Let us neglect the gravity of the liquid in the vessel, and imagine it loaded by a movable piston, on whose surface-unit the pressure p falls. If the piston be displaced a distance $d\,h$, the liquid volume $q \cdot d\,h$ will be discharged. Denoting the density of the liquid by ϱ and its velocity by v, we then shall have

$$q \cdot p \cdot dh = q \cdot dh \cdot \rho \frac{v^2}{2}, \text{ or } v = \sqrt{\frac{2p}{\rho}}.$$

Wherefore, under the same pressure, different liquids are discharged with velocities inversely proportional to the square root of their density. It is generally supposed that this theorem is directly applicable to gases. Its *form*, indeed, is correct; but the deduction frequently employed involves an error, which we shall now expose.

14. Two vessels (Fig. 212) of equal cross-sections are placed side by side and connected with each other by a small aperture in the base of their dividing walls. For the velocity of flow through this aperture we obtain, under the same suppositions as before,

$$q \cdot dh \cdot s (h_1 - h_2) = q \frac{dh \cdot s}{g} \frac{v^2}{2}, \text{ or } v = \sqrt{2g(h_1 - h_2)}.$$

If we neglect the gravity of the liquid and imagine the pressures p_1 and p_2 produced by pistons, we shall similarly have $v = \sqrt{2(p_1 - p_2)/\varrho}$. For example, if the pistons employed be loaded with the weights P and $P/2$, the weight P will sink the distance h and $P/2$ will rise the distance h. The work $(P/2)h$ is thus left, to generate the *vis viva* of the effluent fluid.

A gas under such circumstances would behave differently. Supposing the gas to flow from the vessel containing the load P into that containing the load $P/2$, the first weight will fall a distance h, the second, however, since under half the pressure a gas doubles its volume, will rise a distance $2h$, so that the work $Ph - (P/2)2h = 0$ would be performed. In the case of gases, accordingly, some *additional*

Fig. 212.

work, competent to produce the flow between the vessels must be performed. This work the gas itself performs, by expanding, and by overcoming by its *force of expansion* a pressure. The expansive force p and the volume w of a gas stand to each other in the familiar relation $p w = k$, where k, so long as the temperature of the gas remains unchanged, is a constant. Supposing the volume of the gas to expand under the pressure p by an amount dw, the work done is

$$\int p\,dw = k\int \frac{dw}{w}.$$

For an expansion from w_0 to w, or for an increase of pressure from p_0 to p, we get for the work

$$k\log\left(\frac{w}{w_0}\right) = k\log\left(\frac{p_0}{p}\right).$$

Conceiving by this work a volume of gas w_0 of density ϱ, moved with the velocity v, we obtain

$$v = \sqrt{\frac{2p_0\log\left(\dfrac{p_0}{p}\right)}{\rho}}.$$

The velocity of efflux is, accordingly, in this case also inversely proportional to the square root of the density; its magnitude, however, is not the same as in the case of a liquid.

But even this last view is very defective. Rapid changes of the volumes of gases are always accompanied with changes of temperature, and, consequently also with changes of expansive force. For this reason, questions concerning the motion of gases cannot be dealt with as questions of pure mechanics, but always involve questions of *heat*. [Nor can even a thermodynamical treatment always suffice: it is sometimes necessary to go back to the consideration of molecular motions.]

15. The knowledge that a compressed gas contains stored-up work, naturally suggests the inquiry, whether this is not also true of compressed liquids. As a matter of fact, every liquid under pressure *is* compressed.

To effect compression work is requisite, which reappears the moment the liquid expands. But this work, in the case of the mobile liquids, is very small. Imagine, in Fig. 213, a gas and a mobile liquid of the same volume, measured by OA, subjected to the same pressure, a pressure of one atmosphere, designated by AB. If the pressure be reduced to one-half an atmosphere, the volume of the gas will be doubled, while that of the liquid will be increased by only about 25 millionths. The expansive work of the gas is represented by the surface $A B D C$, that of the liquid by $A B L K$, where

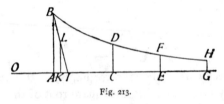

Fig. 213.

$AK = 0.000025\ OA$. If the pressure decreases till it becomes zero, the total work of the liquid is represented by the surface ABI, where $AI = 0.00005 OA$, and the total work of the gas by the surface contained between AB, the infinite straight line $ACEG$ . . . ., and the infinite hyperbola branch $BDFH$ . . . . *Ordinarily,* therefore, the work of expansion of liquids may be neglected. There are however phenomena, for example, the soniferous vibrations of liquids, in which work of this very order plays a principal part. In such cases, the changes of temperature the liquids undergo must also be considered. We thus see that it is only by a fortunate concatenation of circumstances that we are at liberty to consider a phenomenon with any close approximation to the truth as a mere matter of molar mechanics.

16. We now come to the idea which DANIEL BER-
NOULLI sought to apply in his work *Hydrodynamica,
sive de Viribus et Motibus Fluidorum Commentarii*
(1738). When a liquid sinks, the space through which
its center of gravity actually descends (*descensus actu-
alis*) is equal to space through which the center of
gravity of the separated parts affected with the velocities acquired in the fall can ascend (*ascensus poten-
tialis*). This idea, we see at once, is identical with that

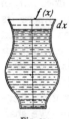

Fig. 214.

employed by Huygens. Imagine a vessel
filled with a liquid (Fig. 214) ; and let its
horizontal cross-section at the distance x
from the plane of the basal orifice, be
called $f(x)$. Let the liquid move and
its surface descend a distance dx. The
center of gravity, then, descends the distance $xf(x) \cdot dx/M$, where $M = \int f(x)
dx$. If k is the space of potential ascent
of the liquid in a cross-section equal to unity, the space
of potential ascent in the cross-section $f(x)$ will be
$k/f(x)^2$, and the space of potential ascent of the center
of gravity will be

$$\frac{k \int \frac{dx}{f(x)}}{M} = k\frac{N}{M},$$

where

$$N = \int \frac{dx}{f(x)}.$$

For the displacement of the liquid's surface through a
distance dx, we get, by the principle assumed, both
N and k changing, the equation

$$-xf(x)dx = Ndk + kdN.$$

This equation was employed by Bernoulli in the solu-

tion of various problems. It will be easily seen, that
Bernoulli's principle can be employed with success
only when the *relative velocities* of the single parts of
the liquid are known. Bernoulli assumes—an assumption apparent in the formulæ—that all particles once
situated in a horizontal plane, continue their motion
in a horizontal plane, and that the velocities in the
different horizontal planes are to each other in the inverse ratio of the sections of the planes. This is the
assumption of *the parallelism of strata*. It does not, in
many cases, agree with the facts, and in others its
agreement is incidental. When the vessel as compared
with the orifice of efflux is very wide, no assumption
concerning the motions within the vessel is necessary,
as we saw in the development of Torricelli's theorem.

17. A few isolated cases of liquid motion were
treated by NEWTON and JOHN BERNOULLI. We shall
consider here one to which a
familiar law is directly applicable. A cylindrical U-tube with
vertical branches is filled with
a liquid (Fig. 215). The length
of the entire liquid column is l.
If in one of the branches the
column be forced a distance x
below the level, the column in

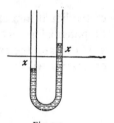

Fig. 215.

the other branch will rise the distance x, and the difference of level corresponding to the excursion x will be
$2x$. If a is the transverse section of the tube and s the
liquid's specific gravity, the force brought into play
when the excursion x is made, will be $2\,a\,s\,x$, which,
since it must move a mass $a\,l\,s/g$ will determine the
acceleration $(2asx)/(als/g) = (2g/l)x$, or, for unit
excursion, the acceleration $2g/l$. We perceive that
pendulum vibrations of the duration

$$T = \pi \sqrt{\frac{l}{2g}}$$

will take place. The liquid column, accordingly, vibrates the same as a simple pendulum of half the length of the column.

A similar, but somewhat more general, problem was treated by John Bernoulli. The two branches of a cylindrical tube (Fig. 216), curved in any manner, make

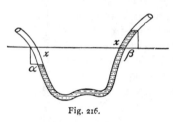

Fig. 216.

with the horizon, at the points at which the surfaces of the liquid move, the angles α and β. Displacing one of the surfaces the distance x, the other surface suffers an equal displacement. A difference of level is thus produced $x(\sin \alpha + \sin \beta)$, and we obtain, by a course of reasoning similar to that of the preceding case, employing the same symbols, the formula

$$T = \pi \sqrt{\frac{l}{g(\sin \alpha + \sin \beta)}}.$$

The laws of the pendulum hold true *exactly* for the liquid pendulum of Fig. 215 (viscosity neglected), even for vibrations of great amplitude; while for the filar pendulum the laws hold only approximately true for small excursions.

18. The center of gravity of a liquid as a whole can rise only as high as it would have to fall to produce its velocities. In every case in which this principle appears to present an exception, it can be shown that the exception is only *apparent*. One example is Hero's fountain. This apparatus, as we know, consists of three

vessels, which may be designated in the descending order as *A, B, C.* The water in the open vessel *A* falls through a tube into the closed vessel *C*; the air displaced in *C* exerts a pressure on the water in the closed vessel *B*, and this pressure forces the water in *B* in a jet above *A* whence it falls back to its original level. The water in *B* rises, it is true, considerably above the level of *B*, but in actuality it merely flows by the circuitous route of the fountain and the vessel *A* to the much lower level of *C*.

Another apparent exception to the principle in question is that of Montgolfier's *hydraulic ram,* in which the liquid by its own gravitational work appears to rise con-siderably above its original level. The liquid flows (Fig. 217) from a cistern *A* through a long pipe *RR* and a valve *V*, which opens inwards, into a vessel *B*. When the current becomes rapid enough, the valve *V* is forced shut, and a liquid mass *m* affected with the

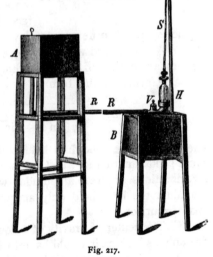

Fig. 217.

velocity *v* is suddenly arrested in *RR*, which must be deprived of its momentum. If this be done in the time *t*, the liquid can exert during this time a pressure

$q = m\,v/t$, to which must be added its hydrostatical pressure p. The liquid, therefore, will be able, during this interval of time, to penetrate with a pressure $p + q$ through a second valve into a *pila Heronis, H,* and in consequence of the circumstances existing there will rise to a higher level in the ascension-tube *SS* than that corresponding to its simple pressure p. It is to be observed here, that a considerable portion of the liquid must first flow off into *B,* before a velocity requisite to close V is produced by the liquid's work in *RR*. A small portion only rises above the original level; the greater portion flows from A into B. If the liquid discharged from *SS* were collected, it could be easily proved that the center of gravity of the quantity thus discharged and of that received in B lay, as the result of various losses, actually *below* the level of A.

The principle of the hydraulic ram, that of the transference of work done by a large liquid mass to a

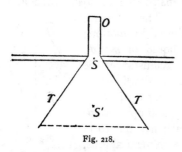

Fig. 218.

smaller one, which thus acquires a great *vis viva,* may be illustrated in the following very simple manner. Close the narrow opening O of a funnel and plunge it, with its wide opening downwards, deep into a large vessel of water. If the finger closing the upper opening be quickly removed, the space inside the funnel will rapidly fill with water, and the surface of the water outside the funnel will sink. The work performed is equivalent to the descent of the contents of the funnel from the center of gravity S of the superficial stratum to the center of gravity S' of the contents of the funnel. If the vessel is sufficiently wide the velocities in it are all very small, and almost the entire *vis viva* is concentrated in the contents of the funnel. If all the parts of the contents had the same velocities, they could all rise to the original level, or the mass as a whole could rise to the height at which its center of gravity was coincident with S. But in the narrower sections of the funnel the velocity of the parts is greater than in the wider sections, and the former therefore contain by far the greater part of the *vis viva*. Consequently, the liquid parts above are violently separated from the parts below and thrown out through the neck of the funnel high above the original surface. The remainder, however, are left considerably below that point, and the center of gravity of the whole never as much as reaches the original level of S.

19. One of the most important achievements of Daniel Bernoulli is his distinction of *hydrostatic* and *hydrodynamic pressure*. The pressure which liquids exert is altered by motion; and the pressure of a liquid *in motion* may, according to the circumstances, be greater or less than that of the liquid *at rest* with the same arrangement of parts. We will illustrate this by a simple example. The vessel A, which has the form of a body of revolution with vertical axis,

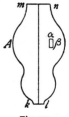

Fig. 219.

is kept constantly filled with a frictionless liquid, so that its surface at $m\,n$ does not change during the discharge at $k\,l$. We will reckon the vertical distance of a particle from the surface $m\,n$ downwards as positive and call it z. Let us follow the course of a prismatic element of volume, whose horizontal base-area is α and height β, in its downward motion, neglecting, on the assumption of the parallelism of strata, all velocities at right angles to z. Let the density of the liquid be ϱ, the velocity of the element v, and the pressure, which is dependent on z, p. If the particle descend the distance $d\,z$, we have by the principle of *vis viva*

$$\alpha\beta\rho\,d\left(\frac{v^2}{2}\right) = \alpha\beta\rho g\,dz - \alpha\frac{dp}{dz}\beta\,dz \quad \ldots \ldots \quad (1)$$

that is, the increase of the *vis viva* of the element is equal to the work of gravity for the displacement in question, less the work of the forces of pressure of the liquid. The pressure on the upper surface of the element is $\alpha\,p$, that on the lower surface is $\alpha[p + (dp/dz)\beta]$. The element sustains, therefore, if the pressure increased downwards, an upward pressure $\alpha(dp/dz)\beta$; and for any displacement dz of the element, the work $\alpha(dp/dz)\beta dz$ must be deducted. Reduced, equation (1) assumes the form

$$\rho \cdot d\left(\frac{v^2}{2}\right) = \rho g\,dz - \frac{dp}{dz}\,dz$$

and, integrated, gives

$$\rho \cdot \frac{v^2}{2} = \rho g z - p + const \quad \ldots \ldots \ldots \quad (2)$$

If we express the velocities in two different horizontal cross-sections a_1 and a_2 at the depths z_1 and z_2 below the surface, by v_1, v_2, and the corresponding pressures by p_1 p_2, we may write equation (2) in the form

$$\frac{\rho}{2} \cdot (v_1^2 - v_2^2) = \rho g (z_1 - z_2) + (p_2 - p_1) \quad \cdot \quad (3)$$

Taking for our cross-section a_1 the surface, $z_1 = 0$, through all cross-sections in the same interval of time, $a_1\,v_1 = a_2\,v_2$. Whence, finally,

$$p_2 = \rho g z_2 + \frac{\rho}{2} v_1^2 \left(\frac{a_2^2 - a_1^2}{a_2^2} \right).$$

The pressure p^2 of the liquid *in motion* (the hydrodynamic pressure) consists of the pressure $\varrho g z_2$ of the liquid *at rest* (the hydrostatic pressure) and of a pressure $(\rho/2) v_1^2 [(a_2^2 - a_1^2)/a_2^2]$ dependent on the density, the velocity of flow, and the cross-sectional areas. In cross-sections larger than the surface of the liquid, the hydrodynamic pressure is greater than the hydrostatic, and *vice versa*.

A clearer idea of the significance of Bernoulli's principle may be obtained by imagining the liquid in the vessel A unacted on by gravity, and its outflow produced by a constant pressure p_1 on the surface. Equation (3) then takes the form

$$p_2 = p_1 + \frac{\rho}{2} (v_1^2 - v_2^2).$$

If we follow the course of a particle thus moving, it will be found that to every increase of the velocity of flow (in the narrower cross-sections) a decrease of pressure corresponds, and to every decrease of the velocity of flow (in the wider cross-sections) an increase of pressure. This, indeed, is evident, wholly aside from mathematical considerations. In the present case every *change* of the velocity of a liquid element must be exclusively produced by the *work of the liquid's forces of pressure*. When, therefore, an element enters into a narrower cross-section, in which a greater velocity of flow prevails, it can acquire this higher velocity only on the condition that a greater pressure acts on its rear surface than on its front surface, that is to say, only when it moves from points of higher to points of lower pressure, or when the pressure decreases in the direction of the motion. If we imagine the pressures in a wide section and in a succeeding narrower section to be for a moment equal, the acceleration of the elements in the narrower section will not take place; the elements will not escape fast enough; they will accumulate before the narrower section; and *at the entrance* to it the requisite augmentation of pressure will be immediately produced. The converse case is obvious.

20. In dealing with more complicated cases, the problems of liquid motion, even though viscosity be

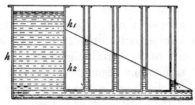

Fig. 220.

neglected, present great difficulties; and when the enormous effects of viscosity are taken into account, anything like a dynamical solution of almost every problem is out of the question. So much so, that although these investigations were begun by Newton, we have, up to the present time, only been able to master a very few of the simplest problems of this class, and that but imperfectly. We shall content ourselves with a simple example. If we cause a liquid contained in a vessel of the pressure-head h to flow, not through an orifice in its base, but through a long cylindrical tube fixed in its side (Fig. 220), the velocity of efflux v will be less than that deducible from Torricelli's law, as a portion of the work is consumed by resistances due to viscosity and perhaps to friction. We find, in fact, that $v = \sqrt{2gh_1}$, where $h_1 < h$. Expressing by h_1 the *velocity*-head, and by h_2 the *resistance*-head, we may put $h = h_1 + h_2$. If to the main cylindrical tube we affix vertical lateral tubes, the liquid will rise in the latter tubes to the heights at which it equilibrates the pressures in the main tube, and will thus indicate at all points the pressures of the main tube. The noticeable fact here is, that the liquid-height at the point of influx of the tube is $= h_2$, and that it diminishes in the direction of the point of outflow, by the law of a straight line, to zero. The elucidation of this phenomenon is the question now presented.

Gravity here does not act *directly* on the liquid in the horizontal tube, but all effects are transmitted to it by the *pressure* of the surrounding parts. If we imagine a prismatic liquid element of basal area α and length β to be displaced in the direction of its length a distance dz, the work done, as in the previous case, is

$$- \alpha \frac{dp}{dz} \beta \, dz = - \alpha \beta \frac{dp}{dz} \, dz.$$

For a finite displacement we have

$$- \alpha \beta \int_{p_2}^{p_1} \frac{dp}{dz} \, dz = - \alpha \beta \, (p_2 - p_1) \quad \cdots \cdots \quad (1)$$

Work is *done* when the element of volume is displaced from a place of *higher* to a place of *lower* pressure. The amount of the work done depends on the size of the element of volume and on the *difference* of pressure at the initial and terminal points of the motion, and not on the length and the form of the path traversed. If the diminution of pressure were twice as rapid in one case as in another, the difference of the pressures on the front and rear surfaces, or the *force* of the work, would be doubled, but the *space* through which the work was done would be halved. The work done would remain the same, whether done through the space $a\,b$ or $a\,c$ of Fig 221.

Through every cross-section q of the horizontal tube the liquid flows with the same velocity v. If, neglecting the differences of velocity in the *same* cross-section, we consider a liquid element which exactly fills the section q and has the length β, the *vis viva* $q\beta\varrho(v^2/2)$ of such an element will persist unchanged throughout its entire course in the tube. This is possible only provided the *vis viva consumed by friction* is replaced by the *work of the liquid's forces of pressure.* Hence, in the direction of the motion of the element the pressure

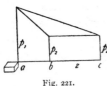

Fig. 221.

must diminish, and for equal distances, to which the same work of friction corresponds, by equal amounts. The total work of gravity on a liquid element $q\,\beta\,\varrho$ issuing from the vessel, is $q\,\beta\,\varrho\,g\,h$. Of this the portion $q\,\beta\,\varrho(v^2/2)$ is the *vis viva* of the element discharged with the velocity v into the mouth of the tube, or, as $v = \sqrt{2gh_1}$, the portion $q\,\beta\,\varrho\,g\,h_1$. The remainder of the work, therefore, $q\,\beta\,\varrho\,g\,h_2$, is consumed in the tube, if owing to the slowness of the motion we neglect the losses within the vessel.

If the pressure-heads respectively obtaining in the vessel, at the mouth, and at the extremity of the tube, are h, h_2, 0, or the pressures are $p = h\,g\,\varrho$, $p_2 = h_2 g\,\varrho$, 0, then by equation (1) of page 513 the work requisite to generate the *vis viva* of the element discharged into the mouth of the tube is

$$q\,\beta\,\rho\,\frac{v^2}{2} = q\,\beta\,(p - p_2) = q\,\beta\,g\,\rho\,(h - h_2) = q\,\beta\,g\,\rho\,h_1,$$

and the work transmitted by the pressure of the liquid to the element traversing the length of the tube, is

$$q\,\beta\,p_2 = q\,\beta\,g\,\rho\,h_2,$$

or the exact amount consumed in the tube.

Let us assume, for the sake of argument, that the pressure does not decrease from p_2 at the mouth to zero at the extremity of the tube by the law of a straight line, but that the distribution of the pressure is different, say, constant throughout the entire tube. The parts in advance then will at once suffer a loss of velocity from the friction, the parts which follow will crowd upon them, and there will thus be produced at the mouth of the tube an augmentation of pressure conditioning a constant velocity throughout its entire length. The pressure at the end of the tube can only be $= 0$ because the liquid at that point is not prevented from yielding to any pressure impressed upon it.

If we imagine the liquid to be a mass of smooth elastic balls, the balls will be most compressed at the bottom of the vessel, they will enter the tube in a state of compression, and will gradually lose that state in the course of their motion. We leave the further development of this simile to the reader.

It is evident, from a previous remark, that the work stored up in the compression of the liquid itself, is very small. The motion of the liquid is due to the work of gravity in the vessel, which by means of the pressure of the compressed liquid is transmitted to the parts in the tube.

An interesting modification of the case just discussed is obtained by causing the liquid to flow through a tube composed of a number of shorter cylindrical tubes of varying widths. The pressure in the direction of outflow then diminishes (Fig. 222) more rapidly in the narrower tubes, in which a greater consumption of work by friction takes place, than in the wider ones. We further note, in every passage of the liquid into a

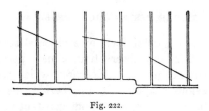

Fig. 222.

wider tube, that is to a *smaller* velocity of flow, an *increase* of pressure (a positive congestion); in every passage into a narrower tube, that is to a *greater* velocity of flow, an abrupt *diminution* of pressure (a negative congestion). The velocity of a liquid element on which no direct forces act can be diminished or increased only by its passing to points of higher or lower pressure.

CHAPTER IV.

THE FORMAL DEVELOPMENT OF MECHANICS.

I.

THE ISOPERIMETRICAL PROBLEMS.

1. When the chief facts of a physical science have once been fixed by observation, a new period of its development begins—the *deductive,* which we treated in the previous chapter. In this period, the facts are reproducible in the mind without constant recourse to observation. Facts of a more general and complex character are copied in thought on the theory that they are made up of simpler and more familiar observational elements. But even after we have deduced from our expressions for the most elementary facts (the principles) expressions for more common and more complex facts (the theorems) and have discovered in all phenomena the same elements, the developmental process of the science is not yet completed. The deductive development of the science is followed by its *formal* development. Here it is sought to put in a clear compendious form, or *system,* the facts to be reproduced, so that each can be reached and mentally pictured with the *least intellectual effort.* Into our rules for the mental reconstruction of facts we strive to incorporate the greatest possible *uniformity,* so that these rules shall be easy of acquisition. It is to be remarked, that the three periods distinguished are not sharply separated from one another, but that the processes of

development referred to frequently go hand in hand, although on the whole the order designated is unmistakable.

2. A powerful influence was exerted on the formal development of mechanics by a particular class of mathematical problems, which, at the close of the seventeenth and the beginning of the eighteenth centuries, engaged the deepest attention of inquirers. These problems, the so-called *isoperimetrical problems,* will now form the subject of our remarks. Certain questions of the greatest and least values of quantities, questions of maxima and minima, were treated by the Greek mathematicians. Pythagoras is said to have taught that the circle, of all plane figures of a given perimeter, has the greatest area. The idea, too, of a certain economy in the processes of nature was not foreign to the ancients. Hero deduced the law of the reflection of light from the theory that light emitted from a point A (Fig. 223) and reflected at M will travel to B by the shortest route.

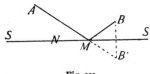

Fig. 223.

Making the plane of the paper the plane of reflection, SS the intersection of the reflecting surface, A the point of departure, B the point of arrival, and M the point of reflection of the ray of light, it will be seen at once that the line AMB', where B' is the reflection of B, is a straight line. The line AMB' is shorter than the line ANB', and therefore also AMB is shorter than ANB. Pappus held similar notions concerning organic nature; he explained, for example, the form of the cells of the honeycomb by the bees' efforts to economize in materials.

These ideas fell, at the time of the revival of the sciences, on not unfruitful soil. They were first taken up by FERMAT and ROBERVAL, who developed a method applicable to such problems. These inquirers observed, as Kepler had already done, that a magnitude y which depends on another magnitude x, generally possesses in the vicinity of its greatest and least values a peculiar property. Let x (Fig. 224) denote abscissas and y ordinates. If, while x increases, y passes through a maximum value, its increase, or rise, will be changed into a decrease, or fall; and if it passes through a minimum value its fall will be changed into a rise. The neighboring values of the maximum or minimum value, consequently, will lie very *near* each other, and

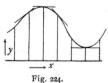

Fig. 224.

the tangents to the curve at the points in question will generally be parallel to the axis of abscissas. Hence, to find the maximum or minimum values of a quantity, we seek the parallel tangents of its curve.

The *method of tangents* may be put in analytical form. For example, it is required to cut off from a given line a a portion x such that the product of the two segments x and $a - x$ shall be as great as possible. Here, the product $x(a - x)$ must be regarded as the quantity y dependent on x. At the maximum value of y any infinitely small variation of x, say a variation ξ, will produce no change in y. Accordingly, the required value of x will be found, by putting

$$x(a - x) = (x + \xi)(a - x - \xi)$$

or

$$ax - x^2 = ax + a\xi - x^2 - x\xi - x\xi - \xi^2$$

or

$$0 = a - 2x - \xi.$$

As ξ may be made as small as we please, we also get

$$0 = a - 2x;$$

whence $x = a/2$.

In this way, the concrete idea of the method of tangents may be translated into the language of algebra; the procedure also contains, as we see, the germ of the *differential calculus*.

Fermat sought to find for the law of the refraction of light an expression analogous to that of Hero for the

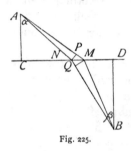

Fig. 225.

law of reflection. He remarked that light, proceeding from a point A, and refracted at a point M, travels to B, not by the shortest route, but in the shortest time. If the path AMB is performed in the shortest time, then a neighboring path ANB, infinitely near the real path, will be described in the same time. If we draw from N on AM and from M on NB the perpendiculars NP and MQ, then the second route, before refraction, is less than the first route by a distance $MP = NM \sin \alpha$, but is larger than it after refraction by the distance $NQ = NM \sin \beta$. On the supposition, therefore, that the velocities in the first and second media are respectively v_1 and v_2, the time required for the path AMB will be a minimum when

$$\frac{NM \sin \alpha}{v_1} - \frac{NM \sin \beta}{v_2} = 0$$

or

$$\frac{v_1}{v_2} = \frac{\sin \alpha}{\sin \beta} = n,$$

where n stands for the index of refraction. Hero's law of reflection, remarks Leibniz, is thus a special case of the law of refraction. For equal velocities ($v_1 = v_2$), the condition of a minimum of *time* is identical with the condition of a minimum of *space*.

Huygens, in his optical investigations, applied and further perfected the ideas of Fermat, considering, not only rectilinear, but also curvilinear motions of light, in media in which the velocity of the light varied continuously from place to place. For these, also, he found that Fermat's law obtained. Accordingly, in all motions of light, an endeavor, so to speak, to produce results in a *minimum* of *time* appeared to be the fundamental tendency.

3. Similar maximal or minimal properties were brought out in the study of mechanical phenomena. As we have already noticed, John Bernoulli knew that a freely suspended chain assumes the form for which its center of gravity lies *lowest*. This idea was, of course, a simple one for the investigator who first recognized the general import of the principle of virtual velocities. Stimulated by these observations, inquirers now began generally to investigate maximal and minimal characters. The movement received its most powerful impulse from a problem propounded by John Bernoulli, in June, 1696[*]—the problem of the *brachistochrone*. In a vertical plane two points are situated, A and B. It is required to assign in this plane the curve by which a falling body will travel from A to B in the *shortest* time. The problem was very ingeniously solved by John Bernoulli himself; and solutions were

* *Acta Eruditorum, Leipzig.*

also supplied by Leibniz, L'Hôpital, Newton, and James Bernoulli.

The most remarkable solution was JOHN BERNOULLI'S own. This inquirer remarks that problems of this class have already been solved, not for the motion of falling bodies, but for the motion of light. He accordingly imagines the motion of a falling body replaced by the motion of a ray of light. (Comp. p. 473.) The two points A and B (Fig. 226) are supposed to be fixed in a medium in which the velocity of light increases in the vertical downward direction by the same law as the velocity of a falling body. The medium is supposed to be constructed of horizontal layers of downwardly decreasing density, such that $v = \sqrt{2gh}$ denotes the velocity of the light in any layer at the distance h below A. A ray of light which travels from A to B under such conditions will describe this distance in the shortest time, and simultaneously trace out the curve of *quickest descent*.

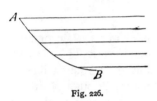

Fig. 226.

Calling the angles made by the element of the curve with the perpendicular, or the normal of the layers, α, α', α''. . . ., and the respective velocities v, v', v''. . . ., we have

$$\frac{\sin \alpha}{v} = \frac{\sin \alpha'}{v'} = \frac{\sin \alpha''}{v''} = \ldots = k = const.$$

or, designating the perpendicular distances below A by x, the horizontal distances from A by y, and the arc of the curve by s,

$$\frac{\left(\dfrac{dy}{ds}\right)}{v} = k.$$

Whence follows

$$d\,y^2 = k^2\,v^2\,d\,s^2 = k^2\,v^2\,(d\,x^2 + d\,y^2)$$

and because $v = \sqrt{2gx}$ also

$$dy = d\,x\sqrt{\frac{x}{a-x}}, \text{ where } a = \frac{1}{2\,g\,k^2}.$$

This is the differential equation of a cycloid, or curve described by a point in the circumference of a circle of radius $r = a/2 = 1/4gk^2$, rolling on a straight line.

To find the cycloid that passes through A and B, it is to be noted that *all* cycloids, inasmuch as they are produced by similar con- structions, are *similar,* and that if generated by the rolling of circles on AD from the point A as origin, are also *similarly situated* with respect to

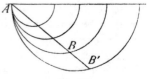

Fig. 227.

the point A. Accordingly, we draw through AB a straight line, and construct any cycloid, cutting the straight line in B' (Fig. 227). The radius of the gen- erating circle is, say, r'. Then the radius of the gener- ating circle of the cyloid sought is $r = r'(AB/AB')$.

This solution of John Bernoulli's, achieved entirely without a method, the outcome of pure geometrical fancy and a skilful use of such knowledge as happened to be at his command, is one of the most remarkable and beautiful performances in the history of physical science. John Bernoulli was an æsthetic genius in this field. His brother James's character was entirely differ- ent. James was the superior of John in critical power, but in originality and imagination was surpassed by the latter. James Bernoulli likewise solved this problem, though in less felicitous form. But, on the other hand, he did not fail to develop, with great thoroughness, a *general* method applicable to such problems. Thus, in these two brothers we find the two fundamental traits of high scientific talent separated from one another—traits, which in the very greatest natural inquirers, in Newton, for example, are combined. We shall soon see those two tendencies, which within one bosom might have fought their battles unnoticed, clashing in open conflict, in the persons of these two brothers.

Vignette to *Leibnitzii et Johannis Bernoullii comercium epistolicum.*
Lausanne and Geneva, Bousquet, 1745.

4. James Bernoulli finds that the chief object of research hitherto had been to find the *values* of a variable quantity, for which a second variable quantity, which is a function of the first, assumes its greatest or its least value. The present problem, however, is to find

from among an *infinite number of curves* one which possesses a certain maximal or minimal property. This, as he correctly remarks, is a problem of an entirely different character from the other and demands a new method.

The principles that James Bernoulli employed in the solution of this problem (*Acta Eruditorum,* May, 1697)* are as follows:

(1) If a curve has a certain property of maximum or minimum, every portion or element of the curve has the same property.

(2) Just as the infinitely adjacent values of the maxima or minima of a quantity in the ordinary problems, for infinitely small changes of the independent variables, are constant, so also is the quantity here to be made a maximum or minimum for the curve sought, for infinitely contiguous curves, constant.

(3) It is finally assumed, for the case of the brachistochrone, that the velocity is $v = \sqrt{2gh}$, where h denotes the height fallen through.

If we picture to ourselves, a very small portion ABC of the curve (Fig. 228), and, imagining a horizontal line drawn through B, cause the portion taken to pass into the infinitely contiguous portion ADC, we shall obtain, by considerations exactly similar to those employed in the treatment of Fermat's law, the well-known relation between the

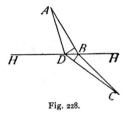

Fig. 228.

sines of the angles made by the curve-elements with the perpendicular and the velocities of descent. In this deduction the following assumptions are made,

* See also his works, Vol. II, p. 768.

(1), that the *part*, or element, ABC is brachistochronous, and (2), that ADC is described in the same time as ABC. Bernoulli's calculation is very prolix; but its essential features are obvious, and the problem is solved* by the above-stated principles.

With the solution of the problem of the brachistochrone, James Bernoulli, in accordance with the practice then prevailing among mathematicians, proposed the following more general "isoperimetrical problem": "Of all isoperimetrical curves (that is, curves of equal perimeters or equal lengths) between the same two fixed points, to find the curve such that the space included (1) by a second curve, each of whose ordinates is a given function of the corresponding ordinate or the corresponding arc of the one sought, (2) by the ordinates of its extreme points, and (3) by the part of the axis of abscissæ lying between those ordinates, shall be a maximum or minimum."

For example. It is required to find the curve BFN, described on the base BN such, that of all curves of

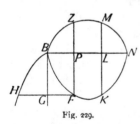

Fig. 229.

the same length on BN, this particular one shall make the area BZN a minimum, where $PZ = (PF)^n$, $LM = (LK)^n$, and so on (Fig. 229). Let the relation between the ordinates of BZN and the corresponding ordinates of BFN be given by the curve BH. To obtain PZ from PF, draw FGH at right angles to BG, where BG is at right angles to BN. By hypothesis, then, $PZ = GH$, and so for the other ordinates. Further, we put $BP = y$, $PF = x$, $PZ = x^n$.

* For the details of this solution and for information generally on the history of this subject, see Woodhouse's *Treatise on Isoperimetrical Problems and the Calculus of Variations*, Cambridge, 1810.—*Trans.*

John Bernoulli gave, forthwith, a solution of this problem, in the form

$$y = \int \frac{x^n \, dx}{\sqrt{a^{2n} - x^{2n}}},$$

where a is an arbitrary constant. For $n = 1$,

$$y = \int \frac{x \, dx}{\sqrt{a^2 - x^2}} = a - \sqrt{a^2 - x^2},$$

that is, BFN is a semicircle on BN as diameter, and the area BZN is equal to the area BFN. For this particular case, the solution, in fact, is correct. But the general formula is not universally valid.

On the publication of John Bernoulli's solution, James Bernoulli openly engaged to do three things: first, to discover his brother's method; second, to point out its contradictions and errors; and, third, to give the true solution. The jealousy and animosity of the two brothers culminated, on this occasion, in a violent and acrimonious controversy, which lasted till James's death. After James's death, John virtually confessed his error and adopted the correct method of his brother.

James Bernoulli surmised, and in all probability correctly, that John, misled by the results of his researches on the catenary and the curve of a sail filled with wind, had again attempted an *indirect* solution, imagining BFN filled with a liquid of variable density and taking the lowest position of the center of gravity as determinative of the curve required. Making the ordinate $PZ = p$, the specific gravity of the liquid in the ordinate $PF = x$ must be p/x, and similarly in every other ordinate. The weight of a vertical filament is then $p \cdot dy/x$, and its moment with respect to BN is

$$\frac{1}{2} x \frac{p \, dy}{x} = \frac{1}{2} p \, dy.$$

Hence, for the lowest position of the center of gravity, $\frac{1}{2} \int p \, dy$, or $\int p \, dy = BZN$, is a maximum. But the fact is here overlooked, remarks James Bernoulli, that with the variation of the *curve BFN* the *weight* of the liquid also is varied. Consequently, in this simple form the deduction is not admissible.

In the solution which he himself gives, James Bernoulli once more assumes that the small portion $F \, F_{,,,}$

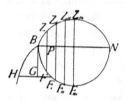

Fig. 230.

of the curve possesses the property which the whole curve possesses. And then taking the four successive points $F \, F_{,} \, F_{,,} \, F_{,,,}$, of which the two extreme ones are fixed, he so varies $F_{,}$ and $F_{,,}$ that the length of the arc $F \, F_{,} \, F_{,,} \, F_{,,,}$ remains *unchanged*,

which is possible, of course, only by a displacement of *two* points. We shall not follow his involved and unwieldy calculations. The principle of the process is clearly indicated in our remarks. Retaining the designations above employed, James Bernoulli, in substance, states that when

$$dy = \frac{p \, dx}{\sqrt{a^2 - p^2}},$$

$\int p \, dy$ is a maximum, and when

$$dy = \frac{(a - p) \, dx}{\sqrt{2 a p - p^2}}$$

$\int p \, dy$ is a minimum.

The dissensions between the two brothers were, we may admit, greatly to be deplored. Yet the genius of the one and the profundity of the other have borne, in the stimulus which Euler and Lagrange received from their several investigations, splendid fruits.

5. Euler (*Problematis Isoperimetrici Solutio Generalis, Com. Acad. Petr.* T. VI, for 1733, published in 1738)* was the first to give a more general method of treating these questions of maxima and minima, or isoperimetrical problems. But even his results were based on prolix geometrical considerations, and not possessed of analytical generality. Euler divides problems of this category, with a clear perception and grasp of their differences, into the following classes:

(1) Required, of *all* curves, that for which a property *A* is a maximum or minimum.

(2) Required, of all curves, equally possessing a property *A*, that for which *B* is a maximum or minimum.

(3) Required, of all curves, equally possessing two properties, *A* and *B*, that for which *C* is a maximum or minimum. And so on.

A problem of the first class is (Fig. 231) the finding of the *shortest* curve through *M* and *N*. A problem of the second class is the finding of a curve through *M* and *N*, which, having the given length *A*, makes the area *MPN* a maximum. A problem of the third class would be: of all curves of the given length *A*, which pass through *M*, *N* and contain the same area *MPN* = *B*, to find one which describes when rotated about *MN* the least surface of revolution. And so on.

---

* Euler's principal contributions to this subject are contained in three memoirs, published in the *Commentaries of Petersburg* for the years 1733, 1736, and 1766, and in the tract *Methodus inveniendi Lineas Curvas Proprietate Maximi Minimive gaudentes*, Lausanne and Geneva, 1744.—*Trans.*

We may observe here, that the finding of an absolute maximum or minimum, without collateral conditions, is meaningless. Thus, all the curves of which in the first example the shortest is sought possess the common property of passing through the points M and N.

The solution of problems of the first class requires the variation of *two* elements of the curve or of *one* point. This is also sufficient. In problems of the second class *three* elements or *two* points must be varied; the reason being, that the varied portion must possess in common with the unvaried portion the property A, and, as B is to be made a maximum or minimum, also the property B, that is, must satisfy *two* conditions. Similarly, the solution of problems of the third class requires the variation of *four* elements. And so on.

Fig. 231.

The solution of a problem of a higher class involves, by implication, the solution of its converse, in all its forms. Thus, in the third class, we vary four elements of the curve, so, that the varied portion of the curve shall share equally with the original portion the values A and B and, as C is to be made a maximum or a minimum, also the value C. But the same conditions must be satisfied, if of all curves possessing equally B and C that for which A is a maximum or minimum is sought, or of all curves possessing A and C that for which B is a maximum or minimum is sought. Thus a circle, to take an example from the second class, contains, of all lines of the same length A, the greatest area B, and the circle, also, of all curves containing the same area B, has the shortest length A. As the condition that the property A shall be possessed in common or shall be a maximum, is expressed in the same manner, Euler saw the possibility of reducing the problems of the higher classes to problems of the first class. If, for example, it is required to find, of all curves having the common property A, that which makes B a maximum, the curve is sought for which $A + mB$ is a maximum, where m is an arbitrary constant. If on any change of the curve, $A + mB$, for any value of m, does not change, this is generally possible only provided the change of A, considered by itself, and that of B, considered by itself, are $= 0$.

6. Euler was the originator of still another important advance. In treating the problem of finding the brachistochrone in a resisting medium, which was investigated by Herrmann and him, the existing methods proved incompetent. For the brachistochrone in a vacuum, the velocity depends solely on the vertical height fallen through. The velocity in one portion of the curve is in no wise dependent on the other portions. In this case, then, we can indeed say, that if the whole curve is brachistochronous, every element of it is also brachistochronous. But in a resisting medium the case is different. The entire length and form of the preceding path enters into the determination of the velocity in the element. The whole curve can be brachistochronous without the separate elements necessarily exhibiting this property. By considerations of this character, Euler perceived, that the principle introduced by James Bernoulli did not hold universally good, but that in cases of the kind referred to, a more detailed treatment was required.

7. The methodical arrangement and the great number of the problems solved, gradually led Euler to sub-

stantially the same methods that Lagrange afterwards developed in a somewhat different form, and which now go by the name of the *Calculus of Variations*. First, John Bernoulli lighted on an *accidental* solution of a problem, by analogy. James Bernoulli developed, for the solution of such problems, a geometrical method. Euler generalized the problems and the geometrical method. And finally, Lagrange, entirely emancipating himself from the consideration of geometrical figures, gave an analytical method. Lagrange remarked, that the increments which functions receive in consequence of a change in their *form* are quite analogous to the increments they receive in consequence of a change of their independent variables. To distinguish the two species of increments, Lagrange denoted the former by δ, the latter by d. By the observation of this analogy Lagrange was enabled to write down at once the equations which solve problems of maxima and minima. Of this idea, which has proved itself a very fertile one, Lagrange never gave a verification; in fact, did not even attempt it. His achievement is in every respect a peculiar one. He saw, with great economical insight, the foundations which in his judgment were sufficiently secure and serviceable to build upon. But the acceptance of these fundamental principles themselves was vindicated only by its results. Instead of employing himself on the demonstration of these principles, he showed with what success they could be employed. (*Essai d'une nouvelle méthode pour déterminer les maxima et minima des formules intégrales indéfinies. Misc. Taur.* 1762.)

The difficulty which Lagrange's contemporaries and successors experienced in grasping his idea clearly, is quite intelligible. Euler sought in vain to clear up the difference between a variation and a differential by imagining constants contained in the function, with the change of which the form of the function changed. The increments of the value of the function arising from the increments of these constants were regarded by him as the variations, while the increments of the function springing from the increments of the independent variables were the differentials. The conception of the Calculus of Variations that springs from such a view is singularly timid, narrow, and illogical, and does not compare with that of Lagrange. Even Lindelöf's modern work, so excellent in other respects, is marred by this defect. The first really competent presentation of Lagrange's idea is, in our opinion, that of JELLETT.* Jellett appears to have said what Lagrange perhaps was unable to say fully, perhaps did not deem it necessary to say.

8. Jellett's view is, briefly, this: The values of some quantities are regarded as *constant,* the values of others as *variable.* Among the latter, one distinguishes independently (or arbitrarily) variable quantities from others that are dependently variable. Analogously, the form of a function may be regarded as *determined* or *indetermined* (variable). If $y = \phi(x)$ and if the form of the function is regarded as variable, then the value of y may change because of an increment dx of the independent variable x as well as because of a change in the form of the function—a transition from ϕ to ϕ_1 The first change is the differential dy, the second, the variation δy. Accordingly,

$$dy = \phi(x + dx) - \phi(x), \text{ and}$$
$$\delta y = \phi_1(x) - \phi(x).$$

----

* An Elementary Treatise on the Calculus of Variations. By the Rev. John Hewitt Jellett, Dublin, 1850.

For example, if we have a plane curve of the indeterminate form $y = \phi(x)$, then the length of its arc between the abscissæ x_0 and x_1 is

$$S = \int_{x_0}^{x_1} \sqrt{1 + \left(\frac{d\varphi(x)}{dx}\right)^2}\, dx = \int_{x_0}^{x_1} \sqrt{1 + \left(\frac{dy}{dx}\right)^2}\, dx,$$

a *determinate* function* of an indeterminate function. For any definite form of the curve or of the function ϕ, the value of S can be determined. For any change of the curve or the form of the function ϕ, the change of the length of arc δS, can be determined. In this example, the arc length S is an integral of a function $u = F\ (dy/dx) = \sqrt{1 + (dy/dx)^2}$, which does not contain $y = \phi(x)$ directly but dy/dx, which depends on y/ϕ.

Now let $u = F(y)$ be a determinate function of an indeterminate function $y = \phi(x)$; then

$$\delta u = F(y + \delta y) - F(y) = \frac{dF(y)}{dy}\, \delta y.$$

Again, let $u = F(y, dy/dx)$ be a determinate function

---

* If with each *number* (belonging to a certain class of numbers) a number is associated, then mathematicians say that a *function* is defined on the said class of numbers. If with each *function* of a certain kind a number is associated, then, following Volterra, mathematicians say that a *functional* is defined for the functions of the said kind. For instance, with each curve $y = \varphi(x)$ that is continuously differentiable between x_0 and x_1 the definition of S pairs the number obtained by integrating from x_0 to x_1 the function $\sqrt{1 + (dy/dx)^2}$. Thus S, the length of curves as defined above, is a functional. On the following pages, a more general functional will be defined. It pairs with $y = \varphi(x)$ the number obtained by integrating from x_0 to x_1 a function $V = F(x, y, dy/dx, \ldots)$. Mach denotes the partial rates of change of V with regard to y, dy/dx, $\ldots$ or the partial derivatives of the function F by $\dfrac{dV}{dy}$, $\dfrac{dV}{d\dfrac{dy}{dx}}$, $\ldots$ Since the publication of the early editions of this book it has become universal practice to denote those partial derivatives by the symbols $\dfrac{\partial F}{\partial y}$, $\dfrac{\partial F}{\partial y'}$, $\ldots$
(Remark by the editor of the 6th American Edition — 1960).

of an indeterminate function, $y = \varphi(x)$. For a change of form of φ, the value of y changes by δy and the value of dy/dx by $\delta(dy/dx)$. The corresponding change in the value of u is

$$\delta u = \frac{dF\left(y, \frac{dy}{dx}\right)}{dy}\, \delta y + \frac{dF\left(y, \frac{dy}{dx}\right)}{d\frac{dy}{dx}}\, \delta\frac{dy}{dx}.$$

The expression $\delta\dfrac{dy}{dx}$ is obtained by our definition from

$$\delta\frac{dy}{dx} = \frac{d(y + \delta y)}{dx} - \frac{dy}{dx} = \frac{d\delta y}{dx}.$$

Similarly, the following results are found:

$$\delta\frac{d^2 y}{dx^2} = \frac{d^2\,\delta y}{dx^2},\ \ \delta\frac{d^3 y}{dx^3} = \frac{d^3\,\delta y}{dx^3},$$

and so forth.

We now proceed to a problem, namely, the determination of the form of the function $y = \varphi(x)$ that will render

$$U = \int_{x_0}^{x_1} V\, dx$$

where

$$V = F\left(x, y, \frac{dy}{dx}, \frac{d^2 y}{dx^2}, \ \dots\right),$$

a maximum or minimum; φ denoting an indeterminate, and F a determinate function. The value of U may be varied (1) by a change of the limits, x_0, x_1. Outside of the limits, the change of the independent variables x, as such, does not affect U; accordingly, if we regard the limits as fixed, this is the only respect in which we need attend to x. The only other way (2) in which the value of U is susceptible of variation is by a change of the *form* of $y = \varphi(x)$. This produces a change of *value* in

$$y, \frac{dy}{dx}, \frac{d^2y}{dx^2}, \ldots$$

amounting to

$$\delta y, \; \delta\frac{dy}{dx}, \; \delta\frac{d^2y}{dx^2} \ldots,$$

and so forth. The total change in U, which we shall call DU, and to express the maximum-minimum condition put $=0$, consists of the differential dU and the variation δU. Accordingly,

$$DU = dU + \delta U = 0.$$

Denoting by $V_1 dx_1$ and $-V_0 dx_0$ the increments of U due to the change of the limits, we then have

$$DU = V_1\, dx_1 - V_0\, dx_0 + \delta \int_{x_0}^{x_1} V dx =$$

$$V_1\, dx_1 - V_0\, dx_0 + \int_{x_0}^{x_1} \delta V.\, dx = 0.$$

But by the principles stated on the preceding page we further get

$$\delta V = \frac{dV}{dy}\delta y + \frac{dV}{d\frac{dy}{dx}}\delta\frac{dy}{dx} + \frac{dV}{d\frac{d^2y}{dx^2}}\delta\frac{d^2y}{dx^2} + \ldots =$$

$$\frac{dV}{dy}\delta y + \frac{dV}{d\frac{dy}{dx}}\frac{d\delta y}{dx} + \frac{dV}{d\frac{d^2y}{dx^2}}\frac{d^2\delta y}{dx^2} + \ldots$$

For the sake of brevity we put

$$\frac{dV}{dy} = N, \frac{dV}{d\frac{dy}{dx}} = P_1, \frac{dV}{d\frac{d^2y}{dx^2}} = P_2, \ldots$$

Then

$$\delta \int_{x_0}^{x_1} V\, dx =$$

$$\int_{x_0}^{x_1} \left(N\delta y + P_1 \frac{d\delta y}{dx} + P_2 \frac{d^2 \delta y}{dx^2} + P_3 \frac{d^3 \delta y}{dx^3} + \ldots \right) dx.$$

One difficulty here is, that not only δy, but also the terms $d\delta y/dx$, $d^2\delta y/dx^2$ . . . . occur in this equation, —terms which are dependent on one another, but not in a directly obvious manner. This drawback can be removed by successive integration by parts, by means of the formula

$$\int u\, dv = uv - \int v\, du.$$

By this method

$$\int P_1 \frac{d\delta y}{dx}\, dx = P_1 \delta y - \int \frac{dP_1}{dx}\, \delta y\, dx,$$

$$\int P_2 \frac{d^2 \delta y}{dx^2}\, dx = P_2 \frac{d\delta y}{dx} - \int \frac{dP_2}{dx} \frac{d\delta y}{dx}\, dx =$$

$$P_2 \frac{d\delta y}{dx} - \frac{dP_2}{dx}\, \delta y + \int \frac{d^2 P_2}{dx^2}\, \delta y\, dx, \text{ and so on.}$$

Performing all these integrations between the limits, we obtain for the condition $DU = 0$ the expression

$$0 = V_1\, dx_1 - V_0\, dx_0$$

$$+ \left(P_1 - \frac{dP_2}{dx} + \cdot\cdot \right)_1 \delta y_1 - \left(P_1 - \frac{dP_2}{dx} + \ldots \right)_0 \delta y_0$$

$$+ \left(P_2 - \frac{dP_3}{dx} + \cdot\cdot \right)_1 \left(\frac{d\delta y}{dx} \right)_1 - \left(P_2 - \frac{dP_3}{dx} + \ldots \right)_0 \left(\frac{d\delta y}{dx} \right)_0$$

$$+ \ldots \ldots \ldots \ldots \ldots \ldots \ldots \ldots \ldots$$

$$+ \int_{x_0}^{x_1} \left(N - \frac{dP_1}{dx} + \frac{d^2 P_2}{dx^2} - \frac{d^3 P_3}{dx^3} + \ldots \right) \delta y \cdot dx,$$

which now contains only δy under the integral sign.

The terms in the first line of this expression are independent of any change in the form of the function and depend solely upon the variation of the limits. The terms of the two following lines depend on the change in the form of the function, for the limiting values of x only; and the indices 1 and 2 state that the actual limiting values are to be put in the place of the general expressions. The terms of the last line, finally, depend on the *general* change in the form of the function. Collecting all the terms, except those in the last line, under one designation $\alpha_1 - \alpha_0$, and calling the expression in parentheses in the last line β, we have

$$0 = \alpha_1 - \alpha_0 + \int_{x_0}^{x_1} \beta \cdot \delta y \cdot dx.$$

But this equation can be satisfied only if

$$\alpha_1 - \alpha_0 = 0 \ldots \ldots \ldots \ldots \ldots (1)$$

and

$$\int_{x_0}^{x_1} \beta \, \delta y \, dx = 0 \ldots \ldots \ldots \ldots \ldots \ldots (2)$$

For if each of the members were not equal to zero, each would be determined by the other. But the integral of an indeterminate function cannot be expressed in terms of its limiting values only. Assuming, therefore, that the equation

$$\int_{x_0}^{x_1} \beta \, \delta y \, dx = 0,$$

holds generally good, its conditions can be satisfied only when $\beta = 0$, since δy is throughout arbitrary and its generality of form cannot be restricted. By the equation

$$N - \frac{dP_1}{dx} + \frac{d^2 P_2}{dx^2} - \frac{d^3 P_3}{dx^3} + \ldots = 0. \ldots (3),$$

therefore, the form of the function $y = \varphi(x)$ that makes the expression U a maximum or minimum is defined. Equation (3) was found by Euler. But Lagrange first showed the application of equation (1) for the determination of a function by the conditions at its limits. By equation (3), which it must satisfy, the *form* of the function $y = \varphi(x)$ is *generally* determined; but this equation contains a number of *arbitrary* constants, whose values are determined solely by the conditions at the limits. With respect to notation, Jellett rightly remarks, that the employment of the symbol δ in the first two terms $V_1 \delta x_1 = V_0 \delta x_0$ of equation (1), the form used by Lagrange, is inconsistent, and he correctly puts for the increments of the independent variables the usual symbols dx_1, dx_0.

9. To illustrate the use to which these equations may be put, let us seek the form of the function that makes

$$\int_{x_0}^{x_1} \sqrt{1 + \left(\frac{dy}{dx}\right)^2}\, dx$$

a minimum—the shortest line. Here

$$V = F\left(\frac{dy}{dx}\right).$$

All expressions except

$$P_1 = \frac{dV}{d\frac{dy}{dx}} = \frac{\frac{dy}{dx}}{\sqrt{1 + \left(\frac{dy}{dx}\right)^2}}$$

vanish in equation (3), and that equation becomes

$dP_1/dx = 0$; which means that P_1, and consequently its only variable, dy/dx, is independent of x. Hence, $dy/dx = a$, and $y = ax + b$, where a and b are constants.

The constants a, b are determined by the values of the limits. If the straight line passes through the points x_0, y_0 and x_1, y_1, then

$$\left. \begin{aligned} y_0 &= ax_0 + b \\ y_1 &= ax_1 + b \end{aligned} \right\} \quad \cdots \cdots \cdots \cdots \quad (m)$$

and as $dx_0 = dx_1 = 0$, $\delta y_0 = \delta y_1 = 0$, equation (1) vanishes. The coefficients $\delta(dy/dx)$, $\delta(d^2y/dx^2)$, ... independently vanish. Hence, the values of a and b are determined by the equations (m) alone.

If the limits x_0, x_1 only are given, but y_0, y_1 are indeterminate, we have $dx_0 = dx_1 = 0$, and equation (1) takes the form

$$\frac{a}{\sqrt{1 + a^2}} (\delta y_1 - \delta y_0) = 0,$$

which, since δy_0 and δy_1 are arbitrary, can only be satisfied if $a = 0$. The straight line is in this case $y = b$, parallel to the axis of abscissæ, and as b is indeterminate, at any distance from it.

It will be noticed, that equation (1) and the subsidiary conditions expressed in equation (m), with respect to the determination of the constants, generally complement each other.

If

$$Z = \int_{x_1}^{x_2} y \sqrt{1 + \left(\frac{dy}{dx}\right)^2} \, dx$$

is to be made a minimum, the integration of the appropriate form of (3) will give

$$y = \frac{c}{2} \left[e^{\frac{x-c'}{c}} + e^{-\frac{x-c}{c}} \right].$$

If Z is a minimum, then $2\pi Z$ also is a minimum, and the curve found will give, by rotation about the axis of abscissæ, the least surface of revolution. Further, the lowest position of the center of gravity of a homogeneously heavy curve of this kind corresponds to a minimum of Z; the curve is therefore a catenary. The determination of the constants c, c' is effected by means of the limiting conditions, as above.

In the treatment of mechanical problems, a distinction is made between the increments of coördinates that *actually* take place in time, namely, $d\,x$, $d\,y$, $d\,z$, and the *possible* displacements δx, δy, δz, considered, for instance, in the application of the principle of virtual velocities. The latter, as a rule, are not variations; that is, are not changes of value that spring from changes in the form of a function. Only when we consider a mechanical system that is a continuum, as for example a string, a flexible surface, an elastic body, or a liquid, are we at liberty to regard δx, δy, δz as indeterminate functions of the coördinates x, y, z, and are we concerned with variations.

It is not our purpose in this work, to develop mathematical theories, but simply to treat the purely physical part of mechanics. But the history of the isoperimetrical problems and of the calculus of variations had to be touched upon, because these researches have exercised a very considerable influence on the development of mechanics. Our sense of the general properties of systems, and of properties of maxima and minima in particular, was much sharpened by these investigations, and properties of the kind referred to were subsequently discovered in mechanical principles with great facility. As a fact, physicists, since Lagrange's time, usually express mechanical principles in a maximal or minimal form. This predilection would be unintelligible without a knowledge of the historical development.

II.

THEOLOGICAL, ANIMISTIC, AND MYSTICAL POINTS OF VIEW IN MECHANICS.

1. If, in entering a parlor in Germany, we happen to hear something said about some man being very pious, without having caught the name, we may fancy that Privy Counsellor X was spoken of, or Herr *von* Y; we should hardly think of a scientific man of our acquaintance. It would, however, be a mistake to suppose that the want of cordiality, occasionally rising to embittered controversy, which has existed in our day between the scientific and the theological faculties, always separated them. A glance at the history of science suffices to prove the contrary.

People talk of the "conflict" of science and theology, or better of science and the church. It is in truth a prolific theme. On the one hand, we have the long catalogue of the sins of the church against progress, on the other side, a "noble army of martyrs," among them no less distinguished figures than Galileo and Giordano Bruno. It was only by good luck that Descartes, pious as he was, escaped the same fate. These things are the commonplaces of history; but it would be a great mistake to suppose that the phrase "warfare of science" is a correct description of its general historic attitude toward religion, that the only repression of intellectual development has come from priests, and that if their hands had been held off, growing science would have shot up with stupendous velocity. No doubt, external opposition did have to be fought; and the battle with it was no child's play. Nor was any means too base for the church to handle in this struggle. She considered nothing but how to conquer; and no temporal policy ever was conducted so selfishly, so unscrupulously, or so cruelly. But investigators have had another struggle on their hands, and by no means an easy one, the struggle with their own preconceived ideas, and especially with the notion that philosophy and science must be founded on theology. It was but slowly that this prejudice little by little was erased.

2. But let the facts speak for themselves, while we introduce the reader to a few historical personages.

Napier, the inventor of logarithms, an austere Puritan, who lived in the sixteenth century, was, in addition to his scientific avocations, a zealous theologian. Napier applied himself to some extremely curious speculations. He wrote an exegetical commentary on the Book of Revelation, with propositions and mathematical demonstrations. Proposition XXVI, for example, maintains that the pope is the Antichrist; proposition XXXVI declares that the locusts are the Turks and Mohammedans; and so forth.

Blaise Pascal (1623-1662), one of the most rounded geniuses to be found among mathematicians and physicists, was extremely orthodox and ascetical. So deep were the convictions of his heart, that despite the gentleness of his character, he once openly denounced at

Rouen an instructor in philosophy as a heretic. The healing of his sister by contact with a relic most seriously impressed him, and he regarded her cure as a miracle. On these facts taken by themselves it might be wrong to lay great stress; for his whole family were much inclined to religious fanaticism. But there are plenty of other instances of his religiosity. Such was his resolve—which was carried out, too—to abandon the pursuits of science altogether and to devote his life solely to the cause of Christianity. Consolation, he used to say, he could find nowhere but in the teachings of Christianity; and all the wisdom of the world availed him not a whit. The sincerity of his desire for the conversion of heretics is shown in his *Lettres provinciales,* where he vigorously declaims against the dreadful subtleties that the doctors of the Sorbonne had devised, expressly to persecute the Jansenists. Very remarkable is Pascal's correspondence with the theologians of his time; and a modern reader is not a little surprised at finding this great "scientist" seriously discussing in one of his letters whether or not the Devil was able to work miracles.

Otto von Guericke, the inventor of the air-pump, occupies himself, at the beginning of his book, now little over two hundred years old, with the miracle of Joshua, which he seeks to harmonize with the ideas of Copernicus. In like manner, we find his researches on the vacuum and the nature of the atmosphere introduced by disquisitions concerning the location of heaven, the location of hell, and so forth. Although Guericke really strives to answer these questions as rationally as he can, still we notice that they give him considerable trouble, questions, be it remembered, that today the theologians themselves would consider absurd.

Yet Guericke was a man who lived after the Reformation!

The giant mind of Newton did not disdain to employ itself on the interpretation of the Apocalypse. On such subjects it was difficult for a sceptic to converse with him. When Halley once indulged in a jest concerning theological questions, he is said to have curtly repulsed him with the remark: "I have studied these things; you have not!"

We need not tarry by Leibniz, the inventor of the best of all possible worlds and of pre-established harmony—inventions which Voltaire disposed of in *Candide,* a humorous novel with a deeply philosophical purpose. But everybody knows that Leibniz was almost if not quite as much a theologian, as a man of science.

Let us turn, however, to the last century. Euler, in his *Letters to a German Princess,* deals with theologico-philosophical problems in the midst of scientific questions. He speaks of the difficulty involved in explaining the interaction of body and mind, due to the total diversity of these two phenomena,—a diversity to his mind undoubted. The system of occasionalism, developed by Descartes and his followers, agreeably to which God executes for every purpose of the soul, (the soul itself not being able to do so,) a corresponding movement of the body, does not quite satisfy him. He derides, also, and not without humor, the doctrine of pre-established harmony, according to which perfect agreement was established from the beginning between the movements of the body and the volitions of the soul, although neither is in any way connected with the other, just as there is harmony between two different but similarly constructed clocks. He remarks, that in this view his own body is as foreign to him as that of a rhinoceros in the midst of Africa, which might just as well be in pre-established harmony with his soul as its own. Let us hear his own words. In his day, Latin was almost universally written. When a German scholar wished to be especially condescending, he wrote in French: "Si dans le cas d'un dérèglement de mon corps Dieu ajustait celui d'un rhinoceros, en sorte que ses mouvements fussent tellement d'accord avec les ordres de mon âme qu'il levât la patte au moment que je voudrais lever la main, et ainsi des autres opérations, ce serait alors mon corps. Je me trouverais subitement dans la forme d'un rhinoceros au milieu de l'Afrique, mais non obstant cela mon âme continuerait les même opérations. J'aurais également l'honneur d'écrire à V. A., mais je ne sais pas comment elle recevrait mes lettres."

One would almost imagine that Euler, here, had been tempted to play Voltaire. And yet, apposite as was his criticism in this vital point, the mutual action of body and soul remained a miracle to him, still. But he extricates himself, however, from the question of the freedom of the will, very sophistically. To give some idea of the kind of questions which a scientist was permitted to treat in those days, it may be remarked that Euler institutes in his physical "Letters" investigations concerning the nature of spirits, the connection between body and soul, the freedom of the will, the influence of that freedom on physical occurrences, prayer, physical and moral evils, the conversion of sinners, and similar topics. All this occurs in a treatise full of clear physical ideas and not devoid of philosophical ones, where the well-known circle-diagrams of logic have their birth-place.

3. Let these examples of religious physicists suffice. We have selected them intentionally from among the foremost of scientific discoverers. The theological pro-

clivities which these men followed, belong wholly to their innermost private life. They tell us openly things which they are not compelled to tell us, things about which they might have remained silent. What they utter are not opinions forced upon them from without; they are their own sincere views. They were not conscious of any theological constraint. In a court which harbored a Lamettrie and a Voltaire, Euler had no reason to conceal his real convictions.

According to the modern notion, these men should at least have seen that the questions they discussed did not belong under the heads where they put them, that they were not questions of science. Still, odd as this contradiction between inherited theological beliefs and independently created scientific convictions seems to us, it is no reason for a diminished admiration of those leaders of scientific thought. Nay, this very fact is a proof of their stupendous mental power: they were able, in spite of the contracted horizon of their age, to which even their own *aperçus* were chiefly limited, to point out the path to an elevation, where our generation has attained a freer point of view.

Every unbiased mind must admit that the age in which the chief development of the science of mechanics took place, was an age of predominantly theological cast. Theological questions were excited by everything, and modified everything. No wonder, then, that mechanics is colored thereby. But the thoroughness with which theological thought thus permeated scientific inquiry, will best be seen by an examination of details.

4. The impulse imparted in antiquity to this direction of thought by Hero and Pappus has been alluded to in the preceding chapter. At the beginning of the seventeenth century we find Galileo occupied with prob-

lems concerning the strength of materials. He shows that hollow tubes offer a greater resistance to flexure than solid rods of the same length and the same quantity of material, and at once applies this discovery to the explanation of the forms of the bones of animals, which are usually hollow and cylindrical in shape. The phenomenon is easily illustrated by the comparison of a flatly folded and a rolled sheet of paper. A horizontal beam fastened at one extremity and loaded at the other may be remodelled so as to be thinner at the loaded end without any loss of stiffness and with a considerable saving of material. Galileo determined the form of a beam of equal resistance at each cross-section. He also remarked that animals of similar geometrical construction but of considerable difference of size would comply in very unequal proportions with the laws of resistance.

The forms of bones, feathers, stalks, and other organic structures, adapted, as they are, in their minutest details to the purposes they serve, are highly calculated to make a profound impression on the thinking beholder, and this fact has again and again been adduced in proof of a supreme wisdom ruling in nature. Let us examine, for instance, the pinion-feather of a bird. The quill is a hollow tube diminishing in thickness as we go towards the end, that is, is a body of equal resistance. Each little blade of the vane repeats in miniature the same construction. It would require considerable technical knowledge even to imitate a feather of this kind, let alone invent it. We should not forget, however, that investigation, and not mere admiration, is the office of science. We know how Darwin sought to solve these problems, by the theory of natural selection. That Darwin's solution is a complete one, may fairly be doubted; Darwin himself questioned it. All external conditions would be powerless if something were not present that *admitted* of variation. But there can be no question that his theory is the first serious attempt to replace mere admiration of the adaptations of organic nature by serious inquiry into the mode of their origin.

Pappus's ideas concerning the cells of honeycombs were the subject of animated discussion as late as the eighteenth century. In a treatise, published in 1865, entitled *Homes Without Hands* (p. 428), Wood substantially relates the following: "Maraldi had been struck with the great regularity of the cells of the honeycomb. He measured the angles of the lozenge-shaped plates, or rhombs, that form the terminal walls of the cells, and found them to be respectively 109° 28' and 70° 32'. Réaumur, convinced that these angles were in some way connected with the economy of the cells, requested the mathematician König to calculate the form of a hexagonal prism terminated by a pyramid composed of three equal and similar rhombs, which would give the greatest amount of space with a given amount of material. The answer was, that the angles should be 109° 26' and 70° 34'. The difference, accordingly, was two minutes. Maclaurin,* dissatisfied with this agreement, repeated Maraldi's measurements, found them correct, and discovered, in going over the calculation, an error in the logarithmic table employed by König. Not the bees, but the mathematicians were wrong, and the bees had helped to detect the error!"

Any one who is acquainted with the method of measuring crystals and has seen the cell of a honeycomb, with its rough and non-reflective surfaces, will question whether the measurement of such cells can be executed

* *Philosophical Transactions* for 1743.—*Trans.*

with a probable error of only two minutes.† So, we must take this story as a sort of pious mathematical fairy-tale, quite apart from the consideration that nothing would follow from it even were it true. Besides, from a mathematical point of view, the problem is too imperfectly formulated to enable us to decide the extent to which the bees have solved it.

The ideas of Hero and Fermat, referred to in the previous chapter, concerning the motion of light, at once received from the hands of Leibniz a theological coloring, and played, as has been mentioned before, a predominant rôle in the development of the calculus of variations. In Leibniz's correspondence with John Bernoulli, theological questions are repeatedly discussed in the very midst of mathematical disquisitions. Their language is not unfrequently couched in biblical pictures. Leibniz, for example, says that the problem of the brachistochrone lured him as the apple had lured Eve.

Maupertuis, the famous president of the Berlin Academy, and a friend of Frederick the Great, gave a new impulse to the theologizing bent of physics by the enunciation of his principle of least action. In the treatise which formulated this obscure principle, and which betrayed in Maupertuis a woeful lack of mathematical accuracy, the author declared his principle to be the one which best accorded with the wisdom of the Creator. Maupertuis was an ingenious man, but not a man of strong, practical sense. This is evidenced by the schemes he was incessantly devising: his bold propositions to found a city in which only Latin should be spoken, to dig a deep hole in the earth to find new substances, to institute psychological investigations by

† But see G. F. Maraldi in the *Mémoirs de l'académie* for 1712. It is, however, now well known the cells vary considerably. See Chauncey Wright, *Philosophical Discussions*, 1877, p. 311.—*Trans.*

means of opium and by the dissection of monkeys, to explain the formation of the embryo by gravitation, and so forth. He was sharply satirized by Voltaire in the *Histoire du docteur Akakia,* a work which led, as we know, to the rupture between Frederick and Voltaire.

Maupertuis's principle would in all probability soon have been forgotten, had Euler not taken up the suggestion. Euler magnanimously left the principle its name, Maupertuis the glory of the invention, and converted it into something new and really serviceable. What Maupertuis meant to convey is very difficult to ascertain. What Euler meant may be easily shown by simple examples. If a body is constrained to move on a rigid surface, for instance, on the surface of the earth, it will describe when an impulse is imparted to it, the shortest path between its initial and terminal positions. Any other path that might be prescribed it, would be longer or would require a greater time. This principle finds an application in the theory of atmospheric and oceanic currents. The theological point of view, Euler retained. He claims it is possible to explain phenomena, not only from their physical *causes,* but also from their *purposes.* "As the construction of the universe is the most perfect possible, being the handiwork of an all-wise Maker, nothing can be met with in the world in which some maximal or minimal property is not displayed. There is, consequently, no doubt but that all the effects of the world can be derived by the method of maxima and minima from their final causes as well as from their efficient ones."*

* "Quum enim mundi universi fabrica sit perfectissima, atque a creatore sapientissimo absoluta, nihil omnino in mundo contingit, in quo non maximi minimive ratio quaedam eluceat: quam ob rem dubium prorsus est nullum, quin omnes mundi effectus ex causis finalibus, ope methodi maximorum et minimorum, aeque feliciter determinari quaeant, atque ex ipsis causis efficientibus." (*Methodus inveniendi lineas curvas maximi minimive proprietate gaudentes.* Lausanne, 1744.)

5. Similarly, the notions of the constancy of the quantity of matter, of the constancy of the quantity of motion, of the indestructibility of work or energy, conceptions which completely dominate modern physics, all arose under the influence of theological ideas. The notions in question had their origin in an utterance of Descartes, before mentioned, in the *Principles of Philosophy,* agreeably to which the quantity of matter and motion originally created in the world—such being the only course compatible with the constancy of the Creator—is always preserved unchanged. The conception of the manner in which this quantity of motion should be calculated was very considerably modified in the progress of the idea from Descartes to Leibniz, and to their successors, and as the outcome of these modifications the doctrine gradually and slowly arose which is now called the "law of the conservation of energy." But the theological background of these ideas only slowly vanished. In fact, at the present day, we still meet with scientists who indulge in self-created mysticisms concerning this law.

During the entire sixteenth and seventeenth centuries, down to the close of the eighteenth, the prevailing inclination of inquirers was, to find in all physical laws some particular disposition of the Creator. But a gradual transformation of these views must strike the attentive observer. Whereas with Descartes and Leibniz physics and theology were still greatly intermingled, in the subsequent period a distinct endeavor is noticeable, not indeed wholly to discard theology, yet to separate it from purely physical questions. Theological disquisitions were put at the beginning or relegated to the end of physical treatises. Theological speculations were restricted, as much as possible, to the question of creation, that, from this point onward, the way might be cleared for physics.

Towards the close of the eighteenth century a remarkable change took place,—a change which was apparently an abrupt departure from the current trend of thought, but in reality was the logical outcome of the development indicated. After an attempt in a youthful work to found mechanics on Euler's principle of least action, Lagrange, in a subsequent treatment of the subject, declared his intention of utterly disregarding theological and metaphysical speculations, as in their nature precarious and foreign to science. He erected a new mechanical system on entirely different foundations, and no one conversant with the subject will dispute its excellencies. All subsequent scientists of eminence accepted Lagrange's view, and the present attitude of physics to theology was thus substantially determined.

6. The idea that theology and physics are two distinct branches of knowledge, thus took, from its first germination in Copernicus till its final promulgation by Lagrange, almost two centuries to attain clearness in the minds of investigators. At the same time it cannot be denied that this truth was always clear to the greatest minds, like Newton. Newton never, despite his profound religiosity, mingled theology with the questions of science. True, even he concludes his *Optics,* whilst on its last pages his clear and luminous intellect still shines, with an exclamation of humble contrition at the vanity of all earthly things. But his optical researches proper, in contrast to those of Leibniz, contain not a trace of theology. The same may be said of Galileo and Huygens. Their writings conform almost absolutely to the point of view of Lagrange, and may be accepted in this respect as class-

ical. But the general views and tendencies of an age must not be judged by its greatest, but by its average, minds.

To comprehend the process here portrayed, the general condition of affairs in these times must be considered. It stands to reason that in a stage of civilization in which religion is almost the sole education, and the only theory of the world, people would naturally look at things from a theological point of view, and that they would believe that this view was possessed of competency in all fields of research. If we transport ourselves back to the time when people played the organ with their fists, when they had to have the multiplication table visibly before them to calculate, when they did so much with their hands that people now-a-days do with their heads, we shall not demand of such a time that it should *critically* put to the test its own views and theories. With the widening of the intellectual horizon through the great geographical, technical, and scientific discoveries and inventions of the fifteenth and sixteenth centuries, with the opening up of provinces in which it was impossible to make any progress with the old conception of things, simply because it had been formed prior to the knowledge of these provinces, this bias of the mind gradually and slowly vanished. The great freedom of thought which appears in isolated cases in the early middle ages, first in poets and then in scientists, will always be hard to understand. The enlightenment of those days must have been the work of a few very extraordinary minds, and can have been bound to the views of the people at large by but very slender threads, more fitted to disturb those views than to reform them. Rationalism does not seem to have gained a broad theater of action till the literature of the eighteenth century. Humanistic, philosophical, historical, and physical science here met and gave each other mutual encouragement. All who have experienced, in part, in its literature, this wonderful emancipation of the human intellect, will feel during their whole lives a deep, elegiacal regret for the eighteenth century.

7. The old point of view, then, is abandoned. Its history is now detectible only in the form of the mechanical principles. And this form will remain strange to us as long as we neglect its origin. The theological conception of things gradually gave way to a more rigid conception; and this was accompanied with a considerable gain in enlightenment, as we shall now briefly indicate.

When we say light travels by the paths of shortest time, we grasp by such an expression many things. But we do not know as yet *why* light prefers paths of shortest time. We forego all further knowledge of the phenomenon, if we find the reason in the Creator's wisdom. We of today know, that light travels by *all* paths, but that only on the paths of shortest time do the waves of light so intensify each other that a perceptible result is produced. Light, accordingly, only *appears* to travel by the paths of shortest time. After the prejudice which prevailed on these questions had been removed, cases were immediately discovered in which by the side of the supposed economy of nature the most striking extravagance was displayed. Cases of this kind have, for example, been pointed out by Jacobi in connection with Euler's principle of least action. A great many natural phenomena accordingly produce the impression of economy, simply because they visibly appear only when by accident an economical accumulation of effects take place. This is the same idea in the province of inorganic nature that Darwin worked out in the domain of organic nature. We facilitate instinctively our comprehension of nature by applying to it the economical ideas with which we are familiar.

Often the phenomena of nature exhibit maximal or minimal properties because when these greatest or least properties have been established the causes of all further alteration are removed. The catenary gives the lowest point of the center of gravity for the simple reason that when that point has been reached all further descent of the system's parts is impossible. Liquids exclusively subjected to the action of molecular forces exhibit a minimum of superficial area, because stable equilibrium can only subsist when the molecular forces are able to effect no further diminution of superficial area. The important thing, therefore, is not the maximum or minimum, but the removal of *work;* work being the factor determinative of the alteration. It sounds much less imposing but is much more elucidatory, much more correct and comprehensive, instead of speaking of the economical tendencies of nature, to say: "So much and so much only occurs as in virtue of the forces and circumstances involved can occur."

The question may now justly be asked: If the point of view of theology which led to the enunciation of the principles of mechanics was utterly wrong, how comes it that the principles themselves are in all substantial points correct? The answer is easy. In the first place, the theological view did not supply the *contents* of the principles, but simply determined their *guise;* their matter was derived from experience. A similar influence would have been exercised by any other dominant type of thought, by a commercial attitude, for instance, such as presumably had its effect on Stevinus's thinking. In the second place, the theological conception of nature itself owes its origin to an endeavor to obtain a more comprehensive view of the world;—the very same endeavor that is at the bottom of physical science. Hence, even admitting that the physical philosophy of theology is a fruitless achievement, a reversion to a lower state of scientific culture, we still need not repudiate the *sound root* from which it has sprung and which is not different from that of true physical inquiry.

In fact, science can accomplish nothing by the consideration of *individual* facts; from time to time it must cast its glance at the world *as a whole*. Galileo's laws of falling bodies, Huygens's principle of *vis viva*, the principle of virtual velocities, nay, even the concept of mass, could not, as we saw, be obtained, except by the alternate consideration of individual facts and of nature as a totality. We may, in our mental reconstruction of mechanical processes, start from the properties of isolated masses (from the elementary or differential laws), and so compose our pictures of the processes; or, we may hold fast to the properties of the system as a whole (abide by the integral laws). Since, however, the properties of one mass always include relations to other masses, (for instance, in velocity and acceleration a relation of time is involved, that is, a connection with the whole world,) it is manifest that *purely* differential, or elementary, laws do not exist. It would be illogical, accordingly, to exclude as less certain this necessary view of the All, or of the more general properties of nature, from our studies. The more general a new principle is and the wider its scope, the *more perfect tests* will, in view of the possibility of error, be demanded of it.

The conception of a will and intelligence active in

nature is by no means the exclusive property of Christian monotheism. On the contrary, this idea is a quite familiar one to paganism and fetishism. Paganism, however, finds this will and intelligence entirely in individual phenomena, while monotheism seeks it in the All. Moreover, a pure monotheism does not exist. The Jewish monotheism of the Bible is by no means free from belief in demons, sorcerers, and witches; and the Christian monotheism of mediæval times is even richer in these pagan conceptions. We shall not speak of the brutal amusement in which church and state indulged in the torture and burning of witches, and which was undoubtedly provoked, in the majority of cases, not by avarice but by the prevalence of the ideas mentioned. In his instructive work on *Primitive Culture* Tylor has studied the sorcery, superstitions, and miracle-belief of savage peoples, and compared them with the opinions current in mediæval times concerning witchcraft. The similarity is indeed striking. The burning of witches, which was so frequent in Europe in the sixteenth and seventeenth centuries, is today vigorously conducted in Central Africa. Even now and in civilized countries and among cultivated people traces of these conditions, as Tylor shows, still exist in a multitude of usages, the sense of which, with our altered point of view, has been forever lost.

8. Physical science rid itself only very slowly of these conceptions. The celebrated work of Giambatista della Porta, *Magia naturalis*, which appeared in 1558, though it announces important physical discoveries, is yet filled with stuff about magic practices and demonological arts of all kinds little better than those of a redskin medicine-man. Not till the appearance of Gilbert's work, *De magnete* (in 1600), was any kind of re-

striction placed on this tendency of thought. When we reflect that even Luther is said to have had personal encounters with the Devil, that Kepler, whose aunt had been burned as a witch and whose mother came near meeting the same fate, said that witchcraft could not be denied, and dreaded to express his real opinion of astrology, we can vividly picture to ourselves the thought of less enlightened minds of those ages.

Modern physical science also shows traces of fetishism, as Tylor well remarks, in its "forces." And the hobgoblin practices of modern spiritualism are ample evidence that the conceptions of paganism have not been overcome even by the cultured society of today.

It is natural that these ideas so obstinately assert themselves. Of the many impulses that rule man with demoniacal power, that nourish, preserve, and propagate him, without his knowledge or supervision, of these impulses of which the middle ages present such great pathological excesses, only the smallest part is accessible to scientific analysis and conceptual knowledge. The fundamental character of all these instincts is the feeling of our oneness and sameness with nature; a feeling that at times can be silenced but never eradicated by absorbing intellectual occupations, and which certainly has a *sound basis,* no matter to what religious absurdities it may have given rise.

9. The French encyclopædists of the eighteenth century imagined they were not far from a final explanation of the world by physical and mechanical principles; Laplace even conceived a mind competent to foretell the progress of nature for all eternity, if but the masses, their positions, and initial velocities were given. In the eighteenth century, this joyful overestimation of the scope of the new physico-mechanical ideas is par-

donable. Indeed, it is a refreshing, noble, and elevating spectacle; and we can deeply sympathize with this expression of intellectual joy, so unique in history.

But now, after a century has elapsed, after our judgment has grown more sober, the world-conception of the encyclopædists appears to us as a *mechanical mythology* in contrast to the *animistic* of the old religions. Both views contain undue and fantastical exaggerations of an incomplete perception. Careful physical research will lead, however, to an analysis of our sensations. We shall then discover that our hunger is not so essentially different from the tendency of sulphuric acid for zinc, and our will not so greatly different from the pressure of a stone, as now appears. We shall again feel ourselves nearer nature, without its being necessary that we should resolve ourselves into a nebulous and mystical mass of molecules, or make nature a haunt of hobgoblins. The direction in which this enlightenment is to be looked for, as the result of long and painstaking research, can of course only be surmised. To *anticipate* the result, or even to attempt to introduce it into any scientific investigation of today, would be mythology, not science.

Physical science does not pretend to be a *complete* view of the world; it simply claims that it is working toward such a complete view in the future. The highest philosophy of the scientific investigator is precisely this *toleration* of an incomplete conception of the world and the preference for it rather than an apparently perfect, but inadequate conception. Our religious opinions are always our own private affair, as long as we do not obtrude them upon others and do not apply them to things which come under the jurisdiction of a different tribunal. Physical inquirers themselves entertain the most diverse opinions on this subject, according to the range of their intellects and their estimation of the consequences.

Physical science makes no investigation at all into things that are absolutely inaccessible to exact investigation, or as yet inaccessible to it. But should provinces ever be thrown open to exact research which are now closed to it, no well-organized man, no one who cherishes honest intentions towards himself and others, will then hesitate any longer to countenance inquiry with a view to exchanging his *opinion* regarding such provinces for positive *knowledge* of them.

When, today, we see society waver, see it change its views on the same question according to its mood and the events of the week, like the register of an organ, when we behold the profound mental anguish which is thus produced, we should know that this is the natural and necessary outcome of the incompleteness and transitional character of our philosophy. A competent view of the world can never be got as a gift; we must acquire it by hard work. And only by granting free sway to reason and experience in the provinces in which they alone are determinative, shall we, to the weal of mankind, approach, slowly, gradually, but surely, to that ideal of a *unified* view of the world which is alone compatible with the economy of a sound mind.

III.

ANALYTICAL MECHANICS.

1. The mechanics of Newton are purely *geometrical.* He deduces his theorems from his initial assumptions entirely by means of geometrical constructions. His procedure is frequently so artificial that, as Laplace

remarked, it is unlikely that the propositions were discovered in that way. We notice, moreover, that the
expositions of Newton are not as candid as those of
Galileo and Huygens. Newton's is the so-called *synthetic* method of the ancient geometers.

When we deduce results from given suppositions,
the procedure is called *synthetic*. When we seek the
conditions of a proposition or of the properties of a figure, the procedure is *analytic*. The practice of the latter
method became usual largely in consequence of the
application of algebra to geometry. It has become
customary, therefore, to call the algebraical method
generally, the analytical. The term "analytical mechanics," which is contrasted with the synthetical, or
geometrical, mechanics of Newton, is the exact equivalent of the phrase "algebraical mechanics."

2. The foundations of analytical mechanics were
laid by EULER (*Mechanica, sive Motus Scientia Analy-
tice Exposita*, St. Petersburg, 1736). But while Euler's
method, in its resolution of curvilinear forces into tangential and normal components, still bears a trace of
the old geometrical modes, the procedure of MACLAU-
RIN (*A Complete System of Fluxions*, Edinburgh,
1742) marks a very important advance. This author
resolves all forces in three fixed directions, and thus
invests the computations of this subject with a high
degree of symmetry and perspicuity.

3. Analytical mechanics, however, was brought to
its highest degree of perfection by LAGRANGE. Lagrange's aim is (*Mécanique analytique*, Paris, 1788) to
dispose *once for all* of the reasoning necessary to resolve
mechanical problems, by embodying as much as possible of it in a single formula. This he did. Every case
that presents itself can now be dealt with by a very

simple, highly symmetrical and perspicuous schema; and whatever reasoning is left is performed by purely mechanical methods. The mechanics of Lagrange is a stupendous contribution to the economy of thought.

In statics, Lagrange starts from the principle of virtual velocities. On a number of material points m_1, m_2, m_3. . . ., definitely connected with one another, are impressed the forces P_1, P_2, P_3. . . . If these points receive any infinitely small displacements p_1, p_2, p_3. . . . compatible with the connections of the system, then for equilibrium $\Sigma P p = 0$; where the well-known exception in which the equality passes into an inequality is left out of account.

Now refer the whole system to a set of rectangular coördinates. Let the coördinates of the material points be x_1, y_1, z_1, x_2, y_2, z_2 . . . . Resolve the forces into the components X_1, Y_1, Z_1, X_2, Y_2, Z_2. . . . parallel to the axes of coördinates, and the displacements into the displacements δx_1, δy_1, δz_1, δx_2, δy_2, δz_2 . . . ., also parallel to the axes. In the determination of the work done only the displacements of the point of application in the direction of each force-component need be considered for that component, and the expression of the principle accordingly is

$$\Sigma(X \delta x + Y \delta y + Z \delta z) = 0 \ldots \ldots (1)$$

where the appropriate indices are to be inserted for the points, and the final expressions summed.

The fundamental formula of dynamics is derived from D'Alembert's principle. On the material points m_1, m_2, m_3 . . . ., having the coördinates x_1, y_1, z_1, x_2, y_2, z_2 . . . . the force-components X_1, Y_1, Z_1, X_2, Y_2, Z_2 . . . . act. But, owing to the connections of the system's parts, the masses undergo accelerations, which are those of the forces.

$$m_1\frac{d^2x_1}{dt^2}, \quad m_1\frac{d^2y_1}{dt^2}, \quad m_1\frac{d^2z_1}{dt^2}\cdots$$

These are called the *effective forces*. But the *impressed forces,* that is, the forces which exist by virtue of the laws of physics, $X, Y, Z\ldots$ and the negative of these effective forces are, owing to the connections of the system, in equilibrium. Applying, accordingly, the principle of virtual velocities, we get

$$\Sigma\left\{\left(X-m\frac{d^2x}{dt^2}\right)\delta x + \left(Y-m\frac{d^2y}{dt^2}\right)\delta y + \left(Z-m\frac{d^2z}{dt^2}\right)\delta z\right\}=0 \quad \ldots\ldots\ldots (2)$$

4. Thus, Lagrange conforms to tradition in making statics precede dynamics. He was by no means compelled to do so. On the contrary, he might, with equal propriety, have started from the proposition that the connections, neglecting their straining, perform no work, or that all the possible work of the system is due to the impressed forces. In the latter case he would have begun with equation (2), which expresses this fact, and which, for equilibrium (or non-accelerated motion) reduces itself to (1) as a particular case. This would have made analytical mechanics, as a system, even more logical.

Equation (1), which for the case of equilibrium makes the element of the work corresponding to the assumed displacement $= 0$, gives readily the results discussed in page 80. If

$$X=\frac{dV}{dx}, \quad Y=\frac{dV}{dy}, \quad Z=\frac{dV}{dz},$$

that is to say, if X, Y, Z are the partial differential co-efficients of one and the same function of the coördi-nates of position, the whole expression under the sign of summation is the total variation, δV, of V. If the latter is $= 0$, V is in general a maximum or a minimum.

5. We will now illustrate the use of equation (1) by a simple example. If all the points of application of the forces are independent of each other, no problem is presented. Each point is then in equilibrium only when the forces impressed on it, and consequently their components, are $= 0$. All the displacements δx, δy, δz . . . . are then wholly arbitrary, and equation (1) can subsist only provided the coefficients of all the displacements δx, δy, δz . . . . are equal to zero.

But if equations obtain between the coördinates of the several points, that is to say, if the points are sub-ject to mutual constraints, the equations so obtaining will be of the form $F(x_1, y_1, z_1, x_2, y_2, z_2 \ldots .) = 0$, or, more briefly, of the form $F = 0$. Then equations also obtain between the displacements, of the form

$$\frac{dF}{dx_1} \delta x_1 + \frac{dF}{dy_1} \delta y_1 + \frac{dF}{dz_1} \delta z_1 + \frac{dF}{dx_2} \delta x_2 + \ldots . = 0,$$

which we shall briefly designate as $DF = 0$. If the system consists of n points, we shall have $3n$ coördi-nates, and equation (1) will contain $3n$ magnitudes δx, δy, δz . . . . If, further, between the coördinates m equations of the form $F = 0$ subsist, then m equa-tions of the form $DF = 0$ will be simultaneously given between the variations δx, δy, δz . . . . By these equations m variations can be expressed in terms of the remainder, and so inserted in equation (1). In (1), therefore, there are left $3n - m$ arbitrary displace-ments, whose coefficients are put $= 0$. There are thus obtained between the forces and the coördinates $3n-m$ equations, to which the m equations $(F = 0)$ must be added. We have, accordingly, in all, $3n$ equations, which are sufficient to determine the $3n$ coördinates of the position of equilibrium, provided the forces are given and only the *form* of the system's equilibrium is sought.

But if the form of the system is given and the *forces* are sought that maintain equilibrium, the question is indeterminate. We have then, to determine $3n$ force-components, only $3n-m$ equations, since the m equations $(F = 0)$ do not contain the force-components.

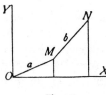

Fig. 232.

As an example of this manner of treatment we shall select a lever $OM = a$, free to rotate about the origin of coördinates in the plane XY, and having at its end a second, similar lever $MN = b$. At M and N, the coördinates of which we shall call x, y and x_1, y_1, the forces X, Y and X_1, Y_1 are applied. Equation (1), then, has the form

$$X\,\delta x + X_1\,\delta x_1 + Y\,\delta y + Y_1\,\delta y_1 = 0 \ldots (3)$$

Of the form $F = 0$ two equations here exist; namely,

$$\left. \begin{array}{l} x^2 + y^2 - a^2 = 0 \\ (x_1 - x)^2 + (y_1 - y)^2 - b^2 = 0 \end{array} \right\} \quad \cdots \cdots \cdots (4)$$

The equations $DF = 0$, accordingly, are

$$\left. \begin{array}{l} x\,\delta x + y\,\delta y = 0 \\ (x_1 - x)\,\delta x_1 - (x_1 - x)\,\delta x + (y_1 - y)\,\delta y_1 - \\ \qquad (y_1 - y)\,\delta y = 0 \end{array} \right\} \quad \cdot (5)$$

Here, two of the variations in (5) can be expressed in terms of the others and introduced in (3). Also for purposes of elimination Lagrange employed a perfectly uniform and systematic procedure, which may be pursued quite mechanically, without reflection. We shall use it here. It consists in multiplying each of the equations (5) by an indeterminate coefficient λ, μ, and adding each in this form to (3). So doing, we obtain

$$\left.\begin{array}{l} [X+\lambda x -\mu(x_1-x)]\delta x + [X_1+\mu(x_1-x)]\delta x_1 \\ [Y+\lambda y -\mu(y_1-y)]\delta y + [Y_1+\mu(y_1-y)]\delta y_1 \end{array}\right\} = 0.$$

The coefficients of the four displacements may now be put directly $= 0$. For two displacements are arbitrary, and the two remaining coefficients may be made equal to zero by the appropriate choice of λ and μ—which is tantamount to an elimination of the two remaining displacements.

We have, therefore, the four equations

$$\left.\begin{array}{l} X + \lambda x - \mu(x_1 - x) = 0 \\ X_1 + \mu(x_1 - x) = 0 \\ Y + \lambda y - \mu(y_1 - y) = 0 \\ Y_1 + \mu(y_1 - y) = 0 \end{array}\right\} \quad \dots \dots \dots (6)$$

We shall first assume that the coördinates are given, and seek the *forces* that maintain equilibrium. The values of λ and μ are each determined by equating to zero two coefficients. We get from the second and fourth equations,

$$\mu = \frac{-X_1}{x_1 - x}, \text{ and } \mu = \frac{-Y_1}{y_1 - y},$$

whence

$$\frac{X_1}{Y_1} = \frac{x_1 - x}{y_1 - y} \quad \dots \dots \dots \dots \dots (7)$$

that is to say, the total component force impressed at N has the direction MN. From the first and third equations we get

$$\lambda = \frac{-X + \mu(x_1 - x)}{x}, \quad \lambda = \frac{-Y + \mu(y_1 - y)}{y},$$

and from these by simple reduction

$$\frac{X + X_1}{Y + Y_1} = \frac{x}{y} \quad \ldots \ldots \ldots \ldots \ldots \ldots \quad (8)$$

that is to say, the resultant of the forces applied at M and N acts in the direction OM.*

The *four* force-components are accordingly subject to only *two* conditions, (7) and (8). The problem, consequently, is an indeterminate one; as it must be from the nature of the case; for equilibrium does not depend upon the absolute magnitudes of the forces, but upon their directions and relations.

If we assume that the forces are given and seek the four *coördinates*, we treat equations (6) in exactly the same manner. Only, we can now make use, in addi-

$$x = \frac{a(X + X_1)}{\sqrt{(X + X_1)^2 + (Y + Y_1)^2}}$$

$$y = \frac{a(Y + Y_1)}{\sqrt{(X + X_1)^2 + (Y + Y_1)^2}}$$

---

* The mechanical interpretation of the indeterminate coefficients λ, μ may be shown as follows. Equations (6) express the equilibrium of two *free* points on which in addition to X, Y, X_1, Y_1 other forces act which answer to the remaining expressions and just destroy X, Y, X_1, Y_1. The point N, for example, is in equilibrium if X_1 is destroyed by a force $\mu(x_1 - x)$, undetermined as yet in magnitude, and Y_1 by a force $\mu(y_1 - y)$. This supplementary force is due to the constraints. Its direction is determined; though its magnitude is not. If we call the angle which it makes with the axis of abscissas a, we shall have

$$\tan a = \frac{\mu(y_1 - y)}{\mu(x_1 - x)} = \frac{y_1 - y}{x_1 - x}$$

that is to say, the force due to the connections acts in the direction of b.

tion, of equations (4). Accordingly, we have, upon the elimination of λ and μ, equations (7) and (8) and two equations (4). From these the following, which fully solve the problem, are readily deduced:

$$x_1 = \frac{a(X + X_1)}{\sqrt{(X + X_1)^2 + (Y + Y_1)^2}} + \frac{b X_1}{\sqrt{X_1^2 + Y_1^2}}$$

$$y_1 = \frac{a(Y + Y_1)}{\sqrt{(X + X_1)^2 + (Y + Y_1)^2}} + \frac{b Y_1}{\sqrt{X_1^2 + Y_1^2}}.$$

Simple as this example is, it is yet sufficient to give us a distinct idea of the character and significance of Lagrange's method. The mechanism of this method is excogitated once for all, and in its application to particular cases scarcely any additional thinking is required. The simplicity of the example here selected being such that it can be solved by a mere glance at the figure, we have, in our study of the method, the advantage of a ready verification at every step.

6. We will now illustrate the application of equation (2), which is Lagrange's form of statement of D'Alembert's principle. There is no problem when the masses move quite independently of one another. Each mass yields to the forces applied to it; the variations $\delta x, \delta y, \delta z \ldots$ are wholly arbitrary, and each coefficient may be singly put $= 0$. For the motion of n masses we thus obtain $3n$ simultaneous differential equations.

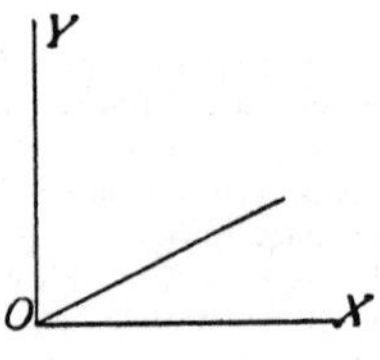

But if equations of condition $(F = 0)$ obtain between the coördinates, these equations will lead to others $(DF = 0)$ between the displacements or variations. With the latter we proceed exactly as in the application of equation (1). Only it must be noted

Fig. 233.

here that the equations $F = 0$ must eventually be employed in their undifferentiated as well as in their differentiated form, as will best be seen from the following example.

A heavy material point m, lying in a vertical plane XY, is free to move on a straight line, $y = a\,x$, inclined at an angle to the horizon. (Fig. 233.) Here equation (2) becomes

$$\left(X - m\frac{d^2 x}{d t^2}\right)\delta x + \left(Y - m\frac{d^2 y}{d t^2}\right)\delta y = 0,$$

and, since $X = 0$, and $Y = -mg$, also

$$\frac{d^2 x}{d t^2}\delta x + \left(g + \frac{d^2 y}{d t^2}\right)\delta y = 0 \quad \dots \dots \dots \text{(9)}$$

The place of $F = 0$ is taken by

$$y = ax \dots \dots \dots \dots \dots \dots \dots \dots \dots \text{(10)}$$

and for $DF = 0$ we have

$$\delta y = a\,\delta x.$$

Equation (9), accordingly, since δy drops out and δx is arbitrary, passes into the form

$$\frac{d^2 x}{d t^2} + \left(g + \frac{d^2 y}{d t^2}\right)a = 0.$$

By the differentiation of (10), or $(F = 0)$, we have

$$\frac{d^2 y}{d t^2} = a\frac{d^2 x}{d t^2},$$

and, consequently,

$$\frac{d^2 x}{d t^2} + a\left(g + a\frac{d^2 x}{d t^2}\right) = 0 \quad \dots \dots \dots \text{(11)}$$

Then, by the integration of (11), we obtain

$$x = \frac{-a}{1 + a^2}g\,\frac{t^2}{2} + bt + c$$

and
$$y = \frac{-a^2}{1 + a^2} g \frac{t^2}{2} + abt + ac,$$

where b and c are constants of integration, determined by the initial position and velocity of m. This result can also be easily found by the direct method.

Some care is necessary in the application of equation (1) if $F = 0$ contains the time. The procedure in such cases may be illustrated by the following example: Imagine in the preceding case the straight line on which m descends to move vertically upwards with the acceleration γ. We start again from equation (9)

$$\frac{d^2 x}{dt^2} \delta x + \left(g + \frac{d^2 y}{dt^2}\right) \delta y = 0.$$

$F = 0$ is here replaced by

$$y = ax + \gamma \frac{t^2}{2} \quad \cdots \cdots \cdots \cdots \cdots \quad (12)$$

To form $DF = 0$, we vary (12) only with respect to x and y, for we are concerned here only with the *possible* displacement of the system in its position *at any given instant,* and not with the displacement that *actually* takes place in time. We put, therefore, as in the previous case,

$$\delta y = a \delta x,$$

and obtain, as before,

$$\frac{d^2 x}{dt^2} + \left(g + \frac{d^2 y}{dt^2}\right) a = 0 \quad \cdots \cdots \cdots \quad (13)$$

But to get an equation in x alone, we have, since x and y are connected in (13) by the *actual* motion, to differentiate (12) with respect to t and employ the resulting equation

$$\frac{d^2 y}{d t^2} = a \frac{d^2 x}{d t^2} + \gamma$$

for substitution in (13). In this way the equation

$$\frac{d^2 x}{d t^2} + \left(g + \gamma + a \frac{d^2 x}{d t^2}\right) a = 0$$

is obtained, which, integrated, gives

$$x = \frac{-a}{1 + a^2} (g + \gamma) \frac{t^2}{2} + b t + c$$

$$y = \left[\gamma - \frac{a^2}{1 + a^2} (g + \gamma)\right] \frac{t^2}{2} + a b t + a c.$$

If a *weightless* body m lies on the moving straight line, we obtain these equations

$$x = \frac{-a}{1 + a^2} \gamma \frac{t^2}{2} + b t + c$$

$$y = \frac{\gamma}{1 + a^2} \frac{t^2}{2} + a b t + a c,$$

—results which are readily understood, when we reflect that, on a straight line moving upwards with the acceleration γ, m behaves as if it were affected with a downward acceleration γ on the straight line at rest.

7. The procedure with equation (12) in the preceding example may be rendered somewhat clearer by the following consideration. Equation (2), D'Alembert's principle, asserts, that all the work that *can* be done in the displacement of a system is done by the impressed forces and not by the connections. This is evident, since the rigidity of the connections allows no changes in the relative positions which would be necessary for any alteration in the potentials of the elastic

Fig. 234.

forces. But this ceases to be true when the connections undergo changes in *time*. In this case, the *changes* of the connections perform work, and we can then apply equation (2) to the displacements that *actually* take place only provided we add to the impressed forces the forces that produce the changes of the connections.

A heavy mass m is free to move on a straight line parallel to OY (Fig. 234.) Let this line be subject to a forced acceleration in the direction of x, such that the equation $F = 0$ becomes

$$x = \gamma \, \frac{t^2}{2}, \quad \ldots \ldots \ldots \ldots \ldots \ldots \quad (14)$$

D'Alembert's principle again gives equation (9). But since from $DF = 0$ it follows here that $\delta x = 0$, this equation reduces itself to

$$\left(g + \frac{d^2 y}{d t^2} \right) \delta y = 0 \quad \ldots \ldots \ldots \ldots \ldots \quad (15)$$

in which δy is wholly arbitrary. Wherefore,

$$g + \frac{d^2 y}{d t^2} = 0$$

and

$$y = \frac{-g \, t^2}{2} + a t + b,$$

to which must be supplied (14) or

$$x = \gamma \, \frac{t^2}{2}.$$

It is patent that (15) does not assign the total work of the displacement that *actually* takes place, but only that of some *possible* displacement on the straight line conceived, for the moment, as fixed.

If we imagine the straight line massless, and cause it to travel parallel to itself in some guiding mechanism moved by a force $m\gamma$, equation (2) will be replaced by

$$\left(m\gamma - m\frac{d^2x}{dt^2}\right)\delta x + \left(-mg - m\frac{d^2y}{dt^2}\right)\delta y = 0,$$

and since δx, δy are wholly arbitrary here, we obtain the two equations

$$\gamma - \frac{d^2x}{dt^2} = 0$$

$$g + \frac{d^2y}{dt^2} = 0,$$

which give the same results as before. The apparently different mode of treatment of these cases is simply the result of a slight inconsistency, springing from the fact that *all* the forces involved are, for reasons facilitating calculation, not included in the consideration at the outset, but a portion is left to be dealt with subsequently.

8. As the different mechanical principles only express different aspects of the same fact, any one of them is easily deducible from any other; as we shall now illustrate by developing the principle of *vis viva* from equation (2) of page 563. Equation (2) refers to instantaneously possible displacements, that is, to "virtual" displacements. But when the connections of a system are independent of the time, the motions *that actually take place* are "virtual" displacements. Consequently the principle may be applied to actual motions. For δx, δy, δz, we may, accordingly, write dx, dy, dz, the displacements which take place in time, and put

$$\Sigma(Xdx + Ydy + Zdz) =$$
$$\Sigma m\left(\frac{d^2x}{dt^2}dx + \frac{d^2y}{dt^2}dy + \frac{d^2z}{dt^2}dz\right).$$

The expression to the right may, by introducing for
$d\,x$, $(d\,x/d\,t)d\,t$ and so forth, and by denoting the velocity by v, also be written

$$\Sigma m\left(\frac{d^2 x}{dt^2}\frac{dx}{dt}dt + \frac{d^2 y}{dt^2}\frac{dy}{dt}dt + \frac{d^2 z}{dt^2}\frac{dz}{dt}dt\right)=$$

$$\tfrac{1}{2}d\Sigma m\left[\left(\frac{dx}{dt}\right)^2 + \left(\frac{dy}{dt}\right)^2 + \left(\frac{dz}{dt}\right)^2\right] = \tfrac{1}{2}d\Sigma mv^2.$$

Also in the expression to the left, $(d\,x/d\,t)\,d\,t$ may be
written for $d\,x$. But this gives

$$\int\Sigma\,(Xdx + Ydy + Zdz) = \Sigma\tfrac{1}{2}m\,(v^2 - v_0^2),$$

where v_0 denotes the velocity at the beginning and v
the velocity at the end of the motion. The integral to the
left can always be found if we can reduce it to a single
variable, that is to say, if we know the course of the
motion in time or the paths which the movable points
describe. If, however, X, Y, Z are the partial differential coefficients of the same function U of coördinates, if, that is to say,

$$X=\frac{dU}{dx},\ \ Y=\frac{dU}{dy},\ \ Z=\frac{dU}{dz},$$

as is always the case when only central forces are involved, this reduction is unnecessary. The entire expression to the left is then a complete differential. And
we have

$$\Sigma\,(U-U_0) = \Sigma\tfrac{1}{2}m\,(v^2 - v_0^2),$$

which is to say, the difference of the force-functions
(or work) at the beginning and the end of the motion
is equal to the difference of the *vires vivæ* at the beginning and the end of the motion. The *vires vivæ* are
in such case also functions of the coördinates.

In the case of a body movable in the plane of X and Y suppose, for example, $X = -y$, $Y = -x$; we then have

$$\int (-y\,dx - x\,dy) = -\int d\,(xy) =$$
$$x_0 y_0 - xy = \tfrac{1}{2} m\,(v^2 - v_0^2).$$

But if $X = -a$, $Y = -x$, the integral to the left is $-\int (a\,dx + x\,d\,y)$. This integral can be assigned the moment we know the path the body has traversed, that is, if y is determined a function of x. If, for example, $y = p\,x^2$, the integral would become

$$-\int (a + 2px^2)\,dx = a\,(x_0 - x) + \frac{2\,p\,(x_0^3 - x^3)}{3}.$$

The difference of these two cases is, that in the first the work is simply a function of coördinates, that a force-function exists, that the element of the work is a complete differential, and the work consequently is determined by the initial and final values of the coördinates, while in the second case it is dependent on the entire path described.

9. These simple examples, in themselves presenting no difficulties, will doubtless suffice to illustrate the general nature of the operations of analytical mechanics. No fundamental light can be expected from this branch of mechanics. On the contrary, the discovery of matters of principle must be substantially completed before we can think of framing analytical mechanics; the sole aim of which is a perfect practical *mastery* of problems. Whosoever mistakes this situation, will never comprehend Lagrange's great performance, which here too is essentially of an *economical* character. Poinsot did not altogether escape this error.

It remains to be mentioned that as the result of the labors of Möbius, Hamilton, Grassmann, and others, a new transformation of mechanics is preparing. These inquirers have developed mathematical conceptions that conform more exactly and directly to our geometrical ideas than do the conceptions of common analytical geometry; and the advantages of analytical generality and direct geometrical insight are thus united. But this transformation, of course, lies, as yet, beyond the limits of an historical exposition.

The *Ausdehnungslehre* of 1844, in which Grassmann expounded his ideas for the first time, is in many respects remarkable. The introduction to it contains epistemological remarks of value. The theory of spatial extension is here developed as a general science, of which geometry is a special tri-dimensional case; and the opportunity is taken on this occasion of submitting the foundations of geometry to a rigorous critique. The new and fruitful concepts of the addition of line-segments, multiplication of line-segments, etc., have also proved to be applicable in mechanics. Grassmann likewise submits the Newtonian principles to criticism, and believes he is able to enunciate them in a single expression as follows: "The total force (or total motion) which is inherent in an aggregate of material particles at any one time is the sum of the total force (or total motion) which has inhered in it at any former time, and all the forces that have been imparted to it from without in the intervening time; provided all forces be conceived as line-segments constant in direction and in length, and be referred to points which have equal masses." By *force* Grassmann understands here the indestructibly impressed velocity. The entire conception is much akin to that of Hertz. The forces (velocities) are represented as line-segments, the moments as surfaces enumerated in definite directions, etc. — a device by means of which every development takes a very concise and perspicuous form. But Grassmann finds the main advantage of his procedure in the fact that every step in the calculation is at the same time the clear expression of every step taken in the thought; whereas in the common method, the latter is forced entirely into the background by the introduction of three arbitrary coördinates. The difference between the analytic and the synthetic method is again done away with, and the advantages of the two are combined. The kindred procedure of Hamilton, which has been illustrated by an example on page 196, will give some idea of these advantages.

IV.

THE ECONOMY OF SCIENCE.[*]

1. It is the object of science to replace, or *save,* experiences, by the reproduction and anticipation of facts in thought. Memory is handier than experience, and often answers the same purpose. This economical office of science, which fills its whole life, is apparent at first glance; and with its full recognition all mysticism in science disappears. Science is communicated by instruction, in order that one man may profit by the experience of another and be spared the trouble of accumulating it for himself; and thus, to spare posterity, the experiences of whole generations are stored up in libraries.

[*] *Cf.* my paper, "Die Leitgedanken meiner naturwissenschaftlichen Erkenntnislehre und ihre Aufnahme durch die Zeitgenossen" (*Scientia: Rivista di Scienza,* vol. vii, 1910, No. 14, 2; or *Physikalische Zeitschrift,* 1910, pp. 599-606).

Language, the instrument of this communication, is itself an economical contrivance. Experiences are analyzed, or broken up, into simpler and more familiar experiences, and then symbolized at some sacrifice of precision. The symbols of speech are as yet restricted in their use within national boundaries, and doubtless will long remain so. But written language is gradually being metamorphosed into an ideal universal character. It is certainly no longer a mere transcript of speech. Numerals, algebraic signs, chemical symbols, musical notes, phonetic alphabets, may be regarded as parts already formed of this universal character of the future; they are, to some extent, decidedly conceptual, and of almost general international use. The analysis of colors, physical and physiological, is already far enough advanced to render an international system of color-signs perfectly practical. In Chinese writing, we have an actual example of a true ideographic language, pronounced diversely in different provinces, yet everywhere carrying the same meaning. Were the system and its signs only of a simpler character, the use of Chinese writing might become universal. The dropping of unmeaning and needless accidents of grammar, as English mostly drops them, would be quite requisite to the adoption of such a system. But universality would not be the sole merit of such a character; since to read it would be to understand it. Our children often read what they do not understand; but that which a Chinaman cannot understand, he is precluded from reading.

2. In the reproduction of facts in thought, we never reproduce the facts in full, but only that side of them which is important to us, moved to this directly or indirectly by a practical interest. Our reproductions

are invariably abstractions. Here again is an economical tendency.

Nature is composed of sensations as its elements. Primitive man, however, first picks out certain compounds of these elements—those namely that are relatively permanent and of greater importance to him. The first and oldest words are names of "things." Even here, there is an abstractive process, an abstraction from the surroundings of the things, and from the continual small changes which these compound sensations undergo, which being practically unimportant are not noticed. No inalterable thing exists. The thing is an abstraction, the name a symbol, for a compound of elements from whose changes we abstract. The reason we assign a single word to a whole compound is that we need to suggest all the constituent sensations at once. When, later, we come to remark the changeableness, we cannot at the same time hold fast to the idea of the thing's permanence, unless we have recourse to the conception of a thing-in-itself, or other such like absurdity. Sensations are not signs of things; but, on the contrary, a thing is a thought-symbol for a compound sensation of relative fixedness. Properly speaking the world is not composed of "things" as its elements, but of colors, tones, pressures, spaces, times, in short what we ordinarily call individual sensations.

The whole operation is a mere affair of economy. In the reproduction of facts, we begin with the more durable and familiar compounds, and supplement these later with the unusual by way of corrections. Thus, we speak of a perforated cylinder, of a cube with beveled edges, expressions involving contradictions, unless we accept the view here taken. All judgments are such amplifications and corrections of ideas already admitted.

3. In speaking of cause and effect we arbitrarily give relief to those elements to whose connection we have to attend in the reproduction of a fact in the respect in which it is important to us. There is no cause nor effect in nature; nature has but an individual existence; nature simply *is*. Recurrences of like cases in which A is always connected with B, that is, like results under like circumstances, that is again, the essence of the connection of cause and effect, exist but in the abstraction which we perform for the purpose of mentally reproducing the facts. Let a fact become familiar, and we no longer require this putting into relief of its connecting marks, our attention is no longer attracted to the new and surprising, and we cease to speak of cause and effect. Heat is said to be the cause of the tension of steam; but when the phenomenon becomes familiar we think of the steam at once with the tension proper to its temperature. Acid is said to be the cause of the reddening of tincture of litmus; but later we think of the reddening as a property of the acid.

Hume first propounded the question: How can a thing A act on another thing B? Hume, in fact, rejects causality and recognizes only a wonted succession in time. Kant correctly remarked that a *necessary* connection between A and B could not be disclosed by simple observation. He assumes an innate idea or category of the mind, a *Verstandesbegriff,* under which the cases of experience are subsumed. Schopenhauer, who adopts substantially the same position, distinguishes four forms of the "principle of sufficient reason" — the logical, physical, and mathematical form, and the law of motivation. But these forms differ only as regards the matter to which they are applied, which may belong either to outward or inward experience.

The natural and common-sense explanation is apparently this. The ideas of cause and effect originally sprang from an endeavor to reproduce facts in thought. At first, the connection of *A* and *B*, of *C* and *D*, of *E* and *F*, and so forth, is regarded as familiar. But after a greater range of experience is acquired and a connection between *M* and *N* is observed, it often turns out that we recognize *M* as *made up of A, C, E,* and *N* of *B, D, F,* the connection of which was before a *familiar* fact and accordingly possesses with us a higher authority. This explains why a person of experience regards a new event with different eyes than the novice. The new experience is illuminated by the mass of old experience. As a fact, then, there really does exist in the mind an "idea" under which fresh experiences are subsumed; but that idea has itself been developed from experience. The notion of the *necessity* of the causal connection is probably created by our voluntary movements in the world and by the changes which these indirectly produce, as Hume supposed but Schopenhauer contested. Much of the authority of the ideas of cause and effect is due to the fact that they are developed *instinctively* and involuntarily, and that we are distinctly sensible of having personally contributed nothing to their formation. We may, indeed, say, that our sense of causality is not acquired by the individual, but has been perfected in the development of the race. Cause and effect, therefore, are things of thought, having an economical office. It cannot be said *why* they arise. For it is precisely by the abstraction of uniformities that we know the question "why."*

* In the text I have employed the term "cause" in the sense in which it is ordinarily used. I may add that with Dr. Carus, following the practice of the German philosophers, I *distinguish* "cause," or *Realgrund*, from *Erkenntnissgrund*. I also agree with Dr. Carus in the statement that "the signification of cause and effect is to a great extent arbitrary and depends much upon

4. In the details of science, its economical character is still more apparent. The so-called descriptive sciences must chiefly remain content with reconstructing individual facts. Where it is possible, the common features of many facts are once for all placed in relief. But in sciences that are more highly developed, rules for the reconstruction of great numbers of facts may be embodied in a *single* expression. Thus, instead of noting individual cases of light-refraction, we can mentally reconstruct all present and future cases, if we know that the incident ray, the refracted ray, and the perpendicular lie in the same plane and that $\sin \alpha / \sin \beta = n$. Here, instead of the numberless cases of refraction in different combinations of matter and under all different angles of incidence, we have simply to note the rule above stated and the values of n, which is much easier. The economical purpose is here unmistakable. In nature there is no *law* of refraction, only different cases of refraction. The law of refraction is a concise compendious rule, devised by us for the mental reconstruction of a fact, and only for its reconstruction in part, that is, on its geometrical side.

5. The sciences most highly developed economically are those whose facts are reducible to a few numerable elements of like nature. Such is the science of mechanics, in which we deal exclusively with spaces, times, and masses. The whole previously established economy of mathematics stands these sciences in stead.

---

the proper tact of the observer." (See his *Fundamental Problems*, pp. 79-91, Chicago: The Open Court Publishing Co., 1891. Also, p. 84.)

The notion of cause possesses significance only as a means of provisional knowledge or orientation. In any exact and profound investigation of an event the inquirer must regard the phenomena as *dependent* on *one another* in the same way that the geometer regards the sides and angles of a triangle as dependent on one another. He will constantly keep before his mind, in this way, all the conditions of fact.

Mathematics may be defined as the economy of counting. Numbers are arrangement-signs which, for the sake of perspicuity and economy, are themselves arranged in a simple system. Numerical operations, it is found, are independent of the kind of objects operated on, and are consequently mastered once for all. When, for the first time, I have occasion to add five objects to seven others, I count the whole collection through, at once; but when I afterwards discover that I can start counting from 5, I save myself part of the trouble; and still later, remembering that 5 and 7 always count up to 12, I dispense with the numeration entirely.

The object of all arithmetical operations is to *save* direct numeration, by utilizing the results of our old operations of counting. Our endeavor is, having done a sum once, to preserve the answer for future use. The first four rules of arithmetic well illustrate this view. Such, too, is the purpose of algebra, which, substituting relations for values, symbolizes and definitively fixes all numerical operations that follow the same rule. For example, we learn from the equation

$$\frac{x^2 - y^2}{x + y} = x - y,$$

that the more complicated numerical operation at the left may always be replaced by the simpler one at the right, whatever numbers x and y stand for. We thus save ourselves the labor of performing in future cases the more complicated operation. Mathematics is the method of replacing in the most comprehensive and *economical* manner possible, *new* numerical operations by old ones done already with known results. It may happen in this procedure that the results of operations are employed which were originally performed centuries ago.

Often operations involving intense mental effort may be replaced by the action of semi-mechanical routine, with great saving of time and avoidance of fatigue. For example, the theory of determinants owes its origin to the remark, that it is not necessary to solve each time anew equations of the form

$$a_1 x + b_1 y + c_1 = 0$$
$$a_2 x + b_2 y + c_2 = 0,$$

from which result

$$x = -\frac{c_1 b_2 - c_2 b_1}{a_1 b_2 - a_2 b_1} = -\frac{P}{N}$$

$$y = -\frac{a_1 c_2 - a_2 c_1}{a_1 b_2 - a_2 b_1} = -\frac{Q}{N},$$

but that the solution may be effected by means of the coefficients, by writing down the coefficients according to a prescribed scheme and operating with them *mechanically*. Thus,

$$\begin{vmatrix} a_1 & b_1 \\ a_2 & b_2 \end{vmatrix} = a_1 b_2 - a_2 b_1 = N$$

and similarly

$$\begin{vmatrix} c_1 & b_1 \\ c_2 & b_2 \end{vmatrix} = P, \text{ and } \begin{vmatrix} a_1 & c_1 \\ a_2 & c_2 \end{vmatrix} = Q.$$

By means of mathematical operations a complete relaxation of the mind can occur. This happens where operations of counting hitherto performed are symbolized by mechanical operations with signs, and our brain energy, instead of being wasted on the repetition of old operations, is spared for more important tasks. The merchant pursues a like economy, when, instead of directly handling his bales of goods, he operates with bills of lading or assignments of them. The drudgery of computation may even be relegated to a machine. Several different types of calculating machines are actually in practical use. The earliest of these (of any complexity) was the difference-engine of Babbage, who was familiar with the ideas here presented.

A numerical result is not always reached by the *actual* solution of the problem; it may also be reached indirectly. It is easy to ascertain, for example, that a curve whose quadrature for the abscissa x has the value x^m, gives an increment $m\,x^{m-1}\,d\,x$ of the quadrature for the increment $d\,x$ of the abscissa. But we then also know that $\int m\,x^{m-1}\,d\,x = x^m$; that is, we recognize the quantity x^m from the increment $m\,x^{m-1}\,d\,x$ as unmistakably as we recognize a fruit by its rind. Results of this kind, accidentally found by simple inversion, or by processes more or less analogous, are very extensively employed in mathematics.

That scientific work should be more useful the more it has been used, while mechanical work is expended in use, may seem strange to us. When a person who daily takes the same walk accidentally finds a shorter cut, and thereafter, remembering that it is shorter, always goes that way, he undoubtedly saves himself the difference of the work. But memory is really not work. It only places at our disposal energy within our present or future possession, which the circumstance of ignorance prevented us from availing ourselves of. This is precisely the case with the application of scientific ideas.

The mathematician who pursues his studies without clear views of this matter, must often have the uncomfortable feeling that his paper and pencil surpass him in intelligence. Mathematics, thus pursued as an object of instruction, is scarcely of more educational value than busying oneself with the Cabala. On the contrary, it induces a tendency toward mystery, which is pretty sure to bear its fruits.

6. The science of physics also furnishes examples of this economy of thought, altogether similar to those we have just examined. A brief reference here will suffice. The moment of inertia saves us the separate consideration of the individual particles of masses. By the force-function we dispense with the separate investigation of individual force-components. The simplicity of reasonings involving force functions springs from the fact that a great amount of mental work had to be performed before the discovery of the properties of the force-functions was possible. Gauss's dioptrics dispenses us from the separate consideration of the single refracting surfaces of a dioptrical system and substitutes for it the principal and nodal points. But a careful consideration of the single surfaces had to precede the discovery of the principal and nodal points. Gauss's dioptrics simply *saves* us the necessity of often repeating this consideration.

We must admit, therefore, that there is no result of science which in point of principle could not have been arrived at wholly without methods. But, as a matter of fact, within the short span of a human life and with man's limited powers of memory, any stock of knowledge worthy of the name is unattainable except by the *greatest* mental economy. Science itself, therefore, may be regarded as a minimal problem, consisting of the completest possible presentment of facts with the *least possible expenditure of thought*.

7. The function of science, as we take it, is to replace experience. Thus, on the one hand, science must remain in the province of experience, but, on the other, must hasten beyond it, constantly expecting con-

firmation, constantly expecting the reverse. Where neither confirmation nor refutation is possible, science is not concerned. Science acts and acts only in the domain of *uncompleted* experience. Exemplars of such branches of science are the theories of elasticity and of the conduction of heat, both of which ascribe to the smallest particles of matter only such properties as observation supplies in the study of the larger portions. The comparison of theory and experience may be farther and farther extended, as our means of observation increase in refinement.

Experience alone, without the ideas that are associated with it, would forever remain strange to us. Those ideas that hold good throughout the widest domains of research and that supplement the greatest amount of experience, are the *most scientific*. The principle of continuity, the use of which everywhere pervades modern inquiry, simply prescribes a mode of conception which conduces in the highest degree to the economy of thought.

8. If a long elastic rod be fastened in a vise, the rod may be made to execute slow vibrations. These are directly observable, can be seen, touched, and graphically recorded. If the rod be shortened, the vibrations will increase in rapidity and cannot be directly seen; the rod will present to the sight a blurred image. This is a new phenomenon. But the sensation of touch is still like that of the previous case; we can still make the rod record its movements; and if we mentally retain the *conception* of vibrations, we can still anticipate the results of experiments. On further shortening the rod the sensation of touch is altered; the rod begins to sound; again a new phenomenon is presented. But the phenomena do not all change at once; only this or that phenomenon changes; conse-

quently the accompanying notion of vibration, which is not confined to any single one, is still serviceable, still economical. Even when the sound has reached so high a pitch and the vibrations have become so small that the previous means of observation are not of avail, we still *advantageously* imagine the sounding rod to perform vibrations, and can predict the vibrations of the dark lines in the spectrum of the polarized light of a rod of glass. If on the rod being further shortened *all* the phenomena suddenly passed into *new* phenomena, the conception of vibration would no longer be serviceable because it would no longer afford us a means of supplementing the new experiences by the previous ones.

When we mentally add to those actions of a human being which we can perceive, sensations and ideas like our own which we cannot perceive, the object of the idea we so form is economical. The idea makes experience intelligible to us; it supplements and supplants experience. This idea is not regarded as a great scientific discovery, only because its formation is so natural that every child conceives it. Now, this is exactly what we do when we imagine a moving body which has just disappeared behind a pillar, or a comet at the moment invisible, as continuing its motion and retaining its previously observed properties. We do this that we may not be surprised by its reappearance. We fill out the gaps in experience by the ideas that experience suggests.

9. Yet not all the prevalent scientific theories originated so naturally and artlessly. Thus, chemical, electrical, and optical phenomena are explained by atoms. But the mental artifice atom was not formed by the principle of continuity; on the contrary, it is a product especially devised for the purpose in view. Atoms

cannot be perceived by the senses; like all substances, they are things of thought. Furthermore, the atoms are invested with properties that absolutely contradict the attributes hitherto observed in bodies. However well fitted atomic theories may be to reproduce certain groups of facts, the physical inquirer who has laid to heart Newton's rules will only admit those theories as *provisional* helps, and will strive to attain, in some more natural way, a satisfactory substitute.

The atomic theory plays a part in physics similar to that of certain auxiliary concepts in mathematics; it is a mathematical *model* for facilitating the mental reproduction of facts. Although we represent vibrations by the harmonic formula, the phenomena of cooling by exponentials, falls by squares of times, etc., no one will fancy that vibrations *in themselves* have anything to do with the circular functions, or the motion of falling bodies with squares. It has simply been observed that the relations between the quantities investigated were similar to certain relations obtaining between familiar mathematical functions, and these *more familiar* ideas are employed as an easy means of supplementing experience. Natural phenomena whose relations are not similar to those of functions with which we are familiar, are at present very difficult to reconstruct. But the progress of mathematics may facilitate the matter.

As mathematical helps of this kind, spaces of more than three dimensions may be used, as I have elsewhere shown. But it is not necessary to regard these, on this account, as anything more than mental artifices.*

* As the outcome of the labors of Lobatchévski, Bolyai, Gauss, and Riemann, the view has gradually obtained currency in the mathematical world, that that which we call *space* is a *particular, actual* case of a more *general,*

This is the case, too, with *all* hypothesis formed for the explanation of new phenomena. Our conceptions of electricity fit in at once with the electrical phenomena, and take almost spontaneously the familiar course, the moment we note that things take place as if attracting and repelling fluids moved on the surface of the conductors. But these mental expedients have nothing whatever to do with the phenomenon *itself*.

---

conceivable case of multiple quantitative manifoldness. The space of sight and touch is a threefold manifoldness; it possesses three dimensions; and every point in it can be defined by three distinct and independent data. But it is possible to conceive of a quadruple or even multiple space-like manifoldness. And the character of the manifoldness may also be differently *conceived* from the manifoldness of actual space. We regard this discovery, which is chiefly due to the labors of Riemann, as a very important one. The properties of actual space are here directly exhibited as objects of *experience*, and the pseudo-theories of geometry that seek to excogitate these properties by metaphysical arguments are overthrown.

A thinking being is supposed to live in the surface of a sphere, with no other kind of space to institute comparisons with. His space will appear to him similarly constituted throughout. He might regard it as infinite, and could only be convinced of the contrary by experience. Starting from any two points of a great circle of the sphere and proceeding at right angles thereto on other great circles, he could hardly expect that the circles last mentioned would intersect. So, also, with respect to the space in which we live, only experience can decide whether it is finite, whether parallel lines intersect in it, or the like. The significance of this elucidation can scarcely be overrated. An enlightenment similar to that which Riemann inaugurated in science was produced in the mind of humanity at large, as regards the surface of the earth, by the discoveries of the first circumnavigators.

The theoretical investigation of the mathematical possibilities above referred to, has, primarily, nothing to do with the question whether things really exist which correspond to these possibilities; and we must not hold mathematicians responsible for the popular absurdities which their investigations have given rise to. The space of sight and touch is *three-dimensional*; that, no one ever yet doubted. If, now, it should be found that bodies vanish from this space, or new bodies get into it, the question might scientifically be discussed whether it would facilitate and promote our insight into things to conceive experiential space as part of a four-dimensional or multi-dimensional space. Yet in such a case, this fourth dimension would, none the less, remain a pure thing of thought, a mental fiction.

But this is not the way matters stand. The phenomena mentioned were not forthcoming until *after* the new views were published, and were then exhibited in the presence of certain persons at spiritualistic *séances*. The fourth dimension was a very opportune discovery for the spiritualists and for theologians who were in a quandary about the location of hell. The use the spirit-

10. My conception of economy of thought was developed out of my experience as a teacher, out of the work of practical instruction. I possessed this conception as early as 1861, when I began my lectures as Privat-Docent, and at the time believed that I was in exclusive possession of the principle—a conviction which will, I think, be found pardonable. I am now, on the contrary, convinced that at least some presentiment of this idea has always, and necessarily must have, been a common possession of all inquirers who have ever made the nature of scientific investigation the subject of their thoughts. The expression of this opinion may assume the most diverse forms; for example, I should most certainly characterize the guiding theme of simplicity and beauty which so distinctly marks the work of Copernicus and Galileo, not only as æsthetical, but also as economical. So, too, New-

---

ualist makes of the fourth dimension is this. It is possible to move out of a finite straight line, without passing the extremities, through the second dimension; out of a finite closed surface through the third; and, analogously, out of a finite closed space, without passing through the enclosing boundaries, through the fourth dimension. Even the tricks that prestidigitateurs, in the old days, harmlessly executed in three dimensions, are now invested with a new halo by the fourth. But the tricks of the spiritualists, the tying or untying of knots in endless strings, the removing of bodies from closed spaces, are all performed in cases where there is absolutely nothing at stake. All is purposeless jugglery. We have not yet found an *accoucheur* who has accomplished parturition through the fourth dimension. If we should, the question would at once become a serious one. Professor Simony's beautiful tricks in rope-tying, which, as the performance of a prestidigitateur, are very admirable, speak against, not for, the spiritualists.

Everyone is free to set up an opinion and to adduce proofs in support of it. Whether, though, a scientist shall find it worth his while to enter into serious investigations of opinions so advanced, is a question which his reason and instinct alone can decide. If these things, in the end, should turn out to be true, I shall not be ashamed of being the last to believe them. What I have seen of them was not calculated to make me less skeptical.

I myself regarded multi-dimensioned space as a mathematico-physical help even prior to the appearance of Riemann's memoir. But I trust that no one will employ what I have thought, said, and written on this subject as a basis for the fabrication of ghost stories. (Compare Mach, *Die Geschichte und die Wurzel des Satzes von der Erhaltung der Arbeit.*)

ton's *Regulæ philosophandi* are substantially influenced by economical considerations, although the economical principle as such is not explicitly mentioned. In an interesting article, "An Episode in the History of Philosophy," published in *The Open Court* for April 4, 1895, Mr. Thomas J. McCormack has shown that the idea of the economy of science was very near to the thought of Adam Smith (*Essays*). In recent times the view in question has been repeatedly though diversely expressed, first by myself in my lecture *Ueber die Erhaltung der Arbeit* (1872), then by Clifford in his *Lectures and Essays* (1872), by Kirchhoff in his *Mechanics* (1874), and by Avenarius (1876). To an oral utterance of the political economist A. Herrmann I have already made reference in my *Erhaltung der Arbeit* (p. 55, note 5); but no work by this author treating especially of this subject is known to me.

11. I should also like to make reference here to the supplementary expositions given in my *Popular Scientific Lectures* (English edition, pages 186 et seq.) and in my *Principles of Heat* (German edition, page 294). In the latter work, the criticisms of Petzoldt (*Vierteljahrsschrift für wissenschaftliche Philosophie*, 1891) are considered. Husserl, in the first part of his work, *Logische Untersuchungen* (1900), has recently made some new animadversions on my theory of mental economy; these are in part answered in my reply to Petzoldt. I believe that the best course is to postpone an exhaustive reply until the work of Husserl is completed, and then see whether some understanding cannot be reached. For the present, however, I should like to premise certain remarks. As a natural inquirer, I am accustomed to begin with some special and definite inquiry, and allow the same to act upon me in all its phases, and to ascend from the special

aspects to more general points of view. I followed this custom also in the investigation of the development of physical knowledge. I was obliged to proceed in this manner for the reason that a theory of theory was too difficult a task for me, being doubly difficult in a province in which a minimum of indisputable, general, and independent truths from which everything can be deduced is not furnished at the start, but must first be sought for. An undertaking of this character would doubtless have more prospect of being successful if one took mathematics as one's subject-matter. I accordingly directed my attention to individual phenomena: the adaptation of ideas to facts, the adaptation of ideas to one another,* mental economy, comparison, intellectual experiment, the constancy and continuity of thought, etc. In this inquiry, I found it helpful and restraining to look upon every-day thinking and science in general, as a biological and organic phenomenon, in which logical thinking assumed the position of an ideal limiting case. I do not doubt for a moment that the investigation can be begun at both ends. I have also described my efforts as epistemological sketches.* It may be seen from this that I am perfectly able to distinguish between psychological and logical questions, as I believe every one else is who has ever felt the necessity of examining logical processes from the psychological side. But it is doubtful if any one who

---

* *Popular Scientific Lectures*, English edition, pp. 244 et seq., where the adaptation of thoughts to one another is described as the object of theory proper. Grassmann appears to me to say pretty much the same in the introduction to his *Ausdehnungslehre* of 1844, page xix: "The first division of all the sciences is that into real and formal, of which the real sciences depict reality in thought as something independent of thought, and find their truth in the agreement of thought with that reality; the formal sciences, on the other hand, have as their object that which has been posited by thought and itself, find their truth in the agreement of the mental processes with one another."

has read carefully even so much as the logical analysis of Newton's enunciations in my *Mechanics,* will have the temerity to say that I have endeavored to erase all distinctions between the "blind" natural thinking of every-day life and logical thinking. Even if the logical analysis of all the sciences were complete, the biologico-psychological investigation of their development would continue to remain a necessity for me, which would not exclude our making a new logical analysis of this last investigation. If my theory of mental economy be conceived merely as a teleological and provisional theme for guidance, such a conception does not exclude its being based on deeper foundations,† but goes toward making it so. Mental economy is, however, quite apart from this, a very clear logical ideal which retains its value even after its logical analysis has been completed. The systematic form of a science can be deduced from the same principles in many different manners, but some one of these deductions will answer to the principle of economy better than the rest, as I have shown in the case of Gauss's dioptrics.‡ So far as I can now see, I do not think that the investigations of Husserl have affected the results of my inquiries. As for the rest, I must wait until the remainder of his work is published, for which I sincerely wish him the best success.

When I discovered that the idea of mental economy had been so frequently emphasized before and after my enunciation of it, my estimation of my personal achievement was necessarily lowered, but the idea itself appeared to me rather to gain in value on this account; and what appears to Husserl as a de-

* *Principles of Heat,* Preface to the first German edition.
† *Analysis of the Sensations,* second German edition, pages 64-65.
‡ *Principles of Heat,* German edition, page 394.

gradation of scientific thought, the association of it with vulgar or "blind" thinking, seemed to me to be precisely an exaltation of it. It has outgrown the scholar's study, being deeply rooted in the life of humanity and reacting powerfully upon it.

CHAPTER V.

THE RELATIONS OF MECHANICS TO OTHER DE-
PARTMENTS OF KNOWLEDGE.

I.

THE RELATIONS OF MECHANICS TO PHYSICS.

1. Purely mechanical phenomena do not exist. The production of mutual accelerations in masses is, to all appearances, a purely dynamical phenomenon. But with these dynamical results are always associated thermal, magnetic, electrical, and chemical phenomena, and the former are always modified in proportion as the latter are asserted. On the other hand, thermal, magnetic, electrical, and chemical conditions also can produce motions. Purely mechanical phenomena, accordingly, are abstractions, made, either intentionally or from necessity, for facilitating our comprehension of things. The same thing is true of the other classes of physical phenomena. Every event belongs, in a strict sense, to all the departments of physics, the latter being separated only by an artificial classification, which is partly conventional, partly physiological, and partly historical.

2. The view that makes mechanics the basis of the remaining branches of physics, and explains all physical phenomena by mechanical ideas, is in our judgment a prejudice. Knowledge which is historically first, is not necessarily the foundation of all that is subsequently

gained. As more and more facts are discovered and classified, entirely new ideas of general scope can be formed. We have no means of knowing, as yet, which of the physical phenomena go *deepest,* whether the mechanical phenomena are perhaps not the most superficial of all, or whether all do not go *equally deep.* Even in mechanics we no longer regard the oldest law, the laws of the lever, as the foundation of all the other principles.

The mechanical theory of nature, is, undoubtedly, in an historical view, both intelligible and pardonable; and it may also, for a time, have been of much value. But, upon the whole, it is an artificial conception. Faithful adherence to the method that led the greatest investigators of nature, Galileo, Newton, Sadi Carnot, Faraday, and J. R. Mayer, to their great results, restricts physics to the expression of *actual facts,* and forbids the construction of hypotheses behind the facts, where nothing tangible and verifiable is found. If this is done, only the simple connection of the motions of masses, of changes of temperature, of changes in the values of the potential function, of chemical changes, and so forth is to be ascertained, and nothing is to be imagined along with these elements except the physical attributes or characteristics directly or indirectly given by observation.

This idea was elsewhere* developed by the author with respect to the phenomena of heat, and indicated, in the same place, with respect to electricity. All hypotheses of fluids or media are eliminated from the theory of electricity as entirely superfluous, when we reflect that electrical conditions are all given by the values of the potential function V and the dielectric

---

* Mach, *Die Geschichte und die Wurzel des Satzes von der Erhaltung der Arbeit.*

constants. If we assume the differences of the values of V to be measured (on the electrometer) by the forces, and regard V and not the quantity of electricity Q as the primary notion, or measurable physical attribute, we shall have, for any simple insulator, for our quantity of electricity

$$Q = \frac{-1}{4\pi} \int \left(\frac{d^2 V}{dx^2} + \frac{d^2 V}{dy^2} + \frac{d^2 V}{dz^2} \right) dv,$$

(where x, y, z denote the coördinates and dv the element of volume,) and for our potential*

$$W = \frac{-1}{8\pi} \int V \left(\frac{d^2 V}{dx^2} + \frac{d^2 V}{dy^2} + \frac{d^2 V}{dz^2} \right) dv.$$

Here Q and W appear as derived notions, in which no conception of fluid or medium is contained. If we work over in a similar manner the entire domain of physics, we shall restrict ourselves wholly to the quantitative conceptual expression of actual facts. All superfluous and futile notions are eliminated, and the imaginary problems to which they have given rise forestalled.

The preceding paragraph, which was written in 1883, met with little response from the majority of physicists, but it will be noticed that physical expositions have since then closely approached to the ideal there indicated. Hertz's "Investigations on the Propagation of Electric Force" (1892) affords a good instance of this description of phenomena by simple differential equations.

The removal of notions whose foundations are historical, conventional, or accidental, can best be furthered by a comparison of the conceptions obtaining

* Using the terminology of Clausius.

in the different departments, and by finding for the conceptions of every department the corresponding conceptions of others. We discover, thus, that temperatures and potential functions correspond to the velocities of mass-motions. A single velocity-value, a single temperature-value, or a single value of potential function, never changes *alone*. But whilst in the case of velocities and potential functions, so far as we yet know, only differences come into consideration, the significance of temperature is not only contained in its difference with respect to other temperatures. Thermal capacities correspond to masses, the potential of an electric charge to quantity of heat, quantity of electricity to entropy, and so on. The pursuit of such resemblances and differences lays the foundation of a *comparative physics,* which shall ultimately render possible the concise expression of extensive groups of facts, without *arbitrary* additions. We shall then possess a homogeneous physics, unmingled with artificial atomic theories.

It will also be perceived, that a real *economy* of scientific thought cannot be attained by mechanical hypotheses. Even if an hypothesis were fully competent to reproduce a given department of natural phenomena, say, the phenomena of heat, we should, by accepting it, only substitute for the actual relations between the mechanical and thermal processes, the hypothesis. The real fundamental facts are replaced by an equally large number of hypotheses, which is certainly no gain. Once an hypothesis has facilitated, as best it can, our view of new facts, by the substitution of more familiar ideas, its powers are exhausted. We err when we expect more enlightenment from an hypothesis than from the facts themselves.

3. The development of the mechanical view was favored by many circumstances. In the first place, a connection of all natural events with mechanical processes is unmistakable, and it is natural, therefore, that we should be led to explain less known phenomena by better known mechanical events. Then again, it was first in the department of mechanics that laws of general and extensive scope were discovered. A law of this kind is the principle of *vis viva* $\Sigma(U_1 - U_0) = \Sigma\frac{1}{2}m(v_1^2 - v_0^2)$, which states that the increase of the *vis viva* of a system in its passage from one position to another is equal to the increment of the force-function, or work, which is expressed as a function of the final and initial positions. If we fix our attention on the work a system can perform and call it with Helmholtz the *Spannkraft*, S,* then the work *actually performed*, U, will appear as a diminution of the *Spannkraft*, K, initially present; accordingly, $S = K - U$, and the principle of *vis viva* takes the form

$$\Sigma S + \tfrac{1}{2} \Sigma m v^2 = const,$$

that is to say, every diminution of the *Spannkraft*, is compensated for by an increase of the *vis viva*. In this form the principle is also called the law of the *Conservation of Energy*, in that the sum of the *Spannkraft* (the potential energy) and the *vis viva* (the kinetic energy) remains constant in the system. But since, in nature, it is possible that *not only vis viva* should appear as the consequence of work performed, but also quantities of heat, or the potential of an electric charge, and so forth, scientists saw in this law the expression of a

---

* Helmholtz used this term in 1847; but it is not found in his subsequent papers; and in 1882 (*Wissenschaftliche Abhandlungen*, II, 965) he expressly discards it in favor of the English "potential energy." He even (p. 968) prefers Clausius's word *Ergal* to *Spannkraft*, which is quite out of agreement with modern terminology.—*Trans.*

mechanical action as the basis of all natural actions. However, nothing is contained in the expression but the fact of an invariable quantitative *connection* between mechanical and other kinds of phenomena.

4. It would be a mistake to suppose that a wide and extensive view of things was first introduced into physical science by mechanics. On the contrary, this insight was possessed at all times by the foremost inquirers and even entered into the construction of mechanics itself, and was, accordingly, not first created by the latter. Galileo and Huygens constantly alternated the consideration of particular details with the consideration of universal aspects, and obtained their results only by a persistent effort for a simple and consistent view. The fact that the velocities of individual bodies and systems are dependent on the spaces descended through, was perceived by Galileo and Huygens only by a very detailed investigation of the motion of descent in particular cases, combined with the consideration of the circumstance that bodies generally, of their own accord, only sink. Huygens especially on this occasion emphasizes the impossibility of a mechanical perpetual motion; he possessed, therefore, the modern point of view. He felt the *incompatibility* of the idea of a perpetual motion with the notions of the natural mechanical processes with which he was familiar.

Take the fictions of Stevinus—say, that of the endless chain on the prism. Here, too, a deep, broad insight is displayed. We have here a mind, disciplined by a multitude of experiences, brought to bear on an individual case. The moving endless chain is to Stevinus a motion of descent that is not a descent, a motion without a purpose, an intentional act that does not answer to the intention, an endeavor for a change which does not produce the change. If motion, generally, is the result of descent, then in the particular case descent is the result of motion. It is a sense of the mutual interdependence of v and h in the equation $v = \sqrt{2gh}$ that is here displayed, though of course in not so definite a form. A contradiction exists in this fiction for Stevinus's exquisite investigative sense that would escape less profound thinkers.

This same breadth of view, which alternates the individual with the universal, is also displayed, only in this instance not restricted to mechanics, in the performances of Sadi Carnot. When Carnot finds that the quantity of heat Q which, for a given amount of work L, has flowed from a higher temperature t to a lower temperature t', can only depend on the temperatures and not on the material constitution of the bodies, he reasons in exact conformity with the method of Galileo. Similarly does J. R. Mayer proceed in the enunciation of the principle of the equivalence of heat and work. In this achievement the mechanical view was quite remote from Mayer's mind; nor had he need of it. They who require the crutch of the mechanical philosophy to understand the doctrine of the equivalence of heat and work, have only half comprehended the progress which it signalizes. Yet, high as we may place Mayer's original achievement, it is not on that account necessary to depreciate the merits of the professional physicists, Joule, Helmholtz, Clausius, and Thomson, who have done very much, perhaps all, towards the detailed *establishment* and *perfection* of the new view. The assumption of a plagiarism of Mayer's ideas is in our opinion gratuitous. They who advance it, are under the obligation to *prove* it. The repeated appearance of the same idea is not new in history. We shall not take up here the discussion of purely personal questions, which thirty years from now will no longer interest students. But it is unfair, from a pretense of justice, to insult men, who if they had accomplished but a third of their actual services, would have lived highly honored and unmolested lives.

In Germany, Mayer's works at first met with a very cool, and in part hostile, reception; even difficulties of publication were encountered; but in England they found more speedy recognition. After they had been almost forgotten there, amid the wealth of new facts being brought to light, attention was again called to them by the lavish praise of Tyndall in his book *Heat: a Mode of Motion* (1863). The consequence of this was a pronounced reaction in Germany, which reached its culminating point in Dühring's work *Robert Mayer, the Galileo of the Nineteenth Century* (1878). It almost appeared as if the injustice that had been done to Mayer was now to be atoned for by injustice towards others. But as in criminal law, so here, the sum of the injustice is only increased in this way, for no algebraic cancelation takes place. An enthusiastic and thoroughly satisfactory estimate of Mayer's performances was given by Popper in an article in *Ausland* (1876, No. 35), which is also very readable from the many interesting epistemological *aperçus* that it contains. I have endeavored (*Principles of Heat*) to give a thoroughly just and sober presentation of the achievements of the different inquirers in the domain of the mechanical theory of heat. It appears from this that each one of the inquirers concerned made some distinctive contribution which expressed their respective intellectual peculiarities. Mayer may be regarded as the philosopher of the theory of heat and energy; Joule, who was also conducted to the principle of energy by philosophical considerations, furnishes the experimental foundation; and Helmholtz gave to it its theoretical physical form. Helmholtz, Clausius, and Thomson form a transition to the views of Carnot, who stands alone in his ideas. Each one of the first-mentioned inquirers could be eliminated. The progress of the development would have been retarded thereby, but it would not have been checked (compare the edition of Mayer's works by Weyrauch, Stuttgart, 1893).

5. We shall now attempt to show that the broad view expressed in the principle of the conservation of energy, is not peculiar to mechanics, but is a condition of logical and sound scientific thought generally. The business of physical science is the reconstruction of facts in thought, or the abstract quantitative expression of facts. The rules which we form for these reconstructions are the laws of nature. In the conviction that such rules are possible lies the law of causality. The law of causality simply asserts that the phenomena of nature are *dependent* on one another. The special emphasis put on space and time in the expression of the law of causality is unnecessary, since the relations of space and time themselves implicitly express that phenomena are dependent on one another.

The laws of nature are equations between the measurable elements $\alpha \beta \gamma \delta \ldots \omega$ of phenomena. As nature is variable, the number of these equations is always less than the number of the elements.

If we know *all* the values of $\alpha \beta \gamma \delta \ldots$, by which, for example, the values of $\lambda \mu \nu \ldots$ are given, we may call the group $\alpha \beta \gamma \delta \ldots$ the cause and the group $\lambda \mu \nu \ldots$ the effect. In this sense we may say that the effect is *uniquely* determined by the cause. The principle of sufficient reason, in the form, for instance, in which Archimedes employed it in the development of the laws of the lever, consequently asserts nothing more than that the effect cannot by any given set of circumstances be at once determined and undetermined.

If two circumstances a and λ are connected, then, supposing all others are constant, a change of λ will be accompanied by a change of a, and as a general rule a change of a by a change of λ. The constant observance of this *mutual* interdependence is met with in Stevinus, Galileo, Huygens, and other great inquirers. The idea is also at the basis of the discovery of *counter-phenomena.* Thus, a change in the volume of a gas due to a change of temperature is supplemented by the counter-phenomenon of a change of temperature on an alteration of volume; Seebeck's phenomenon by Peltier's effect, and so forth. Care must, of course, be exercised, in such inversions, respecting the *form* of the dependence. Figure 235 will render clear how a perceptible alteration of a may always be produced by an alteration of λ, but a change of λ not necessarily by a change of a. The relations between electromagnetic and induction phenomena, discovered by Faraday, are a good instance of this truth.

Fig. 235.

If a set of circumstances $a\,\beta\,\gamma\,\delta\,.\,.\,.$, by which a second set $\lambda\,\mu\,\nu\,.\,.\,.$ is determined, be made to pass from its initial values to the terminal values $a'\,\beta'\,\gamma'\,\delta'.\,.\,.$, then $\lambda\,\mu\,\nu\,.\,.\,.$ also will pass into $\lambda'\,\mu'\,\nu'\,.\,.\,.$ If the first set be brought back to its initial state, also the second set will be brought back to its initial state. This is the meaning of the "equivalence of cause and effect," which Mayer again and again emphasizes.

If the first group suffer only *periodical* changes, the second group also can suffer only periodical changes, not continuous *permanent* ones. The fertile methods of thought of Galileo, Huygens, S. Carnot, Mayer, and their peers, are all reducible to the simple but significant perception, *that purely periodical alterations of one set of circumstances can only constitute the source of similarly periodical alterations of a second set of circumstances, not of continuous and permanent alterations.* Such maxims, as "the effect is equivalent to the cause," "work cannot be created out of nothing," "a perpetual motion is impossible," are particular, less definite, and less evident forms of this perception, which in itself is not especially concerned with mechanics, but is a constituent of scientific thought generally. With the perception of this truth, any metaphysical mysticism that may still adhere to the principle of the conservation of energy* is dissipated.

With these lines, which were written in 1883, compare Petzoldt's remarks on the striving after *stability* in intellectual life ("Maxima, Minima und Ökonomie," *Vierteljahrsschr. für wiss. Philosophie,* 1891).

All ideas of conservation, like the notion of substance, have a solid foundation in the economy of thought. A mere unrelated change, without fixed point of support, or reference, is not comprehensible, not mentally reconstructible. We always inquire, accordingly, what idea can be retained amid all variations as *permanent,* what *law* prevails, what *equation* remains fulfilled, what quantitative *values* remain constant? When we say the refractive index remains constant in all cases of refraction, g remains $= 9 \cdot 810m$ in all cases of the motion of heavy bodies, the energy remains constant in every isolated system, all our assertions have one and the same economical function, namely that of facilitating our mental reconstruction of facts.

---

* When we reflect that the principles of science are all abstractions that presuppose *repetitions* of *similar* cases, the absurd applications of the law of the conservation of forces to the universe as a whole fall to the ground.

The principle of energy is only briefly treated in the text, and I should like to add here a few remarks on the following four treatises, discussing this subject, which have appeared since 1883: *Die physikalischen Grundsätze der elektrischen Kraftübertragung,* by J. Popper, Vienna, 1883; *Die Lehre von der Energie,* by G. Helm, Leipzig, 1887; *Das Princip der Erhaltung der Energie,* by M. Planck, Leipzig, 1887; and *Das Problem der Continuität in der Mathematik und Mechanik,* by F. A. Müller, Marburg, 1886.

The independent works of Popper and Helm are, in the aim they pursue, in perfect accord, and they quite agree in this respect with my own researches, so much so in fact that I have seldom read anything that, without the obliteration of individual differences, appealed in an equal degree to my mind. These two authors especially meet in their attempt to enunciate a general science of energetics; and a *suggestion* of this kind is also found in a note to my treatise *Ueber die Erhaltung der Arbeit (Conservation of Energy,* page 47). Since then "energetics" has been exhaustively treated by Helm, Ostwald, and others.

In 1872, in this same treatise (pp. 42 et seq.), I showed that our belief in the principle of excluded perpetual motion is founded on a more general belief in the *unique* determination of one group of (mechanical) elements, $\alpha\,\beta\,\gamma\,.\,.\,.$, by a group of different elements, $x\,y\,z\,.\,.\,.$ Planck's remarks at pages 99, 138, and 139 of his treatise essentially agree with this; they are different only in form. Again, I have repeatedly remarked that all forms of the law of causality spring from subjective impulses, which nature is by no means compelled to satisfy. In this respect my conception is allied to that of Popper and Helm.

Planck (pp. 21 et seq., 135) and Helm (p. 25 et seq.) mention the "metaphysical" points of view by which Mayer was controlled, and both remark (Planck, p. 26 et seq., and Helm p. 28) that also Joule, though there are no direct expressions to justify the conclusion, must have been guided by similar ideas. To this last I fully assent.

With respect to the so-called "metaphysical" points of view of Mayer, which, according to Helmholtz, are extolled by the devotees of metaphysical speculation as Mayer's highest achievement, but which appear to Helmholtz as the weakest feature of his expositions, I have the following remarks to make. With maxims, such as, "Out of nothing, nothing comes," "The effect is equivalent to the cause," and so forth, one can never convince *another* of anything. How little such empty maxims, which until recently were admitted in science, can accomplish, I have illustrated by examples in my treatise *Die Erhaltung der Arbeit.* But in Mayer's case these maxims are, in my judgment, not weaknesses. On the contrary, they are with him the expression of a *powerful* instinctive yearning, as yet unsettled and unclarified, for a sound, substantial conception of what is now called energy. This desire I should not exactly call metaphysical. We now know that Mayer was not wanting in the conceptual power to give clearness to this desire. Mayer's attitude in this point was in no respect different from that of Galileo, Black, Faraday, and other great inquirers, although perhaps many were more taciturn and cautious than he.

I have touched upon this point before in my *Analysis of the Sensations,* Jena, 1886, English translation, Chicago, 1897, p. 174 et seq. Aside from the fact that I do not share the Kantian point of view, in fact, occupy *no* metaphysical point of view, not even that of Berkeley, as hasty readers of my last-mentioned treatise have assumed, I agree with F. A. Müller's remarks on this question (p. 104 et seq.). For a more exhaustive discussion of the principle of energy see my *Principles of Heat.*

II.

THE RELATIONS OF MECHANICS TO PHYSIOLOGY.

1. All science has its origin in the needs of life. However minutely it may be subdivided by particular vocations or by the restricted tempers and capacities of those who foster it, each branch can attain its full and best development only by a living connection with *the whole.* Through such a union alone can it approach its true maturity, and be insured against lop-sided and monstrous growths.

The division of labor, the restriction of individual inquirers to limited provinces, the investigation of those provinces as a life-work, are the fundamental conditions of a fruitful development of science. Only by such specialization and restriction of work can the economical instruments of thought requisite for the mastery of a special field be perfected. But just here lies a danger—the danger of our overestimating the instruments, with which we are so constantly employed, or even of regarding them as the objective point of science.

2. Now, such a state of affairs has, in our opinion, actually been produced by the disproportionate formal development of physics. The majority of natural inquirers ascribe to the intellectual implements of physics,

to the concepts mass, force, atom, and so forth, whose sole office is to revive economically arranged experiences, a reality beyond and independent of thought. Not only so, but it has even been held that these forces and masses are the real objects of inquiry, and, if once they were fully explored, all the rest would follow from the equilibrium and motion of these masses. A person who knew the world only through the theater, if brought behind the scenes and permitted to view the mechanism of the stage's action, might possibly believe that the real world also was in need of a machine-room, and that if this were once thoroughly explored, we should know all. Similarly, we, too, should beware lest the *intellectual* machinery, employed in the representation of the world on *the stage of thought,* be regarded as the basis of the real world.

3. A philosophy is involved in any correct view of the relations of special knowledge to the great body of knowledge at large—a philosophy that must be demanded of every special investigator. The lack of it is asserted in the formulation of imaginary problems, in the very enunciation of which, whether regarded as soluble or insoluble, flagrant absurdity is involved. Such an overestimation of physics, in contrast to physiology, such a mistaken conception of the true relations of the two sciences, is displayed in the inquiry whether it is possible to *explain* feelings by the motions of atoms.

Let us seek the conditions that could have impelled the mind to formulate so curious a question. We find in the first place that greater *confidence* is placed in our experiences concerning relations of time and space; that we attribute to them a more objective, a more *real* character than to our experiences of colors, sounds, temperatures, and so forth. Yet, if we investigate the matter accurately, we must surely admit that our sensations of time and space are just as much *sensations* as are our sensations of colors, sounds, and odors, only that in our knowledge of the former we are surer and clearer than in that of the latter. Space and time are well-ordered systems of sets of sensations. The quantities stated in mechanical equations are simply ordinal symbols, representing those members of these sets that are to be mentally isolated and emphasized. The equations express the form of interdependence of these ordinal symbols.

A body is a relatively constant sum of touch and sight sensations associated with the same space and time sensations. Mechanical principles, like that, for instance, of the mutually induced accelerations of two masses, give, either directly or indirectly, only some combination of touch, sight, light, and time sensations. They possess intelligible meaning only by virtue of the sensations they involve, the contents of which may of course be very complicated.

It would be equivalent, accordingly, to explaining the more simple and immediate by the more complicated and remote, if we were to attempt to derive sensations from the motions of masses, wholly aside from the consideration that the notions of mechanics are economical implements or expedients perfected to represent *mechanical* and not physiological or *psychological* facts. If the *means* and *aims* of research were properly distinguished, and our expositions were restricted to the presentation of *actual facts,* false problems of this kind could not arise.

4. All physical knowledge can only mentally represent and anticipate compounds of those elements we call sensations. It is concerned with the connection of these elements. Such an element, say the heat of a body

A, is connected, not only with other elements, say with such whose aggregate makes up the flame B, but also with the aggregate of certain elements of our body, say with the aggregate of the elements of a nerve N. As simple object and element N is not essentially, but only conventionally, different from A and B. The connection of A and B is a problem of *physics*, that of A and N a problem of *physiology*. Neither one exists alone; both exist at once. Only provisionally can we neglect either. Processes, thus, that in appearance are purely mechanical, are, in addition to their evident mechanical features, always physiological, and, consequently, also electrical, chemical, and so forth. The science of mechanics does not comprise the foundations, no, nor even a part of the world, but only an *aspect* of it.

At the beginning of this book, the view was expressed that the doctrines of mechanics have developed out of the collected experiences of handicraft by an intellectual process of refinement. In fact, if we consider the matter without prejudice, we see that the savage discoverers of bow and arrows, of the sling, and of the javelin, set up the most important law of modern dynamics—the law of inertia—long before it was misunderstood with thorough-going perversity by Aristotle and his learned commentators. And although first ancient machines for throwing projectiles and catapults and then modern firearms brought this law daily before our eyes, many centuries were needed before the correct theoretical idealization was discovered by the genius of Galileo and Newton. It lay in exactly the opposite direction to that in which the great majority of human beings expected it to lie. Not the conservation, but the decrease of the velocity of projection was to be theoretically explained and justified.

The simple machines—the five mechanical powers —as they are described by Hero of Alexandria in the work of which an Arabian translation came down to the Middle Ages, are without question a product of handicraft. If, now, a child busies himself with mechanical work with quite simple and primitive means— as was the case with my son, Ludwig Mach—the dynamical sensations observed in this connection and the dynamical experiences obtained when adaptive motions are made, make a powerful and lasting impression. If we pay attention to these sensations, we come closer, intellectually speaking, to the instinctive origin of machines. We understand why one prefers the longer lever yielding to the lesser pressure, and why a hammer swung round by its handle through a longer distance can transmit more work or *vis viva*. We understand at once by experiment the transport of loads on rollers, and also how the wheel—the fixed roller—arose. The making of rollers must have gained a great technical importance and have led to the discovery of the lathe. In possession of this, mankind easily discovered the wheel, the wheel and axle, and the pulley. But the primitive lathe is the very ancient fire-drill of savages, which had a bow and cord, though of course this primitive lathe is only fitted for small objects. The Arabians still use it, and, up to quite recent times, it was almost universally in use with our watchmakers. The potter's wheel of the ancient Egyptians was also a kind of lathe. Perhaps these forms served as models for the larger lathe, whose discovery, as well as that of the plumb-line and theodolite, is ascribed to Theodorus of Samos. On it pillars of stone may well have been turned (532 B. C.). Not all knowledge finds immediate use; often it lies fallow for a long time. The ancient Egyptians had wheels on the war-chariots of the king. They actually transported their huge stone monuments, with brazen disregard of the work of men, on sledges. What did the labor of slaves taken as prisoners in war matter to them? The prisoners ought to be thankful that they were not, in the Assyrian manner, impaled, or at least blinded, but only, quite kindly, in comparison with that, used as beasts of burden. Even our noble precursors in civilization—the Greeks—did not think very differently.

But if we suppose even the best will for progress, many discoveries remain hardly comprehensible. The ancient Egyptians were not acquainted with the screw. In the many plates of Rosselini's work no trace of it is to be found. The Greeks ascribed, on doubtful reports, its discovery to Archytas of Tarentum (about 390 B. C.). But with Archimedes (250 B. C.) and with Hero (100 B. C.) we find the screw in very many forms as something well known. Hero can easily say—and even in a way that can be understood by modern schoolmen: "the screw is a winding wedge." But whoever has not yet seen or handled a screw will not by this indication discover one. By analogy with the cases spoken of before, we must suppose that, when an object in the form of a screw—such as a twisted rope or a pair of wires twisted together for ornamental purposes or the spindle-ring of an old fire-drill which had been worn spirally by the cord—fell, by chance, into someone's hand, the thought of construction of a screw lay near to the sensation of the twisting of this thing in and out of the hand. Fundamentally it is chance observations in which the faulty adaptation of human beings to their surroundings expresses itself, and which, when they are once noticed, gives rise to a further adaptation.

My son vividly describes how, in an ethnographical museum, the dynamical experiences of his youth again vividly came to life; how they were awakened again by the perceptible traces of the work on the objects exhibited. May these experiences be used for the finding of a universal genetic technology, and perhaps, by the way, lead a little deeper into the understanding of the primitive history of mechanics.

CHRONOLOGICAL TABLE

OF A FEW

EMINENT INQUIRERS

AND OF

THEIR MORE IMPORTANT MECHANICAL WORKS.

ARCHIMEDES (287-212 B. C.). A complete edition of his works was published, with the commentaries of Eutocius, at Oxford, in 1792; a French translation by F. Peyrard (Paris, 1808); a German translation by Ernst Nizze (Stralsund, 1824).

LEONARDO DA VINCI (1452–1519). Leonardo's scientific manuscripts are substantially embodied in H. Grothe's work, "Leonardo da Vinci als Ingenieur und Philosoph" (Berlin, 1874).

GUIDO UBALDI(o) e Marchionibus Montis (1545-1607). *Mechanicorum Liber* (Pesaro, 1577).

S. STEVINUS (1548-1620). *Beghinselen der Weegkonst* (Leyden, 1585); *Hypomnemata Mathematica* (Leyden, 1608).

GALILEO (1564-1642). *Discorsi e dimostrazioni matematiche* (Leyden, 1638). The first complete edition of Galileo's writings was published at Florence (1842-1856), in fifteen volumes 8vo.

KEPLER (1571-1630). *Astronomia Nova* (Prague, 1609); *Harmonice Mundi* (Linz, 1619); *Stereometria Doliorum* (Linz, 1615). Complete edition by Frisch (Frankfort, 1858).

MARCUS MARCI (1595-1667). *De Proportione Motus* (Prague, 1639).

DESCARTES (1596-1650). *Principia Philosophiæ* (Amsterdam, 1644).

ROBERVAL (1602-1675). *Sur la composition des mouvements. Anc. Mém. de l'Acad. de Paris.* T. VI.

GUERICKE (1602-1686). *Experimenta Nova, ut Vocantur Magdeburgica* (Amsterdam, 1672).

FERMAT (1601–1665). *Varia Opera* (Toulouse, 1679).

TORRICELLI (1608-1647). *Opera Geometrica* (Florence, 1644).

WALLIS (1616-1703). *Mechanica Sive de Motu* (London, 1670).

MARIOTTE (1620-1684). *Œuvres* (Leyden, 1717).

PASCAL (1623-1662). *Récit de la grande expérience de l'équilibre des liqueurs* (Paris, 1648) ; *Traité de l'équilibre des liqueurs et de la pesanteur de la masse de l'air.* (Paris, 1662).

BOYLE (1627-1691). *Experimenta Physico Mechanica* (London, 1660).

HUYGENS (1629-1695). *A Summary Account of the Laws of Motion.* Philos. Trans. 1669; *Horologium Oscillatorium* (Paris, 1673) ; *Opuscula Posthuma* (Leyden, 1703).

WREN (1632-1723). *Lex Naturæ de Collisione Corporum.* Philos. Trans. 1669.

LAMI (1640-1715). *Nouvelle manière de démontrer les principaux théoremes des élémens des mécaniques* (Paris, 1687).

NEWTON (1642-1726). *Philosophiæ Naturalis Principia Mathematica* (London, 1686).

LEIBNIZ (1646-1716). *Acta Eruditorium*, 1686, 1695; *Leibnitzii et Joh. Bernoullii Comercium Epistolicum* (Lausanne and Geneva, 1745).

JAMES BERNOULLI (1654–1705). *Opera Omnia* (Geneva, 1744).

VARIGNON (1654-1722). *Projet d'une nouvelle mécanique* (Paris, 1687).

JOHN BERNOULLI (1667-1748). *Acta Erudit.* 1693; *Opera Omnia* (Lausanne, 1742).

MAUPERTUIS (1698-1759). *Mém. de l'Acad. de Paris,* 1740; *Mém. de l'Acad. de Berlin,* 1745, 1747; *Œuvres* (Paris, 1752).

MACLAURIN (1698-1746). *A Complete System of Fluxions* (Edinburgh, 1742).

DANIEL BERNOULLI (1700-1782). *Comment. Acad. Petrop., T. I. Hydrodynamica* (Strassburg, 1738).

EULER (1707-1783). *Mechanica sive Motus Scientia* (Petersburg, 1736); *Methodus Inveniendi Lineas Curvas* (Lausanne, 1744). Numerous articles in the volumes of the Berlin and St. Petersburg academies.

CLAIRAUT (1713-1765). *Théorie de la figure de la terre* (Paris, 1743).

D'ALEMBERT (1717-1783). *Traité de dynamique* (Paris, 1743).

LAGRANGE (1736-1813). *Essai d'une nouvelle méthode pour déterminer les maxima et minima.* Misc. Taurin. 1762; *Mécanique analytique* (Paris, 1788).

LAPLACE (1749-1827). *Mécanique céleste* (Paris, 1799).

FOURIER (1768-1830). *Théorie analytique de la chaleur* (Paris, 1822).

GAUSS (1777-1855). *De Figura Fluidorum in Statu Æquilibrii. Comment. Societ. Götting.,* 1828; *Neues Princip der Mechanik* (Crelle's Journal, IV, 1829) ; *Intensitas Vis Magneticæ Terrestris ad Mensuram Absolutam Revocata* (1833). Complete works (Göttingen, 1863).

POINSOT (1777-1859). *Éléments de statique* (Paris, 1804).

PONCELET (1788-1867). *Cours de mécanique* (Metz, 1826).

BELANGER (1790-1874). *Cours de mécanique* (Paris, 1847).

MÖBIUS (1790-1867). *Statik* (Leipzig, 1837).

CORIOLIS (1792-1843). *Traité de mécanique* (Paris, 1829).

C. G. J. JACOBI (1804-1851). *Vorlesungen über Dynamik,* herausgegeben von Clebsch (Berlin, 1866).

W. R. HAMILTON (1805-1865). *Lectures on Quaternions,* 1853.—Essays.

GRASSMANN (1809-1877). *Ausdehnungslehre* (Leipzig, 1844).

H. HERTZ (1857-1894). *Principien der Mechanik* (Leipzig, 1894).

INDEX

INDEX

Abendroth, W., 246.

Absolute conservation of energy, 296 et seq.

Absolute space, time, motion, xxvii, xxviii, 271-297, 330, 336-341. See the *nouns*.

Absolute units, 367-375.

Abstractions, 579.

Acceleration, Galileo on, 157 et seq. 174, 188, 189; Newton on, 296; also 266, 282, 288, 303, 306.

Action, least, principle of, 456-474, 549; sphere of, 479.

Action and reaction, 243-245, 301.

Adaptation, in nature, 547; of thoughts to facts, 7, 38-40; 593.

Aërometer, effect of suspended particles on, 254.

Aërostatics, See *Air*.

Affined, 204.

Air, expansive force of, 150; quantitative data of, 147; weight of, 136; pressure of, 137 et seq.; nature of 131 et seq.; function of, with Aristotle and Philoponos, 152-153; Voltaire's ideas on, 135.

Air-pump, 144-150.

Aitken, 184.

Alcohol and water, mixture of, 479 et seq.

Algebra, economy of, 583.

Algebraical mechanics, 561.

All, The, necessity of its consideration in research, 556.

Analytical mechanics, 560-577.

Analytic method, 561.

Anaxagoras, 4, 131.

Anding, A., 293.

Animal, free in space, 379.

Animistic points of view in mechanics, 555 et seq.

Apelt, 334.

Archimedes, on the lever and the center of gravity, 11, 13-17, 19-20, 24-28, 39; on hydrostatics, 107-109; various modes of deduc-tion of his hydrostatic principle, 124; illustration of his principle, 126, 479; also 614.

Archytas, 11-12, 615.

Areas, law of the conservation of, 382-395.

Aristotle, 4, 12-13, 98, 105-106; 131.

Artifices, mental, 588 et seq.

Assyrian monuments, 1.

Atmosphere, See *Air*.

Atoms, mental artifices, 588.

Attraction, 307.

Atwood's Machine, 178, 267.

Automata, 12.

Axes of reference with Galileo and Newton, 264, 280-281, 291.

Axiomatic certainties, 330.

Avenarius, vii, xvii, xix, xxiii, xxvii, 592.

Babbage, on calculating machines, 585.

Babo, von, 180.

Baliani, 170, 238.

Ballistic pendulum, 417.

Balls, elastic, symbolizing pressures in liquids, 418.

Bandbox, rotation of, 390.

Barometer, height of mountains determined by, 139, 140.

Barrow, I., 247.

Base, pressure of liquids on, 110, 120.

Beeckmann, 189.

Belanger, on impulse, 360.

Benedetti, 98, 102, 153-156, 189, 190.

Berkeley, xii, xiii, xix, xx, 609.

Bernoulli, Daniel, his geometrical demonstration of the parallelo-gram of forces, 51-52; criticism of Bernoulli's demonstration, 53-57; on the law of areas, 382; on the principles of *vis viva*, 433, 437-438; on the velocity of liquid ef-flux, 498; his hydrodynamic prin-ciple, 503; on the parallelism of

Date Due
